Computers in Communication

Computers in Communication

Gordon Brebner
Department of Computer Science
University of Edinburgh, Scotland

THE McGRAW-HILL COMPANIES

London • New York • St Louis • San Francisco • Auckland
Bogotá • Caracas • Lisbon • Madrid • Mexico • Milan
Montreal • New Delhi • Panama • Paris • San Juan • São Paulo
Singapore • Sydney • Tokyo • Toronto

Published by
McGraw-Hill Publishing Company
Shoppenhangers Road, Maidenhead, Berkshire, SL6 2QL, England
Telephone 01628 502500
Fax 01628 770224

British Library Cataloguing in Publication Data

The CIP data of this title is available from the British Library, UK

Library of Congress Cataloging-in-Publication Data

The CIP data of this title is available from the Library of Congress, Washington DC, USA

McGraw-Hill
A Division of The **McGraw·Hill** Companies

Typeset by the author

Printed and bound in Great Britain at the University Press, Cambridge

Printed on permanent paper in compliance with ISO Standard 9706

To Rosemary

CONTENTS

List of Trademarks

Advanced Peer-to-Peer Networking (APPN), IBM   International Business Machines Corporation, Old Orchard Road, Armonk, NY 10504, USA

Apple, Appletalk, LocalTalk, Macintosh   Apple Computer Inc., 1 Infinite Loop, Cupertino, CA 95014, USA

AT&T   American Telephone and Telegraph Company, 32 Avenue of the Americas, New York, NY 10013-2412, USA

Banyan VINES   Banyan Systems Inc., 120 Flanders Road, PO Box 5013, Westboro, MA 01581, USA

Cisco, IGRP   Cisco Systems Inc., 170 West Tasman Drive, San Jose, CA 95134-1706, USA

DEC, DECnet, Digital, Digital Equipment Corporation, VT Digital Equipment Corporation, 111 Powdermill Road, Maynard, MA 01754-1418, USA

Ethernet, Xerox   Xerox Corporation, PO Box 1600, 800 Long Ridge Road, Stamford, CT 06904, USA

Java, NFS   Sun Microsystems Inc., 2550 Garcia Avenue, Mountain View, CA 94043-1100, USA

Kerberos   Massachusetts Institute of Technology, Five Cambridge Center, Kendall Square, Room NE25-230, Cambridge, MA 02142-1493, USA

MCI   MCI Communications Corporation, 1801 Pennsylvannia Avenue, NW, Washington, DC 20006, USA

Microsoft, Windows   Microsoft Corporation, One Microsoft Way, Redmond, WA 98052-6399, USA

Motorola   Motorola Inc., 1303 East Algonquin Road, Schaumberg, IL 60196, USA

Netscape Navigator   Netscape Communications Corporation, 501 East Middlefield Road, Mountain View, CA 94043, USA

Novell, NetWare, IntraNetWare   Novell Inc., 1555 North Technology Way, Orem, Utah 8407, USA

Prospero   University of Southern California, 620 West 35th Street, Los Angeles, CA 90089-1333, USA

UNIX   UNIX Systems Laboratories Inc., 190 River Road, Summit, NJ 07901, USA

WAIS   WAIS Inc., 1040 Noel Drive, Menlo Park, CA 94025, USA

X Window System   X Consortium Inc., The Open Group, 11 Cambridge Center, Cambridge, MA 02142-1405, USA

PREFACE

OBJECTIVES

The building of 'information superhighways', in the first instance exemplified by the massive growth of the Internet, has introduced the general public to computers in communication. The end result is to allow the sharing of computer-based resources that are physically located in all parts of the world. From the point of view of most users, the mechanisms by which computers communicate are best kept hidden. However, without these mechanisms, no sharing of resources could happen. This book aims to reveal the basic principles of computer communications, and show how these underpin the practical communication mechanisms that are actually used. It is intended both for readers who will be involved in getting computers to communicate, and also those who just have a general interest in what happens under the tarmac of the information highway.

The book is deliberately focused tightly on matters concerned with communication. It avoids excursions into other areas of computer science, such as computer programming, modelling or simulation, not to mention electronics, photonics or mathematics. Such matters are covered in other specialized texts. The reader requires no background in computer communications, only a basic familiarity with computers in general. A knowledge of computer programming is not required although, in a very few places, a little knowledge would aid full understanding of points made. However, this is not crucial to the main plot. Conversely, the author hopes that the book can also offer something to the reader who does have practical experience of computer communications, by setting this experience in an overall context, and revealing alternative possibilities.

In several respects, this book differs from other general textbooks on computer communications. This is not likely to trouble the beginner, but a little explanation is in order for a reader who is not new to the subject area. First, the general approach is to identify the basic principles of computer communication (strongly motivated by how human communication works), and *then* apply these principles in increasingly practical settings. This contrasts with a more conventional approach that catalogues practical examples of computer communication systems, noting principles (sometimes repeatedly) as the cataloguing proceeds. One important benefit of this book's approach is to give coverage that should stand the test of time. The basic principles underpinning the book have endured through a period when massive changes and developments in practical computer communications have come about.

The second difference concerns the presentation of the material. Most textbooks on computer communications become bogged down, both by technical details and by technical terminology. This obstructs understanding of the important issues. This book tackles the problem in two ways. One is to omit the lowest levels of technical description of virtually all practical mechanisms included. There is sufficient detail for a reader to understand how the mechanisms work, but a reader who needs enough detail to actually implement the mechanisms will need to seek extra specialist information elsewhere. The author feels that the benefits of comprehension for all far outweigh the benefits of encyclopaedic status for a few. The other policy is to use a uniform technical vocabulary throughout the book. Rather than describe practical mechanisms in terms of their particular jargon and acronyms, each is described using consistent terminology that refers back to the basic principles of communication. The author feels that this is desirable both to aid comprehension and also to stress that there are only a few basic ideas underpinning an apparent plethora of different mechanisms.

USING THE BOOK

The book has five main parts:

- Chapter 1: introduction
- Chapters 2–5: basic principles
- Chapters 6–8: computer networking
- Chapters 9–11: three case studies
- Chapter 12: standardization

The recommended way to use the book is to read the five parts in this order, that is, to read the book from beginning to end. Forward and backward cross-references in the text have been designed to be optimal for a sequential reader. However, some readers might want to vary the order, either because they are already familiar with some of the material or because they prefer to see complete practical examples before tackling the basic principles. This is feasible, making use of the index to check up on any unfamiliar terms that are defined in earlier, unread, chapters.

For a reader who wants to start with a large practical example, Chapter 9 is a good place to begin. It contains a case study of accessing a World Wide Web page. This is likely to be familiar from first-hand experience, and studying the communication problems involved gives an introduction to most of the themes of the book. It also allows a look at the workings of the Internet, which is the largest, and best-known, example of computer networking in the world. For a complete beginner, many of the technical details, and general issues, of Chapter 9 can be ignored at a first reading. There will still be enough accessible 'meat' left in the case study.

For a reader who seeks a basic understanding of the techniques of computer communications, without surrounding context, then Chapters 2 to 8 are enough, possibly illuminated by one of the three case study chapters. For a more rapid read, the many examples included in these chapters can be read selectively. To see why Chapters 1 and 12 are dispensible for such a reader, it is worth explaining their purposes.

Chapter 1 is mainly concerned with constraints placed on computer communications:

- by the activities that are using the computer communications; and
- by the physical communication media that are available to link computers.

That is, it presents computer communications as an activity that lies between the physical world of communication media and the logical world of computer users. If these external constraints are taken on trust, the details may be omitted by the reader.

Chapter 12 is also concerned with constraints, but of a different and more specialized type. If one computer is to communicate with one or more others, then there must be agreement on the way in which communication will take place, otherwise the computers will not understand one another. Humans are responsible for putting agreements in place, and this exposes computer communications to the process of international standardization. The result is further constraints on what can be done and, if these are taken on trust, the details may be omitted by the reader.

The material on basic principles contained in Chapters 2, 3 and 5 (Chapter 4 expands on the ideas in Chapter 3) forms the core of the book. Thus, it is essential reading. There is some scope for omission, especially by a reader who has some prior experience of computer communications. Each basic principle introduced is first motivated by examples of the principle occurring in human communications. Some readers may not need this motivation, or will be happy with fewer examples. Also, each basic principle is illustrated with examples of the principle occurring in computer communications. Readers may already be familiar with some examples, or will need fewer examples to understand the point.

The material on computer networking contained in Chapters 6, 7 and 8 introduces the topic in three stages, each stage involving more complex networking techniques. The treatment combines and applies the basic principles of computer communications. It is also essential reading, since networking is absolutely central to modern computer communications. There are many examples in these chapters, some fairly detailed, since the aim is to include all of the important network types. Thus, there are more examples than strictly necessary to illustrate the general principles. A reader may choose to omit some of these examples, perhaps because of prior experience or lack of relevance.

The three case studies in Chapters 9, 10 and 11 are included to show how complete computer communication systems fit together, in order to bridge the gap between physical communication media and the demands of computer users. Readers are recommended to follow at least one of these case studies, otherwise the material in preceding chapters may seem rather disjointed. An alternative, for students aided by an instructor, is to choose a completely different problem that has ready familiarity, and then analyse it in a similar manner to the case studies presented in the book.

EXERCISES AND FURTHER READING

There is an 'exercises and further reading' section at the end of each chapter. In total, there are over 300 exercises in the book. None of these exercises is of the type that is just designed to test whether the reader has read and absorbed the material contained in a chapter. Instead, the exercises are of three main types, intended to encourage the reader to:

- relate topics covered to real-life human communication experience;
- think more deeply about less straightforward topics; and
- find out more information about certain topics.

The exercises vary widely in difficulty, from those that require a few moments' thought, to those that have the potential to become term projects for a group of students. The instructor's guide that is associated with the book gives further guidance on this matter, to assist in the selection of exercises for classroom use.

The exercises, and further reading suggestions, have been carefully chosen to minimize the extra resources that are required by the reader. The author fully appreciates that many readers will not have access to well-stocked libraries containing specialist books, academic journals, technical reports and product documentation. Further, readers might not have access to state of the art computer communication facilities that allow practical experimentation. Therefore, only *one* primary resource is particularly desirable to underpin extra activities of the reader: access to the Internet.

The ability to access the Internet, and its facilities, allows the reader to experiment with many of the technical points covered in the book. Further, it makes a vast collection of information on computer communications readily and freely available to the reader. It seems particularly appropriate for a book on computer communications to make practical use of the technology described as much as possible. The most obvious Internet facility of interest for information gathering is the World Wide Web (WWW). Knowing well that the phrase 'here today, gone tomorrow' could have been invented to describe material on the WWW, the author has avoided giving explicit WWW references, except in a handful of very safe cases. The reader can use one of the WWW searching engines to track down useful material. The technical reasons behind the various frustrations of using the WWW are explained in the case study of Chapter 9.

A key information resource is the ever-enlarging collection of Internet 'Requests for Comments' (RFCs), which can be accessed through the WWW. This is a splendid collection, combining folk history, technical discussion and specification of Internet technology. Many of the end-of-chapter exercises involve looking at an RFC, either to skim or to digest. Of course, this introduces a potential danger of introducing an Internet bias to the material under consideration, but the author believes that the ready availability, and the well-written nature, of the RFC series make it a very worthy source.

Aside from the electronic information resources, the author has been very parsimonious in recommending other printed sources. Fourteen other books are mentioned, covering various areas of computer communications in more depth than is possible (or desirable) in this book. Books still have a place in the electronic information age, to give overall coherence to a body of material culled from numerous sources.

Departing from normal textbook tradition, this book does not include a lengthy section of worthy references to academic papers. This does not reflect laziness on the part of the author — rather, it is the product of experience and consideration. It would have been very easy to include references to the numerous papers consulted while writing the book, not to mention stitching on various 'standard' references. However, not only might many of the referenced papers be inaccessible to the average reader, but also the author knows that most references remain firmly unconsulted even if they are accessible. A further point is that, in this fast-moving subject area, many such references become rapidly out of date.

Having said all this, the book has a few references to particularly relevant journal articles, but these are in journals most likely to be present in a computing library. For general reading, *Computer Communication Review,* published by the ACM SIGCOMM (Special Interest Group on Communication), including the proceedings of the annual SIGCOMM conference, is recommended for the reader who wants to track the major computer communications issues of the day. Other possible medium weight sources on topical communications issues are the IEEE *Network* and *Communications* magazines, or general purpose computing magazines such as *Byte.*

ACKNOWLEDGEMENTS

The basic framework of this book has been in development for many years. A key influence was Mike Padlipsky, Old Network Boy of the Internet, and Scotch whisky buff. On a memorable visit to Edinburgh in 1986, he foresaw the rise of Internet standards and the demise of ISO standards — near-heresy at that time. This helped to reinforce my own feelings of dissatisfaction with the traditional way of teaching computer communications: a combination of 'encyclopaedic catalogue' and 'ISO reference model' (the latter term is explained in Chapter 12). The fourth year computer communications class at Edinburgh University acted as unwitting guinea pigs in 1987, when I tried out a radically different approach — an approach that has largely survived to underpin this book. The students of that class must be thanked for their patience with frequent retuning of the course agenda. Many thanks are due to Fred King, for acting as a sounding board on numerous occasions throughout this period.

My general approach to teaching computer communications has matured since this first experiment, assisted by a fertile environment at Edinburgh University. In the length of a single corridor, one can find expertise ranging from the underlying theory of concurrency to the practical operation of major state of the art computer networks. To single out just two people of many, Robin Milner's work on concurrency has been a significant influence on the basic principles distilled in this book, and Sam Wilson has been an invaluable source of information on the operational details of advanced computer networks.

During the production of this book, various people have been of assistance. At McGraw-Hill, Rupert Knight enticed me into beginning to write, and sustained the tricky early stages with his relentless enthusiasm. Alfred Waller has seen the project into port, sharing his invaluable experience and sage insights into publishing. Throughout, Ros Comer was a continuous source of guidance on the finer points of language and layout. Several anonymous McGraw-Hill reviewers have contributed many helpful suggestions at various stages. At Edinburgh University, Steven Haeck was kind enough to read the near-final manuscript, and make many constructive comments that improved the final manuscript. However, at the end of day, the translation of my thoughts on computer communications into camera-ready copy was my responsibility — aided, abetted and sometimes frustrated by LaTeX— and so I take the blame for anything that is not as it should be.

Finally, but most importantly, huge thanks are due to my wife Rosemary. She encouraged me to embark on the book in the first place, and then put up with a husband who not only disappeared to commune with the computer keyboard most weekends but also insisted on giving her regular progress reports on esoteric topics outside her areas of expertise. This book is dedicated to Rosemary.

Gordon Brebner
Edinburgh
January 1997

ONE

INTRODUCTION

The main topics in this introductory chapter are:

- benefits of communication
- history of human and computer communication mechanisms
- present-day uses for computer communications
- physical underpinning of computer communications
- the main principles of computer communications

1.1 BACKGROUND

Communication is beneficial for the human race. By communicating with one another, information can be shared — past experience, current affairs, predictions of the future — from here, there and everywhere. Also, resources and expertise can be shared, by communicating with the right people. Only hermits are noted for their ability to live satisfactorily in the absence of any communication with other people.

Similar observations apply to computers too. Individual computers are capable of gathering, processing, storing and distributing information, under the direction of humans. They are not only found in distinctive boxes with keyboards, mice and screens attached, or in large cabinets bristling with flashing lights and whirling tape drives, as seen in ageing science fiction films. Their basic information-handling capabilities can be harnessed for controlling other machinery, and so they are also hidden inside things like wristwatches, microwave ovens, central heating systems, factory production line equipment and nuclear power plant safety systems.

There are three main areas where benefits can be expected if one computer is able to communicate with others:

- it can get information that is stored by other computers;
- it can get other computers to do specialized work; and
- it can communicate with humans that use other computers.

The benefits need not only be in one direction — this computer can also export its own information, its specialized abilities and access to its human users. The outcome is a beneficial sharing of resources.

At present, it is usually the case that computers inside personal items like wristwatches, or inside domestic equipment like microwave ovens, do not communicate with others. However, this situation is on the point of changing, given that communication between the more recognizable types of computers has proved to be very useful, and that appropriate communication technologies are becoming available. It might also seem that a conventional home computer, or a single computer tucked into a musty office, is island-like, cut off from the world community of computers. However, this is a delusion, since such computers usually have an obliging communication mechanism: human beings transferring the latest fruits of the computing trade on floppy disk or compact disk.

This book is concerned with computer communications where there are no human middlemen, so that computers can converse directly with one another. Although human participation in the role of intermediary is being eliminated, it should not be forgotten that computers only communicate because humans have instructed them to and, moreover, that this is possible because humans have instructed the computers how to communicate.

The techniques covered arise from communications between conventional types of computer. However, they are equally applicable to the world of the future, where there will be things like intelligent houses with communication not just between domestic appliances but also with the fabric of the building itself.

The act of communication is not always easy for humans. For example, it is not feasible for every person in the world to communicate effectively with any other person whenever desired. Differences in culture, availability and physical location cause problems. Things get even tougher if one broadens communication to include other species of animal or plant, never mind any alien life forms that might visit or be visited.

Similar problems affect communication between computers. Because of this, the solutions used are rather similar to those developed by humans over the millenia, and so are usually familar to the neophyte from normal human experience. This is good, since it means that explanations of techniques used can be well motivated by appropriate examples from human communications.

Aside from reconciling differences between communicating parties, a further problem is how communication is physically achieved. For humans, there is a mixture of movement and media. People may move to make communication easier, or even possible at all. For example, one may climb a mountain to consult a guru, travel to work or attend a concert. Once the communicating parties are in suitable positions, communication might take place using sound through the air, visible sign language, telephones, television or sending items through the postal system. For computers, there is usually no movement involved in communication, robotics still being in its relative infancy. Because of this, computers communicate using media that physically reach the computers, and that are capable of transmitting computer conversations. Of course, this does not rule out the possibility of humans moving the computers to places where they can be reached by media.

The main themes of the book are concerned with these communication problems. The next two sections of this introductory chapter provide the enclosing framework for these themes. First, Section 1.2 surveys the main practical uses made of communications between computers. Such uses provide the rationale for indulging in computer communications in the first place, and indicate what are the main requirements of computer communications. After this, Section 1.3 surveys the physical media that can be used for the transmission of information between computers. This indicates the physical limits that apply to the process of communication and which, ultimately, affect what can be offered to the users of computer communication facilities.

Following Sections 1.2 and 1.3, Section 1.4 introduces the main themes of the book. The features of communications are classified under three main headings:

- **information:** the type of information that is communicated;
- **time:** when, and how quickly, a communication takes place; and
- **space:** which computers, and inter-connecting channels between computers, are involved in a communication.

There are two main problems associated with communication: achieving agreement between computers on the nature of communications; and implementing the required communications using available physical communication media. The processes of achieving agreement and implementation require a lot of human decision making prior to communications being possible. This entails human communications of a rather specialist kind. The section ends by introducing the ways in which these technical and political human communications have a major influence on what is possible and not possible in the computer world.

1.2 USES OF COMPUTER COMMUNICATIONS

Present-day uses for computer communications arise from a convergence between two different worlds. The first is a computer-centred world, where computers existed, and then it became convenient to inter-connect them. The other is a human-oriented world, where communication facilities existed, and then it became convenient to computerize these facilities. In the latter world, there was also convergence between telecommunications facilities largely used for inter-personal communication, such as the telephone, and broadcasting facilities largely used for entertainment, such as television.

One term often used to describe the fruits of this general convergence is the **global village.** It is interesting to note that, when this phrase was coined by Marshall McLuhan in 1964, it was based on an extrapolation of existing human-oriented facilities, and did not envisage the future involvement of computer communications. However, it is still equally apt to embrace computer-based facilities. The technology shift is captured in another, more modern, term: the **information superhighway**. This refers to the collection of communication technology and information technology that will be used to underpin the global village of the future. Just as roads underpin the movement of people and goods, so the information superhighway will underpin the movement of information.

To understand the current uses made of computer communications, and to point ahead to future developments, it is useful to conduct a brief historical survey of how different communication systems emerged and then converged. This follows in the next three sub-sections, which cover developments on the computer front, on the telecommunications front, and on the broadcasting front, respectively. A central concern is the demands that are made by these differing types of systems if they are to be realized using computer communication facilities.

To quantify these demands, time can be measured as usual, in seconds, or in fractions or multiples of seconds. Information can be measured using **bits** (short for 'binary digits'). The quantity of information available in one bit is that required to distinguish between two possible values. For example, one bit of information is enough to distinguish between 'on' and 'off', 'black' and 'white' or 'yes' and 'no'. A more precise definition of both information, and the bit as a measure of information, is given in Chapter 2. The bit is like any other unit, so it is convenient to talk of the kilobit (kbit), which is 1000 bits, and the megabit (Mbit), which is 1 000 000 bits. Note that these units are powers of ten, rather than the powers of two (2^{10} and 2^{20}) sometimes used in computing circles.

1.2.1 Computer-oriented communication

In the earliest days of computers, the roomful of boxes containing the component parts of the computer was the centre of attention. Information was supplied to a computer, and retrieved from a computer, by people who were physically present in the computer room. This is illustrated schematically in Figure 1.1(a). In the

(a) computer with only directly-connected input/output devices

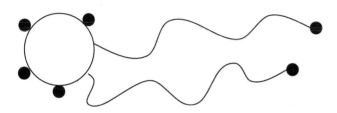

(b) computer with remotely-connected input/output terminals

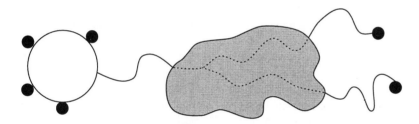

(c) computer with terminals connected via communication service

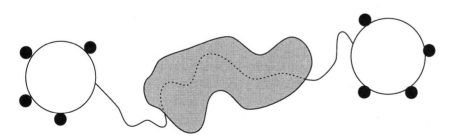

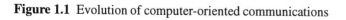

(d) computers connected via communication service

Figure 1.1 Evolution of computer-oriented communications

unusual event that there was any communication between computers at all, this was done by carrying paper tapes or magnetic tapes produced by one computer along to be read by another computer. Thus, people were the servants of computers as much as computers were the servants of people.

The first developments in communications were designed to make life easier for people, by eliminating the need to be physically adjacent to the computer when interacting with it. One of these was the introduction of a remote terminal — a typewriter-like device with a keyboard for input and a teleprinter for output — connected by a cable to the computer. This is illustrated in Figure 1.1(b). In essence, terminals were just extra peripheral devices for the computer, the only difference being that they were a longer distance away. Because of this, and the fact that terminals were electro-mechanical devices, they operated at very slow speeds in computer terms: a few characters input per second and around 30 characters output per second.

This basic capability evolved into a model of interaction with computers that survives to the present day. There are two major improvements. One is in terminal technology, with video screens replacing printing, giving a consequent large increase in output speed, and also with terminals containing computers as their controllers (or, indeed, computers just emulating the behaviour of terminals). Beyond terminals, full WIMP (Window, Icon, Menu and Pointer) interfaces can also be used, if the computer is sophisticated enough to interact via such an interface.

The other improvement is the one of interest to computer communications. This is that the direct physical cable between terminal and computer can be replaced by any communication channel capable of passing information between a terminal and a computer. Useable communication channels include links through the conventional telephone system or through specialized computer networks. This is illustrated in Figure 1.1(c). The **Internet** has become the best-known specialized mechanism for providing links between computers located in all continents of the world. In 1997, it was estimated that around 20 000 000 computers could make use of the Internet. The overall effect presented to a human user is still of a direct physical link between terminal and computer, but this is an illusion.

The increased distancing of terminal from computer presents two agreement problems that must be solved to enable communications. The first is that a wide variety of terminals (or computers masquarading as terminals) may be used, and each terminal's behaviour must be reconciled with that expected by the computer. This issue, and its resolution, is discussed more fully on page 44 in Chapter 2.

The second problem concerns the human using the terminal. This person issues commands to the computer, or inputs information to the computer, through the keyboard. Given that the terminal is behaving as though it is a real peripheral device of the computer, its user must know exactly how to interact with that particular type of computer. This is acceptable if only one computer is ever used via the terminal, but the whole point is that communications advances now allow all sorts of computers, all over the world, to be used. No single person can be expected to know how to interact with all of the different computers that are accessible.

A solution to this problem is for a particular user just to interact with his or her own familiar computer, and then that computer interacts with any other computers of interest on the user's behalf. This introduces some automation of the task faced by the human terminal user. In early versions of such facilities, the user's computer just pretended to be a human user of the distant computers. That is, it transmitted information that appeared to come from a terminal keyboard (albeit with a rather faster typist than normal) and then received back information by pretending to be a terminal printer or screen. This sort of interaction allowed files to be sent to, or fetched from, distant computers, electronic mail to be sent or received, or processing jobs to be given to distant computers and their results to be retrieved. These operations were directed by the human user.

In time, such procedures became refined, to eliminate the unnecessary humanization of the dialogue between computers. That is, new agreements to specify appropriate direct interactions between computers replaced the more verbose means used by humans. The resulting situation is illustrated in Figure 1.1(d). The new agreements covered things like transferring files, sending electronic mail and submitting jobs. These matters are discussed in more detail in Section 2.2.2. A further effect of such advances is that users gradually got more insulated from the vagaries of particular types of computer. For example, the same user interface could be used for something like electronic mail, regardless of what type of computers were involved in the mail transmission. Further extensions of the insulation process led to **distributed systems,** where the existence of a collection of computers is completely hidden from a user. Thus, it appears that a service is being provided by one single uniform computer system.

In distributed systems, it is fairly common for the computers involved to take on **server** and **client** roles. Servers have special capabilities, for example, storing particular information, performing particular processing of information, or having particular input or output devices attached. Clients can make use of servers by issuing requests, and by getting responses back. Overall, this is similar to the ways that humans directly interact with computers. However, in a distributed system, the computer client-server relationships are hidden from human users. The World Wide Web (WWW), considered in detail in Chapter 9, is an example familiar to many, in which the illusion of a world of information is presented to a user. In fact, the user's computer acts as a client, and computers throughout the world that store particular WWW pages of information act as servers for this client.

In summary, the evolution of computer-oriented communications has gone from a situation where a terminal was connected to a computer via a physical cable to a situation where a computer can be connected to numerous other computers via indirect channels supplied by complex communication systems. The types of information shared are more complex and the quantities are potentially huge. The speeds are related to computer rates rather than human rates. Early terminal links required the communication of only 300 bits per second. Two high speed computers might be able to cope with the communication of 1 gigabit (1 000 000 000 bits) per second nowadays. When pushed to one extreme, a collection of communicating computers can be placed into one box, with high speed electronic interconnections, to form a **supercomputer**: a new, high speed computer consisting of many individual computing elements. This is the point at which computer communications meets the subject area of parallel computation.

1.2.2 Telecommunications

Human communication via a medium is long-established. In the mid-nineteenth century, the **telegraph** became an economical method for the electrical transmission of information represented in Morse code, or similar, over long distances. In style, Morse code is not far removed from the bit-focused approach to information followed by computers. The difference is that information is expressed in a three-valued form (dot, dash and pause) rather than a two-valued form. The major problem with the telegraph was the need for an expensive physical cable between the communicating parties. The development of telegraph systems often occurred in parallel with railways, with telegraph cables being located alongside the railway tracks. The medium was used fairly wastefully, because the speed of transmission was limited to the speed at which people could press Morse keys to send information and the speed at which people could listen to the transmission to decode the information. Thus, the restriction came from human frailty, rather than any fundamental physical limitation of the cabling.

Around the turn of the twentieth century, the **telephone** became available. Like the telegraph, a large investment in cabling was necessary. However, unlike the telegraph, the telephone was an analogue device, in that human speech was directly converted to electrical waveforms for transmission. Experience of using the telephone until recent times confirms the main problem of analogue transmission: dubious effects on the electrical signal are reflected as audible crackles, whines, etc. on reception. Nowadays, most telephone systems employ digital transmission systems, with speech being represented using a series of bits. This is because it is rather easier to reconstruct a series of bits from a damaged electrical signal than directly encoded speech. A complex collection of automated national and international telephone exchanges provides a near-worldwide telephone communication system, going a fair way towards a goal of allowing every person in the world to speak to any other. Transmission is not exclusively via cabling, with radio transmission becoming increasingly used by mobile telephones.

The telephone system can be used as a vehicle for computer-oriented communication, for example, as a way of connecting a terminal to a computer. This does not involve computers speaking to one another in a human way over the telephone. Instead, electrical waveforms of a similar style to those used to encode speech are transmitted, but they in fact encode series of bits of computer-style information.

The modern telephone system has evolved to be like a distributed computer system. The digital exchanges are just special-purpose computers, and the links between them are similar to links between computers. To an extent, a fully digital service is offered to end users of the telephone system, through the **Integrated Services Digital Network** (ISDN) service, and other offerings. However, in the main, the typical subscriber is still supplied with a traditional analogue-style service. This is implemented using the underlying digital facilities. Thus, if such a subscriber uses the service for computer communication, the information is needlessly passed through an analogue form before being transmitted digitally.

One extension to the conventional telephone is to make it into a videophone, where a picture of the caller is transmitted as well as the speech of the caller. It is feasible to supply a videophone service over a normal telephone line, but the picture quality is fairly crude. A further catch is that few people possess videophone equipment. However, the main problem is that much more information must be transmitted to encode pictures and so, to achieve better quality, a better communication facility is needed. For speech alone, it is enough to transmit about 10 kbits of information per second. However, for video, around 60 kbits of information per second is a lowest limit, with 1500 kbits per second needed for decent quality.

Other telecommunication services have been developed in addition to the telephone, geared to transmitting textual-style information rather than direct human communications. One of these is **fax** (facsimile), which can be used to transmit an image of a text page from one telephone user to another. This involves a digital representation of the page in terms of bits of information being transmitted between two fax machines, encoded as a telephone-style electrical signal. The fax machines are really just special-purpose computers, which conduct a conversation over the analogue telephone system. Many computers can now emulate the behaviour of fax machines, so that they can transmit page images to other computers or fax machines. The number of bits of information needed to represent a faxed page varies according to the pattern of black on white on the page. However, on average around 200 kbits are sufficient for one page. More details of fax representation are given on page 53 in Chapter 2.

A further family of information transmission services are those based on **telex**. These are fairly close relatives of computer-oriented communications that involve textual information being sent between computers and terminals or between computers. Telex itself is a long-established telegraph service that allows textual messages to be transmitted between subscribers. There are about 1.2 million subscribers, mostly businesses, around the world. The messages can be composed of upper case letters, numerical digits and some punctuation symbols. They are transmitted in a digital form, but the transmission rate is only 50 bits of information per second, which equates to only 10 characters per second. A more modern offering is **teletex,** which allows a much larger character set, including graphical symbols and word processor style facilities for composing pages. It has a much faster transmission rate of 2400 bits per second, but this is still slow in modern computer communications terms.

Both telex and teletex are simple mechanisms for transmitting digital information between two specialized machines. **Videotex** is a variation that is akin to the client-server model for distributed computer systems. It is used over the normal telephone system. A user of videotex has a special video terminal attached to the telephone line, and this acts as the client. It is possible to telephone computer databases worldwide, and then conduct searches for information. The databases are the servers. One place where videotex is often seen is in travel agents, where video terminals are used to interrogate holiday and travel booking systems over the telephone. Videotex is an example of a **value-added service** operated by telecommunications providers. In addition to a raw communication facility, extra services are provided on top. This type of service illustrates a convergence of telecommunications with computer-oriented communications.

1.2.3 Radio and television broadcasting

Around the turn of the twentieth century, at the same time as the telephone was becoming available, a free worldwide communication medium was being harnessed: the electro-magnetic spectrum. This allowed the use of physical phenomena such as radio waves, rather than expensive cabling, for transmission. The first application was a wireless version of the telegraph. Although it seems strange with hindsight, the obvious use of this medium — broadcasting — was not realized until after the First World War. To an extent, this was due to an attitude that communication technology was only relevant for closed commercial and military applications, rather than the general public. The average person was still expected to travel in order to communicate directly. A further reason was that the cost of transmitting and receiving equipment was high, although there were no wiring charges. Since these early days, of course, there has been a massive growth in public radio and television broadcasting services.

Until the last years of the twentieth century, the sound and picture information of radio and television has been transmitted as an analogue signal. These signals are received using aerials, and then decoded. Digital transmission, with sound and pictures encoded using series of bits was only in its infancy in 1997. However, this is undoubtably the transmission style of the future, marking a convergence of radio and television with computer-oriented communications.

Compared with the transmission demands of the telephone and videophone, those of radio and television are much higher. In the case of radio, this is because the whole range of audible sounds must be transmittable, rather than just human speech. In addition, extras like stereo sound are also desirable. The information rate needed for high fidelity audio transmission is about 100 kbits per second — ten times the rate for telephony. In the case of television, pictures are larger, contain more detail and involve more motion over time. The information rate needed for television-quality video transmission is about 15 Mbits per second — again, ten times the rate for decent-quality video telephony. Note that these rates are for digital transmission, where extensive computer processing of the information can compress it to fit these rates. Without such processing, the required rate for raw video transmission is very much higher.

Just as videotex allows the retrieval of digital information using the telephone system, so **teletext** is a similar facility associated with television. Note the subtle naming difference between the television world's *teletext* and the telecommunication world's *teletex*. Teletext has far less scope for interaction, compared with videotex. All of the available information is continuously broadcast in sequence, and then the television receiver has to filter out the information of no interest. This is in contrast to videotex, where a user explicitly initiates searches for information of interest. However, like videotex, teletext illustrates a convergence with computer-oriented communications.

The rise of cable television illustrates a convergence between television and telecommunications. The same physical technology can be used to deliver both telephone and television services to an end user. It can also be used to deliver computer-oriented communication services, thus completing a convergence of communication facilities with rather different histories. Although analogue transmission is still used for cable television, in time, with the advance of digital television, the raw physical cable will become the bearer of a digital information bit carrying service, which is then used to support different value-added communication services. This will represent each user's slip road to the information superhighway.

1.2.4 Summary of uses of computer communications

From the point of view of a user, the interface to the converged world of computer communications, telecommunications, radio and television should allow easy access to information, communication facilities and entertainment. These will be underpinned by the techniques associated with computer communications today. Indeed, already, devices such as telephones and television sets are really just special-purpose computers from a technical communications point of view.

The blending of computers and everyday devices will continue, to a point where, from a user point of view, explictly visible computers are not likely to be present. A user will only see 'intelligent' devices; these will be internally underpinned by computers. The devices might be examples of **wearable computing,** embedded in clothes or spectacles (or even in the body), or of **ubiquitous computing,** embedded in numerous objects all around the user. One particular new type of device will be a **knowledge robot** — a device that can speculatively gather information that is likely to be of interest to its user, an improvement on conventional facilities that only gather specific information when ordered to.

One key component of a converged information interface is the use of **multimedia,** where a blend of sound, still pictures, video sequences and computer-encoded text can be used to present information. Another key component is in-

Table 1.1 Information sizes (in bits)

Type of information	Number of bits
Teletext page	7000
One second of speech	10 000
Computer-stored page of text	12 000
Character-based video screenful	15 000
One second of hi-fi audio	100 000
Typical faxed page	200 000
One second of quality videophone	1 500 000
Colour photograph	4 000 000
One second of entertainment video	15 000 000

teraction, where the user is not just a passive recipient as for radio and television. As a first step, **video on demand** is seen as an extremely important facility of the future. This allows users to select films for immediate viewing — akin to going to a local video library — rather than having to fit in with the whims of broadcasters. Later, it is possible to imagine that interactive films will be produced, allowing viewers to participate. Such facilities are rather akin to the interactive multi-player computer games, already available through computer communication facilities.

The demands placed on computer communication facilities will be great. From an information point of view, a rich variety of types of information must be dealt with, with some of these types being very large in terms of the number of bits of information required. Table 1.1 summarizes some approximate minimum information sizes, many of them quoted earlier in this section. From a time point of view, ideally communications must happen quickly enough that there is no unacceptable delay to a user (for example, within half a second for speech over a telephone) and then continue at a fast enough rate to ensure high quality. From a space point of view, ideally any computer-based device in the world should be able to communicate with any other, and adequate channels must be available to support these communications. Most of this is technically feasible, but only at extortionate cost. Therefore, the practical decision is on what facilities can be provided at a reasonable cost.

1.3 PHYSICAL LINKS

Just as human communication devices like the telephone and television are underpinned by physical media, including electrical cables and broadcasts in the electro-magnetic spectrum, so are computer communications. This book is not concerned with the physical details of how such media are put to work, but only with their observable behaviour: essentially, how good they are for transferring information. This encompasses issues like transfer rate, reliability and, of course, cost.

For any particular medium, its communication abilities can be categorized under the headings of information, time and space. The assumption for information is that any medium used is capable of transmitting bits, i.e., two-valued quantities, either individually or in a sequence. This service must be offered as a basis for computer communications, and nothing more complex is necessary. Therefore, the bit forms the atomic unit of information for communications. The details of how bits are transmitted, as signals, waveforms or whatever, are not relevant here. The service need not necessarily be completely reliable, in that some transmitted bits may be lost completely, changed in value or duplicated. However, it must be sufficiently useable that it can form a basis for reliable bit transmission.

From the point of view of time, there are two main measures of interest. The first is the **latency** of the medium, that is, the time taken between a bit being transmitted and being received. The fundamental physical limitation is the speed of light, which is approximately 3×10^8 metres per second. For example, this means that information cannot travel a distance of 1 km in less than 3.3 microseconds. Even if the raw medium has a latency as low as the limit imposed by the speed of light, the actual latency is likely to be increased by delays in the interfaces used to convert bits of information into signals for a medium.

The other important time measure is the **rate** of the medium, that is, the rate at which individual bits in a sequence can be transmitted. The subject of *discrete* information theory is covered in some detail in Section 2.3.1. Fundamental results from *continuous* information theory, a topic not covered in this book, can be used to derive limits on the maximum transmission rate possible over particular media, given things like the frequency range of waveforms used and inherent error rates.

Finally, from the point of view of space, physical media must act as channels for information being sent between computers. A fundamental property is the area or distance that can be spanned by a particular medium. Most media introduce a deterioration in the quality of tranmission as distance increases, to the point where practical bit transmission is no longer possible. Very often, this effect is related to the transmission rate — roughly, the faster the rate, the shorter the distance that can be spanned by most media.

Aside from area and distance, two distinctions can be drawn between the inherent properties of media. One distinction is between:

- **unicast**; and
- **broadcast**

media. The former allow one computer to send information to one other computer. This is the style of transmission familiar from the telephone system. The latter allow one computer to send information to a whole collection of other computers, some of which may choose to receive the information and others may not. This style is familiar from broadcast radio and television.

This first distinction should not be confused with a different distinction. It is between:

- **guided**; and
- **unguided**

media. The former involve some type of cabling that is used to physically guide transmitted bits along their channel. The latter involve transmission 'through the air' with the effects spreading over a wider area, which includes the destination(s) of the channel. Examples of the less obvious cases are radio telephones for unicast, unguided transmission, and cable television for broadcast, guided transmission. The term **wireless** is also used to refer to unguided media, reflecting the fact that no wires are involved. This category includes radio, which traditionally was synonymous with the word 'wireless'.

Section 1.3.1 below contains a brief survey of the most common types of raw physical transmission media, focusing on their bit transmission abilities. However, this is not the full story for real-life computer communications. In practice, there are legal and technical restrictions on the ways that media can be used. For example, there are no general rights that allow guided media to be laid across private property. In most countries, only national telecommunications organizations have the right to lay long distance cabling, or cabling along public roads. As another example, there are wide restrictions on transmission in the electro-magnetic spectrum. Different frequencies are allocated to radio and television companies, the military and other specialized uses. As a result, for communications outside small privately owned areas, services offered by approved operators must form the basis for computer communications, rather than direct use of raw media. Such services are discussed in Section 1.3.2 below.

1.3.1 Physical media and their properties

The basic characteristic of practical physical media is that, when in use, they involve the continuous transmission of a signal from one point to other points. This signal is used to encode sequences of one or more bits that form the information to be communicated. A distinction can be drawn between:

- **digital**; and
- **analogue**

signals. A digital signal has a small number of different possible values, for example, two different values for a binary digital signal. Over time, the value of the signal changes in discrete steps. An analogue signal has values drawn from a continuous range and, over time, the value of the signal varies over this range.

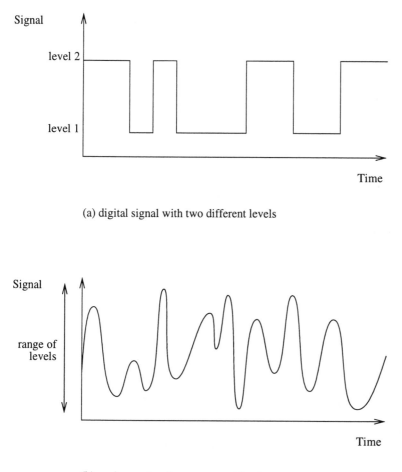

(a) digital signal with two different levels

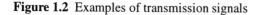

(b) analogue signal over range of levels

Figure 1.2 Examples of transmission signals

Figure 1.2 shows examples of these two types of signal viewed over a period of time.

While 'digital' and 'analogue' are familiar terms in the computer world, two alternative terms are often used loosely in the computer communications world: **baseband** and **broadband.** The looseness of usage comes from the fact that these terms have precise meanings in the telecommunications world, referring to the way in which frequency bands are used. However, this is not an immediate concern here, although a little more discussion of this point is contained on page 65 in Chapter 2. One further caveat is that the term 'broadband' is now often just used as an adjective meaning 'high speed', in contrast to another technical term, 'narrowband', which is now often used just to mean 'low speed'.

The most-used media for computer communications are examined below in increasing order of *potential* (rather than typical) transmission rate. This has the side-effect of presenting unguided media first, then guided media. The main benefit of unguided media over guided media is one of mobility. The positioning of computers is not constrained by the need to be adjacent to inter-connecting cabling. If the equipment needed for transmitting and receiving over an unguided medium is small and light enough, then mobility translates into true portability.

There are few suprises below, since all of the media should be familiar to the reader from everyday human communication systems. The presentation omits two more special-case suggestions that have arisen from the Internet community. One is Waitzman's 'avian carrier' system, proposed on 1 April 1990, which involves bits being transferred on a roll of paper attached to one leg of an avian carrier. The other is Eriksson's 'accoustical transmission medium' system, proposed on 1 April 1996, which involves bits being transmitted using Morse code on sound waves emitted by the beeper of a personal computer.

Radio

Radio waves are an unguided, broadcast, analogue medium. The two different bit values are represented by differing transmitted radio waveforms. In the past, radio has been a Cinderella of the computer communications trade, to a large extent restricted to recreational activities by amateur radio enthusiasts. One notable exception was the pioneering use of radio to link computers distributed over various Hawaiian islands, which has impacted on several general aspects of computer communications. However, radio is now 'going to the ball', with the advent of portable computers and mobile communications. Radio offers a free medium, and flexibility of attachment for computers.

Radio transmission is used over fairly modest distances, for example, within a town or city, or over a smallish rural area. Over such distances, transmission rates are slow — as little as 50 kbits per second. However, over much shorter distances like 100 metres, rates of 10 Mbits per second are feasible. This allows a little flexibility, combined with respectable (i.e., computer-like) speeds. Each transmitted bit arrives at the speed of light over a direct path, which means the shortest latency possible. The problem with radio over any distance is interference from geographical features like mountains and lakes, not to mention amateur radio enthusiasts, and this can lead to rather high probabilities of bits arriving at a receiver damaged. Whole sequences of bits might be lost in unfavourable atmospheric conditions. This is far less of a problem over short distances indoors.

Infra-red

Infra-red transmission involves higher frequency waveforms than radio, and is useful for communication over short distances. In the home, it is a familiar medium for things like remote television controls. A main difference from radio is that infra-red waves do not pass through solid objects, and so its use is confined to within rooms. Infra-red is not normally usable outdoors, since the sun's rays interfere. One advantage of the in-room use is that infra-red transmissions are not regulated, unlike radio, which involves the central assignment of radio frequencies to users.

Infra-red is an unguided medium that is naturally unicast, but can be used in either unicast or broadcast modes. For unicasts, an infra-red beam is just directly pointed at a receiver. For broadcasts, a diffuser is used to spread out the infra-red transmission so that it can be received throughout a room.

Digital transmission is used for slower infra-red transmissions, up to about 2 Mbits per second. This essentially involves switching the infra-red beam off and on to represent the two different bit values. For faster transmission rates, around 10 Mbits per second, analogue transmission has to be used, as for radio. Different waveforms are used to represent the different bit values. The main reliability issue involved in infra-red transmission is ensuring that the beam is not interrupted by a person or object standing in the way. This has the effect of causing the complete loss of a bit stream for the duration of the obstruction.

Microwave

Microwaves reside between radio and infra-red in the electro-magnetic spectrum. While infra-red is an attractive alternative to radio over very short distances, microwaves are usually used over long distances. However, some use is made of microwaves in the **Industrial Scientific and Medical** (ISM) waveband over short distances. One waveband is allocated to ISM worldwide, and transmissions can be made locally with no further regulation.

Microwaves travel in a straight line, and are narrowly focused, and so are an unguided unicast medium. Analogue transmission is used. Operating directly between two points, microwaves can be used for terrestial communications over distances of 50 kilometres or more. Until the advent of fibre-optic cables, discussed below, microwave transmission formed the backbone of the long-distance telephone system for several decades.

Microwaves are also used for communication between the earth and **geostationary satellites.** By bouncing signals off satellites, global distances can be covered. Direct transmission is not possible because of the curvature of the earth. When using such a satellite, most of the microwave's voyage is outside the atmosphere, and so interference is reduced and reliability is improved. Within the atmosphere, microwaves are prone to disruption, for example, by passing through rainfall. Geostationary satellites are positioned about 36 000 km above the equator, and revolve at the same rate as the earth, which means that they appear stationary from the ground. Positioning of such satellites is closely regulated.

In the 1990s, the first proposals for collections of **low-orbit satellites** were made. These orbit at an altitude of only 750 km, and so would not be geostationary. Instead, a sufficiently large number of satellites are in orbit that at least one is over any particular area at all times. The Iridium low-orbit satellite project, proposed by Motorola, and targeted at mobile telephones, involves 66 satellites (and six spares). The first three satellites were launched at the beginning of 1997, and the system was scheduled to come into service in 1998. The Teledesic project, proposed by Microsoft and McCaw Cellular Communications, and targeted at high speed computer communications, involves 840 satellites (and 84 spares), and is scheduled to supply a service by 2001.

Transmission rates of around 50 Mbits per second are possible with microwaves over long distances. However, when a geostationary satellite is involved as an intermediary, the latency in receiving each transmitted bit is around 250 milliseconds because of the round-trip distance of 72 000 km to and from the satellite. This latency is distinctly high in computer terms. It is much reduced if a low-orbit satellite is used instead, since the round-trip distance is reduced by a factor of around 50.

Copper cable

In contrast to the above unguided media, copper cable is the traditional, and most widespread, guided transmission medium. This transmits electrical signals that represent bits. Either digital or analogue transmission can be used, with the terms 'baseband' and 'broadband' normally being used for this medium. Digital transmission is used over relatively short distances, say up to 1 km, at which point interference effects impact on reliability. Analogue transmission can be used over longer distances, up to 100 km, since anologue signalling is more robust to interference.

Copper cables are packaged in two main forms, both familiar in the typical home. The first is **twisted pair,** where two wires are twisted together in a helix like a DNA molecule — for example, this is used to connect telephones to the exchange. The other is **coaxial cable,** where a woven cylindrical wire surrounds a stiff core wire — for example, this is used to connect television sets to aerials.

The most common type of twisted pair cable is **unshielded twisted pair** (UTP). This cable comes in various quality grades, the two most common being Grade 3 and Grade 5, which determine the feasible transmission rates and distances. Usually, four twisted pair cables are bundled together, a practice originating from telephone wiring. This is exploited in some computer wiring configurations, for example, a 100 Mbits per second transmission rate over 100 metres can be obtained by using three parallel Grade 3 UTP cables. The same parameters can be achieved using a single Grade 5 UTP cable. The contrast to UTP is **shielded twisted pair** (STP), where the twisted pairs are enclosed within a protective shield to reduce interference. STP cable has better transmission characteristics than UTP, but much greater care has to be taken when installing it.

Twisted pair is cheaper than coaxial cable. However, while both types offer 100 Mbit per second transmission rates over short distances, coaxial cable allows higher transmission rates than twisted pair over longer distances. For example, only about 4 Mbits per second over a distance of 1 km is feasible with twisted pair, whereas 100 Mbits per second is still feasible over this distance with coaxial cable. The electrical signals travel at about two-thirds the speed of light, so the latency suffered for each bit is not as low as possible, but is still acceptable. Reliability is significantly better than for the various unguided media above, as long as the signals are not subject to any external electrical interference.

Fibre optic cable

Fibre optic cable, which transmits light waves used to represent bits, is the guided technology that is becoming the norm for computer communications. The cable consists of a glass fibre, within a plastic coating to shield it from external light sources. Bits are transmitted digitally, with the absence or presence of a pulse of light used to represent bits of information. Fibre optic cable is now used extensively within the long-distance telephone system, replacing microwave transmission over long distances. It is also finding its way into many streets, albeit stopping short of individual homes, as cable television companies go about their business. Fibre optic cable is very cheap to manufacture, and has significant advantages over electrical cable. It can be used over distances of up to 100 km, and is almost totally reliable since the light signals are not affected by any external electrical or magnetic influences.

In 1997, rates of around 100 Mbits per second were commonplace, with 1000 Mbit per second transmission rates also being used in practice. In theory, rates of millions of Mbits per second are feasible. Each bit arrives at around two-thirds the speed of light, a similar latency as for electrical cable. The feasible transmission rates pose problems for conventional computers because transmission is slowed by the conversion between the electronic signals used by computers and the light signals used for transmission. One possibility under investigation is optical computing, where photonics are used as the internal basis of computers, rather than electronics.

Discussion

The basic properties of the media described above are summarized in Table 1.2, which gives typical present-day parameters, selected from wide possible ranges. The high cost assessment for satellite microwaves comes from the cost of launching the satellite. Once a satellite is in orbit, using it for bouncing microwaves is free. Note that assessing the unguided media as 'free' refers to the media themselves, and does not include the cost of transmitting and receiving equipment.

There is a place for all of these media types to support modern computer communications. For the future, radio and infra-red are likely to dominate mobile computer communications, operating over relatively short distances to allow connectivity between computers and fixed base stations connected to guided media. Fibre optic cabling is already becoming the dominant medium for longer distance communications, and is increasingly used for shorter distance guided communication as well. It has the advantages of speed and reliability over electrical cabling,

Table 1.2 Properties of raw communication media

Medium	Area	Mbits /second	Latency	Reliability	Cost
Radio	100 m 20 km	10 0.05	Minimal	Variable	Free
Infra-red	room	2	Minimal	Good	Free
Terrestial μwave	50 km	50	Minimal	Variable	Free
Satellite μwave	global	50	High	Good	High
Twisted pair	100 m 1 km	100 4	Low	Good	Cheap
Coaxial cable	1 km	100	Low	Good	Moderate
Fibre optic	10 km	1000	Low	Excellent	Cheap

as well as being cheaper, smaller and lighter. Although fibre optic cable has taken over some of the roles originally envisaged for satellites, there will still be a case for using satellites for use in terrain where cabling is hard, or where mobility or broadcasting over a wide area is a desirable feature.

1.3.2 Physical communication services

For communications outside a privately owned area, it is normally necessary to use physical links provided by a public telecommunications carrier (also known as a 'bearer', a 'common carrier' or a 'public carrier'. The organizations that act as carriers are known by various generic names, including **TELCO** (short for 'telecommunications company'), **PNO** (short for 'public network operator') and **PTT** (short for 'post, telegraph and telephone' administration). Until the mid-1980s, it was the case that a single carrier dominated in each country. In the USA, this was AT&T (operators of the Bell telephone system). In other countries, this was usually a branch of the government concerned with all types of communications (as the PTT name suggests).

In 1984, the telecommunications market in the USA was transformed, when AT&T was divided into many smaller component companies which, in turn, encouraged the growth of realistic competitors. Two examples, which have grown into serious long-distance carriers in the USA are MCI and SPRINT. MCI's name stood for 'Microwave Communications Inc' originally, reflecting its use of microwave transmission — however, it has now switched to fibre optic cable. The 'SPR' in SPRINT's name stands for 'Southern Pacific Railway', reflecting its use of cables run alongside existing railway tracks on railway land. In other countries, privatization and deregulation have also led to the growth of a competitive market. This includes telecommunication services being offered by companies involved in providing cable television.

Below, standard facilities that are made available by public carriers are outlined. These range from the normal telephone, through Integrated Services Digital
Network (ISDN) services as an advance on the telephone, to dedicated digital
transmission lines leased from the carriers. All of these are guided unicast media
from a computer communications point of view. The telephone involves analogue
transmission, and the others involve digital transmission. Note that this section
only includes raw transmission capabilities provided by public carriers. Other,
value added, services are described at other points in the book.

Telephone system

The worldwide telephone system, now sometimes known as the **Plain Old Telephone System** (POTS), was designed to transfer human speech. It allows a telephone user to make a call to the huge number of other telephones located around
the world. This includes mobile telephones that are linked to the main wiring
infrastructure of the POTS using radio transmission.

The POTS transmits analogue electrical waveforms representing speech, with
wave frequencies in the range 400 to 3400 hertz being allowed to pass through
the system. Although the user interface is analogue, long distance links now use
digital transmission, as do many local parts of the system. This aspect of the
telephone system is outlined below. There is more discussion of the POTS itself
in Section 7.5.1.

Here, the interest is in how the analogue user interface can be used to support
computer communications. In fact, this is little different from directly using media
such as radio or broadband coaxial cable. Different signal waveforms are used
to represent the two bit values. A device called a **modem** (short for 'modulator
demodulator') is used to perform the conversion. In the case of the telephone, a
waveform representing a sequence of bits must always have its frequency within
the allowable 3000 hertz range, otherwise it will not pass through the POTS intact.
This fact places fundamental limitations on the rate at which information can be
transmitted.

For typical parameters of the telephone system, information theory indicates
that the maximum rate is around 35 000 bits per second. Modems able to operate at
a 33 600 bit per second rate are available. This is essentially a practical upper limit
for modem technology over analogue telephone links and, in fact, such modems
may need to operate at lower rates (e.g., 28 800 bits per second) over typical
telephone links. Modems capable of a 56 000 bit per second rate began appearing
at the end of 1996, but these rely on there being an all-digital telephone link;
such links have a 64 000 bit per second theoretical upper limit. Other modems are
advertized with higher rates than information theory allows, but these are achieved
by compressing the bit stream before it is actually transmitted through the POTS.

Another feature of the POTS as a computer communication medium is that it
has a fairly high error rate. This arises from the fact that perfection is not deemed
necessary in order for humans to make sense of speech transmitted over the

telephone. People can ignore hisses, crackles and the loss of occasional fragments of speech. In bit terms, a telephone link might have an error rate as bad as 1 in 10 000 bits sent being damaged, although modern digital links are far better. This compares with a figure of 1 in 10^{12} for many direct media, such as coaxial or fibre optic cable. Thus, computers must take considerably more precautions for error detection and correction when using a telephone link. Appropriate techniques are described in Section 3.3.3.

Integrated Services Digital Network (ISDN)

As mentioned above, the internal workings of the telephone system are now predominately digital in nature. When this is extended to provide a digital interface to the end user, the resulting service is called the **Integrated Services Digital Network** (ISDN). The name reflects the fact that the service can be used to transmit not just speech, but also other types of information. Anything that can be encoded using digital bits can be transmitted using ISDN. As with the telephone system, calls can be made to any other subscriber that is connected to the ISDN system. A further benefit of all-digital transmission is that the reliability of bit transmission is dramatically increased.

Design work on the original ISDN service began in 1984, with the aim of establishing a worldwide system by the beginning of the 21st century. This happened before it was clear that there would be a huge demand for both high speed computer communications and real-time video communications in the future. Thus, the service was rather modest in terms of bit transmission rate. It is now known as **Narrowband ISDN** (N-ISDN), to reflect its modesty. The N-ISDN basic service, designed to replace a normal telephone link, offers two 64 Kbit per second digital channels for voice or data transmission, together with a 16 Kbit per second control channel. Each 64 Kbit per second channel is suitable for transmitting speech in digital form, so essentially N-ISDN offers the equivalent of two telephone lines instead of one.

N-ISDN services are now offered in many countries. To allow use of N-ISDN when a remote computer supports it, and the normal telephone service otherwise, devices called **hybrid modems** are available. These allow a computer to use a single piece of equipment to interface with the telephone system, and to gain speed benefits from N-ISDN when possible.

Unfortunately, because of the slow rate at which the details of N-ISDN were agreed, it had already become technically obsolete before a significant number of products had become available. Not only are the bits rates too slow for computer and video communications, but they are also far higher than necessary for speech communication, given rapid advances in the art of speech compression.

The result is that **Broadband ISDN** (B-ISDN) has been developed, with the intention of producing a new system that is robust against future technological developments. The basic standard B-ISDN rate is to be 155 Mbits per second, with higher rates possible. This gives a service that is comparable to most of the raw media mentioned in the previous pages. The major underpinning technology for B-ISDN is described in detail in Section 7.5.6. Its worldwide implementation presents major technical challenges — in contrast to N-ISDN, which used fairly safe and well-proved technology. In the first instance, B-ISDN services will only be available to fixed-position users, but work is underway towards a high speed mobile B-ISDN service by 2005. This involves extension of the technology used for mobile telephones.

Leased lines

The telephone system and N-ISDN offer slow information transmission rates to the end user, with B-ISDN offering the prospect of a dramatically improved service in the future. If a user requires faster transmission rates at present, then the norm is to make use of a **leased line** provided by a public carrier. This gives a direct link between two fixed points, with transmission rates of up to 45 Mbits per second being available in North America and Japan, and up to 34 Mbits per second being available elsewhere in the world. The difference in rates is due to the slightly different digital transmission mechanisms used in the two different parts of the world. These will be described in outline here.

The first point to note is that the digital transmission channels used by the public carriers operate at bit rates far higher than those seen by individual end users. The basic idea is that a collection of end user calls share the same transmission channel. User calls are given transmission rights in rotation, each one transmitting for a very short period before allowing the next one in turn to transmit. After each one has had a turn, the cycle repeats again. The process is designed so that each user call can transmit at its required bit rate. For example, if 100 users share a 10 Mbits per second channel, then each user is able to transmit up to 100 kbits per second.

The benefit of this approach for the public carrier is that it needs only to provide a total bit rate high enough to support the maximum number of simultaneous user calls that can be active at one time. It does not use a separate physical channel for each different user, with a large proportion of these channels being idle at any given time.

In North America and Japan, the basic unit of digital transmission is known as the **T1 carrier**, which operates at a rate of 1.544 Mbits per second. For telephones, this is used to support 24 separate calls. The value of the analogue signal for each call is sampled 8000 times per second, and each value is represented by seven bits of digital information. This allows 128 signal values to be differentiated — in essence, the analogue signal is turned into a 128-level digital signal with level changes allowed 8000 times per second. There is a little more discussion of how this sampling is done on page 70 in Chapter 2. An extra bit of control information is added to each sample value, giving eight bits per call. The total of 192 ($= 24 \times 8$) bits of call information, plus one extra control bit to give 193 bits in total, is

Table 1.3 Standard carrier channel types

Carrier name	Mbits per second	Shared use
T1	1.544	–
T2	6.312	$4 \times$ T1
T3	44.736	$6 \times$ T2
T4	274.176	$6 \times$ T3
E1	2.048	–
E2	8.848	$4 \times$ E1
E3	34.368	$4 \times$ E2
E4	139.264	$4 \times$ E3
E5	565.148	$4 \times$ E4

transmitted 8000 times per second, i.e., every 125 microseconds. This leads to the overall 1.544 Mbit per second rate.

In other parts of the world, the basic unit is known as the **E1 carrier**. This is broadly similar, except that there are 32 channels, rather than 24, and eight bits of sample information are used rather than seven. The overall transmission rate is 2.048 Mbits per second.

In turn, the basic T1 and E1 carrier channels can be combined to share faster channels. Table 1.3 shows other members of the Tn and En families. Note that the higher members of the families have rates that are a little larger than the exact total of the rates that can share these members. This is to allow for the transmission of extra bits of control information. Apart from T1 and E1 themselves, T3 and E4 are the most common types found in practice for leased lines.

These families, where higher transmission rate channels are shared by lower transmission rate channels, are known as the **Plesiochronous Data Hierarchy** (PDH). The term 'plesiochronous' means 'nearly synchronous'. This refers to the fact that the exact bit transmission rates on each of the lower rate channels might differ fractionally, perhaps one is a tiny bit faster than another or a tiny bit slower. The transmission mechanisms used for the PDH carriers make allowances for these clocking variations. Such differences in transmission rate arise from the fact that a different clock is used on each channel. The problem of reconciling different views on absolute time is one of the major topics examined in Chapter 3.

The problem of inconsistent clocks in PDH stems from the fact that it was an evolutionary technology, which had to take into account existing channels with separate clocks. For modern transmission over fibre optic cabling, it was possible to install new systems that have a single master clock. This gives the **Synchronous Data Hierarchy** (SDH) instead. The work on SDH originated from a standard developed by Bellcore in the USA called **SONET** (pronounced 'sonnet', and short

Table 1.4 SONET carrier channel types

Carrier name	Mbits per second
STS-1	51.84
STS-3	155.52
STS-9	466.56
STS-12	622.08
STS-18	933.12
STS-24	1244.16
STS-36	1866.24
STS-48	2488.32

for Synchronous Optical NETwork). This more poetic name is usually applied to the technology, rather than 'SDH', since the two only differ in minor ways.

The basic SONET STS-1 channel operates at 51.84 Mbits per second. This allows the transmission of 810 8-bit values every 125 microseconds. It is sufficient to carry one T3 channel, with some extra padding, or up to 32 separate T1 channels. There is then a hierarchy of channel rates based on STS-1, with a member named 'STS-n' being able to carry n different STS-1 channels. For example, an STS-3 channel can carry three STS-1 channels, then an STS-12 channel can carry four STS-3 channels. The defined SONET rates are shown in Table 1.4. The different rates chosen cover a range that includes values desirable to different constituencies in Europe, Japan and the USA, reflecting the fact that SONET and SDH were designed to be unifying standards for high speed digital transmission. For each defined STS-n bit channel, there is a corresponding OC-n optical channel that is used for the actual transmission. Thus, for example, an OC-3 optical channel is used to implement an STS-3 bit channel.

For SDH, the defined channels and their rates are the same, except that the hierarchy starts at the STS-3 rate, rather than the STS-1 rate. Thus, the basic SDH STM-1 channel operates at 155.52 Mbits per second, and then the other members of the family corresponding to those shown in Table 1.4 are STM-3, STM-4, STM-6, STM-8, STM-12 and STM-16. Note that the STM-1 basic rate is the same as the standard rate chosen for the B-ISDN service — this is no coincidence. However, as will be explained in detail in Section 7.5.6 and in Chapter 11, B-ISDN is not implemented directly using SDH channels. The 'STM' name stands for **Synchronous Transfer Mode,** that is, continuous transmission at fixed time periods. However, an alternative — **Asynchronous Transfer Mode** (ATM) — lies between the B-ISDN service and the physical STM transmission system. This adds a considerable amount of flexibility.

1.4 HOW COMPUTERS COMMUNICATE

From Sections 1.2 and 1.3, it might seem that there is not much to say about about what lies between the requirements of users and the physical transmission services available. However, this is certainly not the case, and indeed forms the subject matter of the rest of this book. If computers are not involved, then it is reasonably practicable to supply user services directly using physical media, as illustrated by the traditional telephone system and radio and television systems. However, when computers are introduced into the picture, there is much more scope for introducing subtlety into the communication system, and also into what can be done using the communication system. This turns the resulting provision into a distributed information processing system.

In this section, there is a short summary of the main concepts that underpin the material in the rest of the book. The material is not organized following an approach that is traditional in almost all books on computer communications: begin with the capabilities of the physical transmission medium, and then show how to add successive layers of enrichment, so that eventually it is possible to construct a service that is adequate to directly support the requirements of a distributed information processing system. Such an approach is founded on two practical considerations. One is the practical engineering of computer communication systems. The other is a conceptual model that is often used by people when they are agreeing on how computer communications should work — a topic covered in detail in Chapter 12.

These practical considerations are both important. However, this book is designed to focus on the more fundamental principles of the subject. That is, the nature of computer communications, the general problems that arise, and then the ways in which the problems can be solved. The book does not start from the solutions, and then try to find some principles. A beneficial side-effect of this approach for the beginner is that most of the principles have close analogies from human communications, and so are familiar from everyday conversation and interaction.

The three main strands of the book are introduced in Sections 1.4.1, 1.4.2 and 1.4.3 below. The first is concerned with the nature of communications. The second is concerned with how communication can be achieved. The third is concerned with human influences on the first two strands.

1.4.1 Information, time and space

A first strand of the book is the identification of three separate components of a communication between computers: **information, time** and **space**. These have already been referred to in this chapter, and correspond to the 'what', 'when' and 'where' of communication.

Information has the starring role, in that the sharing of information is the whole purpose of communication. The nature of information is the topic of Chapter 2.

Much of the content of Chapter 2 is not just of interest from a communications perspective. It is also relevant to a general study of how computers process and store information. Another way of putting this is that information can have an existence independent from any act of communication.

Time refers to how a communication proceeds over the course of a time period. This includes issues like when a communication time period begins and ends, whether communication is continuous over the time period, the time taken for information to travel between computers, and the rate at which information is communicated. Chapter 3 is concerned with the fundamental problems associated with time. Then, Chapter 4 introduces the most common ways in which the temporal behaviour of computer communications is organized. This involves combining various issues considered independently in Chapter 3. Note that time is a concept completely dependent on communications, unlike information and space. It captures the nature of particular acts of communication.

Space refers to the computers involved in a communication, and the channel used to connect them. This first requires some means of uniquely identifying computers and channels. There is a natural measure for time — seconds etc. — but, for space, a natural measure like physical position is not very appropriate since computers are likely to move around. The role of the channel in any communication is to supply a means by which information can flow between computers, in the way demanded by the pattern of information sharing in the communication. For example, in the simplest case, with only two computers sharing information with each other, a channel must allow information to be transmitted in each direction between the computers. Chapter 5 is concerned with the fundamental problems associated with space. Like information, spaces can exist independently of particular acts of communication.

It is convenient to imagine an external observer of a communication. After the communication has happened, the observer will know both the static components of the communication — the information communicated and the space it was communicated over — and the dynamic component — the timing of the act of communication. This particular frame of reference is important. First, the exact nature of a communication might not be known before it happens or during its happening, since many communications are evolutionary in nature. Second, the exact nature of a communication might not be known to any of the computers involved at any stage, since many communications are distributed in nature.

1.4.2 Agreement and implementation

For the three components of communications, there are two issues that are important to each:

- **agreement**; and
- **implementation**.

Agreement is concerned with reconciling differences of viewpoint between computers on the components of communications. This characterizes the distinctive feature of computer communications, that separates the area from the study of individual computer systems. There are two basic ways in which agreement can be achieved: absolute and relative. With absolute agreement, communicating parties operate within some pre-agreed context. With relative agreement, communicating parties reach agreement when an act of communication actually takes place. For example, when a communication begins, one party may just agree to cope with the style of communication being used by another party. However, note that all relative agreement arrangements have some sort of enclosing absolute agreement. This specifies exactly how the relative arrangements will work.

For information, it includes establishing agreement on the type of information that is being communicated. For time, it includes agreement on the time periods for communications and the rates at which communication takes place. For space, it includes agreement on identifiers for computers and channels, and the sets of computers involved in particular communications. These matters are discussed in Chapters 2, 3 and 5 for information, time and space respectively. The meta-problem of how humans agree on how computers should communicate is covered in the final chapter, Chapter 12.

Implementation is concerned with how the desired information, time and space properties of a communication can be realized using some other type of communication, usually something simpler and more physically practical. This is not just the issue of how the communication can be implemented at all, but is also concerned with the *quality* of the implementation. This encompasses issues like speed, latency, reliability, security and cost. Implementation problems are, of course, familiar from the study of computer systems in general. There, a reasonably attractive interface is provided to a human user. However, ultimately, this must be implemented using the very simple electronic logic circuits within the computer. In computer communications, there is a similar process, in that a communication service that is attractive to a user must be implemented in terms of a raw capability that can move bits of information between computers.

For information, two main implementation issues are how complex information types can be represented in terms of simpler types, and how information can be transformed to raise the quality of its communication. These points are addressed in Chapter 2.

For time, the fundamental implementation issue is concerned with how complex communications proceed over time. It may be possible to implement a single communication using a series of simpler communications over separate time periods. Conversely, it may be possible to combine several separate communications into one simple communication that is easier to carry out. Another implementation technique is to adjust the actual timing of a communication to improve its quality. These matters are discussed in Chapter 3, and then practical examples of common implementation packages are described in Chapter 4.

For space, the main implementation problem is concerned with producing channels with the required behaviour, given that physical media have a limited repertoire. This leads to the major area of **computer networking**, where collections of computers and channels act in combination to implement required communication channels. Another implementation technique is to combine several channels so that they share the same channel or, alternatively, to split one channel so that it uses several channels in parallel. The basic space implementation issues are discussed in Chapter 5.

The important space-related topic of computer networking is covered in considerable depth in this book. The extent to which the computers in a network must play an active part in the implementation of communications varies. Chapters 6, 7 and 8 introduce networks in three stages, each requiring a larger contribution from computers as network components, not just as network users. These are accompanied by examples covering the most common types of network occurring in practice. The most famous example included in Chapter 8 is the **Internet**.

Throughout Chapters 2 to 8, implementation is concerned with small steps from the more ideal to the more practical. The first four of these chapters deal with implementation steps for only one of the information, time or space components, as appropriate. The latter three chapters involves a synthesis of information, time and space implementation steps. The ultimate goal of implementation is to completely realize a required user communication in terms of physical communication facilities. To show how this can be done, using small implementation steps, Chapters 9, 10 and 11 contain three case studies, each illustrating a complete implementation process. The details of the case studies have been chosen to reflect applications of contemporary interest, implementation techniques that are of the most frequently encountered types, and physical facilities characteristic of the present day (or future).

1.4.3 Human influences

Chapters 2 to 5 present the fundamental principles of computer communications, then Chapters 6 to 8 introduce somewhat more practical issues related to computer networking, and finally Chapters 9 to 11 describe complete engineering solutions to specific computer communications problems. Underpinning the practical solutions to the fundamental problems is another process: human communication about how to achieve agreement and implementation within computer communications.

Chapter 12 is concerned with this process, which is an example of **standardization**. For many everyday objects, it is useful to have national and international agreements on standards, e.g., for screw sizes or for car tyres. Computer communications is an important special case in that, without standardization of some kind, computers cannot be made to converse unless they are completely identical in all respects.

The main products of standardization are **communication protocols**. These are collections of rules that govern the ways in which computers interact with one another when communicating. The reason for using protocols is to allow communications with particular requirements to be implemented using communications that are more practically realizable. Thus, a protocol is a collection of prior agreements that is designed to assist in the implementation process. These agreements may involve information, time and/or space matters. The word 'protocol' is used because it matches the human usage, where protocols may be official formulae or bodies of diplomatic etiquette, that is, formalized rules of behaviour between people. This is exactly what is needed for communicating computers.

Before Chapter 12, which is the final chapter of the book, is reached, most chapters make reference to techniques and technologies that have been standardized. In order to avoid confusion, the names of the four most important standardization forces will be mentioned here:

- the Institute of Electrical and Electronics Engineers (IEEE);
- the International Organization for Standardization (ISO);
- the International Telecommunication Union, Telecommunications sector (ITU-T); and
- the Internet community.

These bodies crop up frequently in the text, but the names of some of the other bodies discussed in Chapter 12 also appear. In the case of the Internet community, there are also frequent references to the Internet Request For Comments (RFC) documents, which constitute an easily accessible, and very interesting, collection of literature on computer communications.

Finally, in this introduction, a word of warning is necessary over terminology. Human agreements on how computers should converse have to be precise, otherwise they would be unsuitable for computer use. However, the vocabulary used by humans to discuss computer communications is rather less than precise. Often, different words are used for the same concept. Sometimes, the same word is used for different concepts. This can be very confusing, not just for the novice but also for the expert.

It is common for textbooks to discuss the concepts of a particular area of computer communications using the vocabulary normally used within the area. This book takes an opposite line. The same term is used for a concept, wherever it crops up. Potentially, there is a problem, in that the reader may have vocabulary problems when going on to read specialist literature for particular areas. However, it is felt that the benefits of consistency and uniformity throughout the book far outweigh this concern.

One example will serve to illustrate this point. In this book, the word **message** is used to denote any unit of information that is shared between computers. For example, at the simplest level, a message might just be a single bit of information. In the literature, at least eight other words or terms are used in various different contexts: cell, datagram, data unit, frame, packet, segment, slot and transfer unit. It is interesting that the term 'data unit' originated in an international standardization effort that sought to invent a neutral term, different from all those in use already. However, it is a bit of a mouthful, so 'message' has been preferred as a more natural term to use here.

1.5 CHAPTER SUMMARY

Communication is beneficial to the human race, and can also be beneficial for computers. It enables the sharing of information, and of information processing facilities, as well as giving access to human computer users.

Future usage of computer communications is centred around the notion of a worldwide information superhighway. This has its roots in three existing areas: computer-oriented communications, which arose as a means of connecting computers and their human users; telecommunications, which arose as a means of connecting human users; and radio and television, which arose as a broadcast education, information and entertainment service. A convergence of the three technologies has occurred, as a result of the omnipresence of computers inside everyday objects and devices. The resulting provision, incorporating interactive, multimedia access to information, makes far higher demands on computer communication systems than has been the case in the past.

Computer communications are supported by physical media linking the computers. These may be guided or unguided, unicast or broadcast. The main media types are radio, infra-red, microwave (including satellites), copper cable and fibre optic cable. These have differing properties, in terms of information transfer rate, distance spanned and reliability. For communications outside a small, privately owned area, the services of public carriers are usually required. These include the telephone system or more computer-oriented services such as ISDN and leased lines. For the future, Broadband ISDN is set to become the standard digital information transmission provision.

To bridge the gap between the required user information service and the physical media, the techniques of computer communications are required. The key components of any particular communication are the information shared, the time period of the communication, and the space of computers and channels involved. Carrying out the communication requires agreement on the nature of the communication between the computers and channels, and the implementation of the required communication in terms of communications that can be physically realized. Communication protocols are central to this — these are rules for how communications proceed, agreed by human designers. The construction and selection of protocols, and other areas of agreement on computer communications, involves standardization processes. These impose constraints on what is possible and not possible in practical computer communications.

1.6 EXERCISES AND FURTHER READING

1.1 Survey the ways in which you communicate with other people, noting the differing purposes of communications with different people.

1.2 Make a list of all of the computers that you use at home, at work, or elsewhere. Remember to include the computers hidden inside everyday objects. Consider any benefits that would arise if any of the computers was able to communicate with any of the others.

1.3 Suppose that you wish to communicate with someone who is standing beside you, but that neither of you speaks the same language. How might communication take place?

1.4 Discuss the ways in which technologies of all kinds have made the notion of a 'global village' possible.

1.5 Collect references to the 'information superhighway' from newspapers and magazines. What do these popular media usually mean when they use this phrase?

1.6 If you have access to the Internet, compare remote `telnet` access to a computer with direct WIMP access to the same computer. Why is `telnet` access still in existence?

1.7 Discuss possible benefits of client-server distributed computer systems, compared with stand-alone computer systems.

1.8 Fax has become a very popular telecommunications service. Why has this happened, given that there are alternatives such as teletex or computer-based electronic mail?

1.9 If a cable television company operates in your area, find out the range of services that are offered. Do these extend beyond television?

1.10 If you have access to the World Wide Web, search for extra information on any topics in this chapter that are of interest to you. How easy do you find such searching, and how useful are the results?

1.11 Outline the facilities that you would want your own personal knowledge robot to have. How easy do you think it would be to implement these facilities?

1.12 Around 65 000 films have been made since the introduction of cinema. Do you think it is feasible to operate a video on demand service that allows any user to see any film that they want at any time?

1.13 Radio waves, infra-red and microwaves are three examples of electromagnetic phenomena used for bit transmission. How useful do you think visible light would be as an unguided transmission medium?

1.14 Obtain information about the various low-orbit satellite proposals, in particular find an explanation of why the Iridium project involves only 66 satellites, whereas the Teledesic project involves 840.

1.15 Compare the information sizes given in Table 1.1 and the information transfer rates for media given in Table 1.2. Comment on the types of multimedia user facilities that could be provided if the raw media were directly used to transfer information between users.

1.16 Find out who the main public telecommunications carriers are in your area. What services do they offer to the end user, and how much do these services cost? In particular, is an ISDN service available to the domestic telephone user?

1.17 If you have access to a modem, look at the list of facilities and options it offers to its user. How many of these are comprehensible to a person who knows little about the details of computer communications? What is the maximum information transfer rate possible?

1.18 From Table 1.3, work out the number of bits per second that are overheads for control purposes when (a) 24 T1 channels share a T3 channel; and (b) 64 E1 channels share an E4 channel.

1.19 The OC-3 carrier is shared by three OC-1 carriers. There is a variant called 'OC-3c' that is used as an unshared carrier. The total user information rate of OC-3 is 148.608 Mbits per second, whereas the user information rate of OC-3c is 149.760 Mbits per second. This compares with the raw bit rate of 155.520 Mbits per second. Calculate the number of control information bits that are used in each 125 microsecond cycle by (a) OC-3; and (b) OC-3c.

1.20 Look at everyday objects to see if they have been subject to national or international standardization. What compromises do you think were made in order to standardize any particular object?

Further reading

At this stage, there is no need for further reading on the core areas covered by the rest of the book. However, some readers may be interested in finding out more about the surrounding context: the uses of computer communications, and/or the physical support for computer communications.

For the reader who is interested in the uses of computer communications, there is a variety of possible sources. First, on computer-oriented communication, the wide range of books about the Internet and, specifically, about the World Wide Web, convey a reasonable impression of how computer communication facilities are presented to users currently. However, any reader who is exposed to a computing environment with communication facilities should encounter specific technical descriptions targeted at users. For general issues involving the telephone system, radio and television, and the convergence of these technologies with computer technologies, more specialized reading is available. *Multimedia Networking* by Agnew and Kellerman (Addison-Wesley 1996) deals with this convergence, and the emergence of interactive, digital, multimedia communications.

For the reader who is interested in the physical underpinning of computer communications, there is a great deal of technical literature. This book deliberately skims over the details of how bits can be transmitted; other books focus entirely on this issue. Basic technologies, such as electronics, photonics and radio, are covered in the electrical engineering or physics literature. *Telecommunication System Engineering* by Freeman (Wiley 1996) covers the technological bases for telecommunications and computer communications.

For more details of the techniques used for transmitting binary information using digital and analogue signals, and in fact for more technical details of many topics introduced in this book, two general purpose textbooks of interest are *Data Communications, Computer Networks and Open Systems* by Halsall (Addison-Wesley 1996) and *Data and Computer Communications* by Stallings (Prentice-Hall 1997).

CHAPTER TWO

INFORMATION

The main topics in this chapter about information are:

- agreement between computers on information types
- examples of communicated information types
- overview of information theory
- transformation of information to improve quality: compression, redundancy for error detection, and encryption
- examples of representing complex types using simpler types

2.1 INTRODUCTION

The purpose of communication between computers is to share information. The act of communicating can be viewed as an act of *information sharing*. That is, after communication, the computers involved have increased the amount of mutually shared information. The idea is illustrated in Figure 2.1. This simple model will

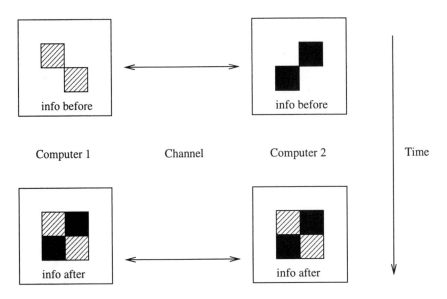

Figure 2.1 Communication model involving information sharing

be extended twice, in Chapters 3 and 5, first to address timing matters, and then to include more than two computers.

The figure is rather idealized, since it suggests that there are always two collections of information beforehand, and then a single shared collection of information afterwards. For example, it may be the case that some of the shared information is only generated during the course of the communication, as a result of information already shared. Alternatively, it may be the case that, during or after communication, one computer destroys its copy of the information, since it is transferring its information to the other. However, for any communication, it is possible for an external observer to identify the information that a computer has gained from the communication taking place. A desirable feature of a communication system is that the communicating parties themselves can also be certain of the results of a communication.

The idea of an **external observer** of communications is important to understanding the concepts described in this chapter, and subsequent chapters. To fully understand the nature of communications, and the effects of communications, it is necessary to stand back and observe all parties involved in the communication.

To allow communication, a **channel** is necessary. This connects the communicating computers together, and allows flows of information to take place between the computers. For this chapter, a simple model of a communication system will be used. It consists of two computers, connected by a channel. A **communication** is an act of information sharing between the two computers via the channel. The channel can be regarded as the component that carries out the functions of the communication system.

In general, a channel must allow information to flow in either direction between the computers, at the same time if necessary. This is important if it is to be used for true *sharing* of information. However, to illustrate many points as simply as possible, communications will just involve a unidirectional information flow, so that always one computer is sending information to the other. For such communications, a simpler channel can be used.

Note that taking sharing of information, rather than just transferring of information, as the basic model is important. This is because it is not possible to model all types of sharing in terms of two directions of transferring: if there are any relationships between the two directional information flows, they cannot be accounted for. Further, the sharing model extends more naturally when more than two computers are taken into account later.

To achieve communication between two computers, there are two fundamental issues that need to be addressed. First, it is necessary to have agreement on the type of information that is being shared. Second, it is necessary to have implementation that allows required types of information to be shared using channels that allow different — usually simpler — types of information to be communicated. The implementation work may not just be concerned with achieving communication, but also with achieving it at some particular level of quality.

2.2 AGREEMENT ON INFORMATION TYPES

A first question to answer is: 'what is information?' The term 'information' is used in preference to 'data' throughout this book for a good reason. Loosely speaking, information is more interesting than data. That is, whereas data might be seen just as a random collection of values drawn from some data type, information is a collection of values with some observable structure or properties. The structure or properties will be of interest to the communicating computers. More importantly, they can usually be exploited in the communication system to aid agreement or implementation.

The subject area of **information theory**, founded by Claude Shannon in 1949, provides a formal definition of information. This can be used to underpin some of the issues that arise in communications. However, in general, the term 'information' is used in this book with a rather less formal meaning. A brief introduction to discrete information theory is given in Section 2.3.1, where it is shown that some of its most important results have implications for what it is possible to implement in communication systems.

Given philosophical agreement on the nature of information, a more practical concern is for the two communicating computers to agree on the particular type of information that is being shared. As stated in Chapter 1, for the purposes of this book, there is only one type that is assumed to be universally understood by all computers and channels. This is the **bit** type, i.e., a type that has two possible values. These two values will be denoted here by **0** and **1**. The use of any other type requires an agreement to be established between the two computers, whether it is just a string of bits or is a higher level type such as an integer or a structure. If the two computers are functionally identical, the agreement is easier, as long as the channel is appropriately equipped to deal with the range of types that is supported by the computers.

2.2.1 Absolute and relative information types

In some case, agreements on the type of information communicated are static. That is, they are fixed in advance of the communication, so both communicators know exactly what type of information is involved. This sort of absolute agreement is fairly common at all levels of communication.

The alternative is for agreement on information to be established dynamically when a communication actually takes place. Then, the beginning of a communication determines the type of information involved. There are two possibilities:

- a recipient recognizes an information type from what is received, and copes with it; or
- an initial part of the communication is concerned with negotiation of an agreement between the communicators.

In either event, this agreement relative to the communication must rely on some enclosing prior agreement on the mechanism that is being used. Thus, there is always some absolute agreement prior to communication. This agreement is an inherent feature of the communication system, fixed by its human designers.

2.2.2 Examples of communicated information types

Bits, bytes and words

As mentioned above, the **bit** will be regarded as a universal data type. If strings of bits are communicated, then there is scope for variability. Most computers have the **byte** built in as a standard type. This is a string of eight bits. The word 'octet' is sometimes used as a more internationally neutral term for the byte, to stress the point that there are eight bits. There are also other, larger units, including the **word** and larger variants such as 'longwords' and 'quadwords'. These are strings of 16, 32, 64, or perhaps more, bits. The first agreement problem is on the lengths of words, if these are to be used as a standard type. Usually, a word is the standard size of value that can be processed by a computer, so naturally this varies between computers.

The second agreement problem is on the ordering of bits within bytes or words. This is the so-called **bit sex** problem. In the case of bits in a string, computers usually have a notion of a 'most significant bit' and a 'least significant bit'. These might be at the beginning and end, respectively, of a communicated string, or vice versa. If two computers differ on their interpretation of bit strings, then an agreement is necessary. In a similar way, when a string of bytes is communicated, there is a **byte sex** problem. This involves reconciling forward or reverse orderings of bytes within words.

Characters

While bits and bytes are the basic data types for computers to operate on, a **character set** is the basic data type for interactions between computers and humans. For interactions in languages such as English, an appropriate character set contains letters ('A', 'B', . . .), digits ('0', '1', . . .) and punctuation symbols (e.g., '.', ';', '('). A widespread character set is ASCII (American Standard Code for Information Interchange), also known as IA5 (International Alphabet 5). This contains 128 characters, which include those appropriate to English, as well as 33 special characters which were designed for computers to read, rather than humans. Many of these computer control characters are now of historic significance only.

An older alternative to ASCII is EBCDIC (Extended Binary Coded Decimal Interchange Code) which has more than 128 characters, but is not a strict superset of ASCII. Thus, if two computers are to communicate characters, but use different character sets, they must establish an agreement on which character set to use or, alternatively, agree that the communication system is responsible for converting one family of characters to the other as best as possible. ASCII and EBCDIC are both targeted at English language applications.

There are extended versions of ASCII that include characters from other, related alphabets, for example, letters with accents such as 'ü'. The most ambitious character set is given by the International Organization for Standardization (ISO) standard 10646-1, which is a universal character set that encompasses symbols needed to represent most known human languages. This set is also known as **Unicode,** since it results from the convergence of the efforts of ISO and an organization called the Unicode Consortium. The character set contains well over 30 000 characters, arranged into different alphabets.

Data types

Bits and bytes are the low level data types that are handled by most computer hardware. However, at the instruction set level or higher, a richer range of data types is supported. These include integers and reals, as well as means of constructing new data types, such as arrays, records and structures. The actual range of data types varies, both between different computers' instruction sets, and between high-level programming languages that are used for programming the processing of information on computers. To allow communication of information expressed in terms of higher level data types, agreement is necessary on a shared standard.

External Data Representation (XDR)

This standard was introduced by Sun Microsystems, and is used by some kinds of Internet communications. This standard is heavily influenced by the fact that most of the communicating computers will be executing programs written in the

C programming language, and so its range of data types is very similar. The basic types are:

- 32-bit signed integer in range $[-2147483648, 2147483647]$
- 32-bit unsigned integer in range $[0, 4294967295]$
- enumeration: this is a subset of signed integers
- boolean: this has the value FALSE or TRUE
- 64-bit signed or unsigned 'hyper' integers
- 32-bit floating point
- 64-bit double-precision floating point
- fixed-length opaque: this is a fixed-length byte string
- variable-length opaque: this is a variable-length byte string
- string: this is a variable-length ASCII character string.

More complex types can be constructed using the basic types. The constructors are:

- fixed-length array of homogeneous elements
- variable-length array of homogeneous elements
- structure: ordered collection of heterogeneous elements
- discriminated union: integer selector followed by alternative types
- void: a 'no data' type.

A reader familiar with C, or any similar high-level programming language, such as Pascal, will recognize these basic types and constructors easily.

Abstract Syntax Notation 1 (ASN.1)

This international standard is a more ambitious alternative to XDR. It originated as a component of a telecommunications message handling standard, but now has much wider use by a variety of communication protocols. The overall style of the data types supported is similar to that of XDR. However, the types are defined in a way that is less tied to a particular programming language or computer instruction set. There are also some extra data types. The basic types are:

- integer: any length of integer
- enumeration: a subset of integers
- boolean: with value FALSE or TRUE
- real: any length of floating point
- bitstring: variable-length bit string
- octetstring: variable-length byte string.

There are also some pre-defined types:

- character string: variable-length string of characters from 12 possible character sets, including IA5 and ISO 10646-1
- generalized time: date, time and time differential in ISO 8601 format
- universal time: date, time and time differential in Coordinated Universal Time (UTC) format.

UTC is mentioned on page 84 in Chapter 3. ISO 8601 uses the same time system, but has a slightly different format for expressing times as strings. For example, in the simplest case, 15.40 (3.40 pm) Greenwich Mean Time (GMT) on Tuesday 27 February 1997 could be represented as '199702271540Z' in ISO 8601 format, and as '9702271540Z' in UTC format. For times given in other time zones, the 'Z' is replaced by a time differential that shows the difference, in hours and minutes, of the time zone from GMT. For example, a time in a time zone that is six hours behind GMT would be followed by '-0600'.

A characteristic of all these ASN.1 types is that they have variable lengths. Unlike many of the XDR types, they are not restricted to particular sizes. More complex types can be constructed using the basic and pre-defined types. The constructors are:

- sequence: ordered collection of heterogeneous elements
- sequence of: ordered collection of homogeneous elements
- set: unordered collection of heterogeneous elements
- set of: unordered collection of homogeneous elements
- choice: range of alternative types
- null: a 'no data' type.

Here, sequence is similar to XDR's variable-length array, and sequence-of is similar to XDR's structure. A reader familiar with any high-level programming language should be able to see how that language's data types might be mapped onto those of ASN.1.

Programs

XDR and ASN.1 are standards for high-level language data types. Until 1995, there were no standards for high-level language programs, because there was no agreed common denominator of the wide range of different programming languages in use. However, the Java language, devised by Sun Microsystems, became an overnight *de facto* standard, in order to allow programs to be obtained from the World Wide Web. Java is an object-oriented programming language that has C++ as an ancestor. Small Java programs called **applets** can be executed on any computer that has an interpreter for the Java language. These Java interpreters provide the means for programs to be independent of the underlying computer system. The details of Java are a topic for a computer programming text, rather than a communications text, so are omitted here. However, there is a little more discussion of Java on page 317 in Chapter 9, which contains the World Wide Web case study.

Communication information

When computers communicate, there is normally an exchange of **messages**. As stated in Chapter 1, the term 'message' is used throughout the book to mean any unit of information that is sent from one computer to another. In different, more specialized, contexts, other terms, such as 'cell', 'frame', 'packet' or 'data unit' are used. Before communication can take place, there must be agreement on the format of the messages to be exchanged. This might either be a permanent absolute agreement, or might be agreed at the beginning of the communication. At more physical levels of communication, a message is likely just to be a string of bits. At higher levels, it can usually be regarded as a structure, to use the XDR term. That is, a message consists of a series of components, each component being a simple data type. The range of types and constructors provided by ASN.1 is sufficient to describe sorts of messages used in virtually all communication protocols.

There are some classic message formats, which will be encountered frequently at various points in this book. For now, only the names will be noted. Details of binary representations of these formats appear in Section 2.3.2, and the full pictures emerge later in the book, as the purpose of the messages becomes clear. The classics include:

- the ATM 'cell', used at a low level in high speed communications;
- the HDLC 'frame', used as a low-level message that is independent of underlying physical communication media;
- the IP 'datagram', used as the basis for information transfer in the Internet; and
- the combined TCP/IP 'segment', used as the basis for reliable information transfer in the Internet.

One further classic is the IEEE 802 'frame', but discussion of it is deferred until Chapter 6.

There are other kinds of information concerned with computer communications that have to be agreed upon. They include time-related information, such as absolute time measurements or time difference measurements. They also include space-related information, such as absolute physical positions or computer naming conventions. These topics will be introduced in context later in the book.

Computer system information

As explained in Chapter 1, the aim of computer communications is not just to share information, but also to share control. That is, to allow one computer to harness capabilities of another computer in a natural way. To share information and control between computer systems, it is necessary to agree on which components of computer systems are of interest, and then to agree on the characteristics of these components. Components include not just the information processing capabilities of computer systems and the information storage capabilities of computer systems, but also peripheral devices that allow interaction with humans and the outside world.

Usually, one or more standard models of particular components exist. A model of a component is referred to as a **virtual** component. That is, it does not actually exist, but it has the characteristics of a real component. Thus, operations on a virtual component can be implemented using the operations available from a real component. Later, in Section 7.2, an example specific to computer communications is mentioned: the **virtual circuit**. This is a model of an idealized channel connecting two computers. Here, the examples are applicable to individual computer systems.

Shared information processing

Batch job processing is an old-fashioned model of how computer systems process information. A job is input to the system, which then cogitates, and finally outputs results. The job is formulated as a sequence of commands of the kind a user might issue to a command line interpreter (a shell, in Unix parlance). The ISO standard Job Transfer and Manipulation (JTM) service, which provides facilities for submission of jobs and collection of results in a system-independent manner, is an example of a standard that might be used as an agreement between two computers that are to interact in a batch processing style.

A more modern model, which has exactly the same sort of basic operation but at a finer grain, is based on the idea of a procedure call. A procedure is called with some parameters, is then executed, and finally returns results. The ISO standard Remote Operations Service (ROS) service or the Sun Microsystems' Remote Procedure Call (RPC) standard are examples of possible agreed mechanisms for two computers to interact in a procedure-calling style. Note that in both cases — jobs and procedures — the agreements just cover the delivery of input/parameters and output/results. The actual range of computations performed is defined by the executing computer, and the requesting computer just makes use of these as appropriate.

Shared information storage

For information storage, all computer systems have some kind of file system. One possible basis for agreements would just be for a computer to advertise operations on its file system along with any other computational operations it offers to the outside world. However, given the pervasiveness of file systems, there are standardized virtual file systems that are computer system independent. Examples include the ISO standard File Transfer, Access and Management (FTAM) service and the Internet File Transfer Protocol (FTP) standard. These include standard operations such as file copying, deletion and renaming. Communicating computers can agree to use such standards, and then must map the virtual file system operations on to their own file systems. These standards are based on the notion that each computer system has its own independent file system.

A more modern approach is to establish an agreement that makes the separate file systems all appear to be part of one overarching file system. Thus, communicating computers must map the virtual file system on to their own file system, and map their own file system into the overarching file system. An example is Sun Microsystems' Network File System (NFS), which has a virtual file system that is heavily influenced by the Unix file system. This type of service is beginning to get away from the normal domain of computer communications, and is heading towards the realm of distributed systems.

Shared peripheral devices

Almost all computer systems have some kind of peripheral devices that allow humans to interact with the computer system. Most also have other devices that allow interaction with the outside world in various ways. As described in Chapter 1, the evolution of computer-oriented communications began with the connection of user terminals to computers. Because of this, one of the most standardized types of computer system application, along with the virtual file system, has been the virtual terminal. This is a system independent model of a human interaction device.

Communicating computer systems must map the operations of a virtual terminal on to their own interactive terminal operations. The earliest virtual terminals, for example the ITU-T (International Telecommunication Union, Telecommunications sector) XXX ('triple X') standard or the Internet TELNET standard, had operations to model the behaviour of a simple text line oriented terminal. The more advanced ISO standard Virtual Terminal (VT) service has a more sophisticated model, with a two-dimensional screen oriented terminal allowing cursor movement operations.

The Digital (DEC) VT100 terminal specification has also become a *de facto* standard for screen oriented terminals. Related to this type of terminal are form oriented terminals, which allow fields to be defined on screen for form-filling. The IBM 3270 terminal has become a *de facto* standard virtual terminal of this type.

Moving beyond simple terminals, WIMP (Window, Icon, Menu and Pointer) systems are now the norm. An example of a standard virtual WIMP system is the X-Window system. Such a standard is massively more complex than a simple virtual terminal standard. In practice, it is very hard for a computer using a different WIMP system to map faithfully all of the X-Window operations on to the different system. However, it is feasible, as evidenced for example by the existence of X-Window emulators for computers running the Microsoft Windows system.

Apart from terminals, there is not a large number of standards for other types of peripheral device. Printers are one example of a pervasive type, and the Internet LPD (Line Printer Daemon) defines a modest virtual printer model. This is mainly based on the traditional notion that lines of text characters are sent to a printer, as one would expect from the use of the term 'line printer'. However, it does include a small range of specialized information formats for printers, including the concept of printing arbitrary patterns that are described in text using the Postscript language. This makes the standard more useful for modern dot matrix printers or laser printers.

An important class of devices are those used in Computer Integrated Manufacturing (CIM), for example, numerically controlled tools, automatic guided vehicles and robots. The ISO Manufacturing Message Standard (MMS) defines the Virtual Manufacturing Device (VMD), which is an abstraction of any factory-floor device, and includes operations for things such as starting and stopping, and status checking. There are also refinements of this model for specific categories of device. For example, there is a virtual robot model, which includes operations for controlling robot arms and manipulating robot tools. MMS and the VMD are discussed in more detail in Chapter 10, which contains a case study concerned with factory automation.

2.3 IMPLEMENTATION OF REQUIRED INFORMATION TYPES

Agreement on information is only concerned with how the communicators agree on the type of information that is being shared. However, it is usually the case that the information has no direct physical representation which is suitable for real communication. Thus, some implementation work must be carried out. It may be necessary to transform the information in some way to ensure that its communication can be carried out correctly and efficiently. This matter is covered in Section 2.3.1. Also, it is usually necessary to represent complex types of information in terms of simpler types. Ultimately, information is represented by bits, which are regarded as the atomic unit of information here. Changes in representation are covered in Section 2.3.2. Sometimes, transformation of information and change of representation are combined, since the transformation necessitates the use of a different representation.

Virtually all of the implementation methods described here also have applications to information processing and storage, as well as to communications. This reflects the fact that information is a static component of communications, which can lead an independent existence in different circumstances.

2.3.1 Transformation of information

When information is transformed to aid implementation, an essential feature of the transformation must be that it preserves the information content. There is no point in managing to implement a communication if, in fact, a computer is receiving the wrong information. The meaning of 'information' will vary, depending on the context of a communication. Before looking at transformations that are of practical use, it is worth having a brief look at the topic of discrete **information**

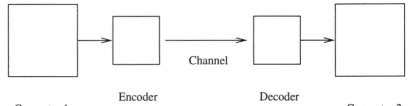

Channel

Encoder   Decoder

Computer 1   Computer 2

Figure 2.2 Coding and communication

theory. This theory yields some general results about the nature of information, and the transformations that can be applied to it.

Information theory

In discrete information theory, the notion of a **code** is central. A transformation is a mapping from strings in one alphabet to strings in another (possibly the same) alphabet. A transformation that is done before communication takes place is an **encoding**. The set of strings that can be targets of the mapping is called the code. A transformation to restore the original information after communication is a **decoding**. The overall process is illustrated in Figure 2.2.

Three particular types of transformation are of interest in discrete information theory. First, transformations that compress information, and so reduce the quantity of communication. Second, transformations that add redundancy to information, and so allow communication over error-prone channels. Third, transformations that scramble information, and so allow secure communication over insecure channels. Fundamental results from discrete information theory give absolute limits on the extent to which such transformations can be done. For example, these rule out compression techniques that compress information so much that more than one value is mapped to the same communicated value. This would cause confusion for a recipient of the information.

To understand these results, it is first necessary to have a formal definition of information. Suppose that there are n different values that a piece of information might have. Further, suppose that the probability that the i-th value (for $1 \leq i \leq n$) is actually communicated on any particular occasion is p_i. Then, the *information*, measured in bits, provided by the occurrence of the i-th value when a communication takes place is:

$$\log_2 \frac{1}{p_i}$$

('$\log_2$' denotes the base-two logarithm function). Here, it is assumed that $p_i > 0$ — otherwise, there is no point in including this i-th value.

As an example of how this definition works, consider the communication of a single bit. If the value **0** always occurs, its probability is equal to 1, so the information provided by its occurrence is equal to zero. This is saying that, because the information communicated is always the same, there is no actual information content. Alternatively, suppose that the values **0** and **1** occur randomly. Then, the probability of a **0** is equal to 0.5, so the information provided by its occurrence is equal to one bit. This is saying that, because the information communicated is always completely unpredictable, every bit matters in terms of information content.

The idea of **entropy** is central to information theory. Entropy can be regarded as a measure of uncertainty about values that are communicated or, alternatively, as a measure of the randomness of values communicated. Precisely, entropy is the expected amount of information conveyed by a communicated value. It is equal to

$$\sum_{i=1}^{n} p_i \log_2 \frac{1}{p_i}.$$

It is possible to show that the entropy is maximal when all of the possible values occur with equal probability, that is, when communicated values are as random as possible. The effect of communication is to reduce entropy — when information is shared, recipients have less uncertainty about the information. Of course, if entropy is zero, there is no need for communication because there is no uncertainty to be resolved.

Discrete information theory uses the idea of entropy to establish fundamental limits on what is possible in terms of compression, error correction and encryption. The most interesting point made by these results is that the limits actually exist. In practical circumstances, the results are usually not applicable, for two main reasons. First, it is often not possible to derive a reliable probability distribution for the occurrence of different values. Second, the theory shows that the limits are achievable, but it does not yield practical coding mechanisms to do this.

The main result relevant to compression is the **source coding theorem**. Loosely stated, this result says that, if the entropy of a data type is e, then e transmitted bits, and no fewer, are enough to communicate each value. This places a limit on the degree of compression that is possible while still retaining error-free communication. As an example, suppose that the data type is the letters of the Roman alphabet: 'A', ..., 'Z'. Also, suppose that letters occur with the probabilities that are usual in normal English text. For example, the probability of an 'E' is 0.1305 but the probability of 'Z' is only 0.0009. The entropy in this case is equal to around 4.15, and so this number of bits is necessary and sufficient to communicate each letter.

A little interpretation of this fact is necessary. Clearly, it is not possible to communicate 4.15 bits (or, indeed, any non-integral number of bits). However, it would be possible to communicate 100 consecutive letters using around 415 bits of information. In fact, the source coding theorem relies on encoding a large block of values as one entity, so that the encoded result has an integral length. This gives the appropriate average compression for each individual encoded value. Later in this section, a practical compression technique that can approach this limit is described.

The main result relevant to error correction is the **channel coding theorem**. It is based on a measure known as the **channel capacity**. For any channel, this is the maximum reduction in entropy that can be achieved by communication. That is, before communication, the receiver is uncertain about what values will be sent and, after communication, it should be less uncertain. If the channel is completely reliable, then there will be no remaining uncertainty, of course. Loosely stated, the theorem says that, if the channel capacity is c, then each value transmitted can communicate c bits of information, and no more. This places a limit on the amount of redundancy that is necessary to ensure error-free communication. The redundancy is the difference between the number of bits transmitted per value and the number of bits of information communicated per value.

The exact definition of channel capacity is not included here. It depends upon the characteristics of the channel, in terms of probabilities of values being changed when traversing the channel. There is no real practical point in computing its value. Rather, it is just necessary to be aware of its existence as a theoretical limit. This is because, first, the model of how errors occur — values are independently damaged — is often not realistic and, second, practical error correction codes seldom approach the theoretical limit.

Finally, information theory has contributed to an understanding of encryption to obtain security of communication. Roughly, the aim is to maximize uncertainty for any eavesdroppers on communication. The model used by Shannon is now somewhat dated. It was predicated on a **ciphertext only** attack, that is, the eavesdropper only has access to encrypted information, not to pairs of original and encrypted information. A central result was that, if information is perfectly compressed, then knowledge of the overall distribution of communicated information, together with possession of sample encrypted information, does not reduce the uncertainty about the encryption method used. Nowadays, the security of encryption schemes hinges upon making cryptanalysis by eavesdroppers very much harder than normal decryption by legitimate receivers. Shannon was aware of this idea, and his thoughts on it have influenced present-day approaches to security.

Compression

The role of compression is to reduce the amount of communication required in order share information. The assumption is that there is some redundancy within the information, and that this can be removed before transmission. There are limits on how far information can be compressed without confusion being introduced for the recipient, and Shannon's source coding theorem illuminates this. In practice, however, a distinction is made between **lossless** and **lossy** compression. The former ensures that information is received intact. The latter allows a small amount of confusion to occur, assuming that the receiver is able to cope with some loss of information. As a general rule, lossless compression is needed for computer-oriented communication, whereas lossy compression is sometimes acceptable for human-oriented communication.

Most of the practical compression techniques used are not just applicable to the communication of information. They are also appropriate for the storage of information, to reduce the amount of space required. The opposite is not entirely true, since compression techniques for storage may be too time-consuming for use in communication. (Of course, if one wants to generalize, a storage medium can just be viewed as a type of communication channel with an arbitrarily long delay, in which case communication and storage are the same thing.)

Compression techniques tend to be specific to particular data types, taking advantage of special characteristics of the data type or of known distributions of values from these data types. A general theme is that some values occur more frequently than others. Then, roughly speaking, the information is transformed so that the most common values are mapped to short representations, and the less common values are mapped to longer representations. This means that, with very high probability, compression will be achieved. When all values occur with about the same frequency, compression is not likely to yield benefits.

Three general techniques are described below: Huffman coding, run length encoding and difference encoding. Then, there are three examples of how information compression is carried out in a special-case way for particular applications: fax transmission, video transmission and Internet message transmission. For these examples, only a bare outline of the full compression method is given, since it is necessarily application specific. However, there is enough detail to illustrate how the various general-purpose compression methods are used in combination. More details of these methods are explored in three exercises at the end of this chapter, respectively, Exercises 2.21, 2.22 and 2.23.

Huffman coding

Huffman coding gives a general-purpose way of deriving variable-length codes based on occurrence probabilities of different values. The probabilities might be known in advance, for example, the occurrence probabilities of the different letters of the alphabet in English text are well-documented. Alternatively, an estimate of the probabilities might be obtained by measurement of occurrence frequencies in communication that has taken place in the past. This idea can be extended to **dynamic Huffman coding**, where the code is changed as communication takes place, reflecting the changing frequencies of values transmitted.

The construction of a Huffman code involves building a tree which has leaves corresponding to different values in the information data type. The leaves of the tree are labelled with three items:

1. the value that the leaf represents;
2. the occurrence probability of the value;
3. the binary code to be used for the value.

The nodes of the tree are labelled with an occurrence probability, which is the total probability of all leaves that are below the node. Thus, the root node is labelled with the total probability for all values, and this will be equal to one.

The first step of the construction algorithm is to create a set of one-leaf trees, one for each value in the data type. Each leaf has its value filled in, and also the known probability of that value occurring. It has an empty bit string attached as a code at this initial stage.

Then, there is a sequence of steps, each one combining two trees into a larger tree. This continues until there is only one tree that contains all of the leaves. If there are n different values in the data type, there will be $n - 1$ combining steps.

Each combining step involves finding the two trees in the current set that have roots labelled with the two smallest (or equal smallest) probabilities. These are then combined into a new tree, which consists of a node with the two chosen trees as left and right children. The new node is labelled with a probability that is the sum of the two child trees' root node probabilities. Each leaf in the left child tree has a **0** prefix added to its binary code. Each leaf in the right child tree has a **1** prefix added to its binary code.

As an example of this process, consider a data type that has six different values: A, B, C, D, E and F. Suppose that these values occur with probabilities 0.05, 0.48, 0.15, 0.10, 0.09 and 0.13. Figure 2.3 illustrates how the tree is formed. The effect of the construction process is that each value is allocated a binary code that has length equal to the depth of the value's leaf in the final tree. The leaves for values that have low probabilities are added to the final tree at an earlier stage than those with higher probabilities, and so are placed more deeply in the tree. Thus, values with lower probabilities are given longer codes than those with higher probabilities, as required.

In fact, it can be proved that Huffman coding gives the best possible variable-length code, in the sense that the expected number of bits used to represent each value is as close to the entropy of the data type as possible. As seen above, the source coding theorem derives the entropy as the absolute lower limit on possible compression. Huffman coding also ensures a property essential to variable-length coding: no valid code is a prefix of another valid code. Thus, a bit sequence formed from a series of binary codes can be unambiguously interpreted. For example, using the code from Figure 2.3, the bit sequence **10100000** unambiguously represents BAD.

Run length encoding

A second general compression theme is that of repetition. This is where a particular value occurs repeated two or more times in succession. Note that this is not necessarily the same thing as a value occurring very frequently. Run length encoding is a particular compression technique that takes advantage of repetition. As a first example, consider information which is a sequence of bits, containing alternating runs of **0**s and **1**s. Instead of transmitting the actual bits, an alternative

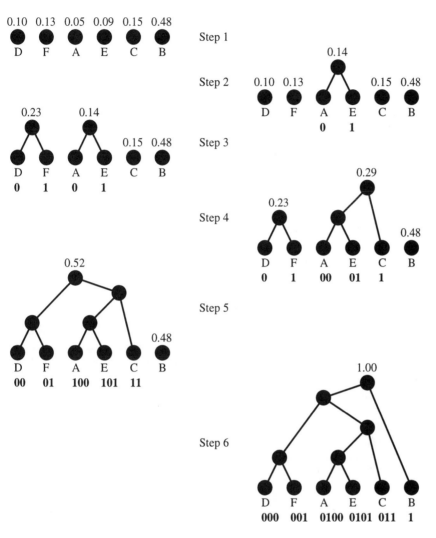

Figure 2.3 Building a tree using Huffman's algorithm

is to transmit the binary representation of integers that give the run lengths. For example, if the information is:

00000000001111111111111110101111110000

then the run lengths are 10, 15, 1, 1, 1, 5 and 4, so a binary encoding might be:

1010 1111 0001 0001 0001 0101 0100

which is a saving of about 28% in terms of the number of bits required. This example assumes that the runs occur in the order zeros, ones, zeros, ones, etc.

As a different example, consider information which is a sequence of characters. In this case, as for any data type with more than two elements, it is not enough just to transmit run lengths. It is necessary also to indicate which character is in each run. For example, if the information is:

zzzzzzzzzzzzzzzzzzz!!!!..........xxxxxxxx

then a possible run length encoding consists of a sequence of pairs:

(19,'z'), (4,'!'), (10,'.'), (8,'x')

which is rather more compact.

Difference encoding

Rather than strict repetition, there might be near-repetition, where one value is followed by another value that is not much different (assuming that there is some measure that allows values to be compared). This also lends scope for compression, using a difference encoding method. Rather than transmitting a sequence of values, a sequence of the differences between adjacent values is transmitted. The compression comes from the fact that the differences should have smaller representations than the values themselves. To do this, it is necessary to communicate, or agree in advance, some starting value, so that the first difference makes sense. One problem with this approach is that if, for any reason, a difference value gets lost, then the whole of the rest of the sequence becomes wrong. This is not the case if the actual values are communicated.

As an example, consider the communication of a sequence of integers. If the information is:

135, 142, 131, 127, 128, 136, 139, 126, 126, 134

the difference coding might be:

7, -11, -4, 1, 8, 3, -13, 0, 8

which is more compact. If an initial value of zero, say, had been pre-agreed, then a difference of 135 would have to be transmitted first. Another example is the communication of a video sequence. Normally, video films do not differ very much from frame to frame, with major differences only occurring when the whole scene changes. Thus, instead of transmitting a complete picture for every frame, only the differences between adjacent frames need be transmitted, with considerable compression gains.

Example: Fax compression

Fax (facsimile) transmission involves the communication of representations of sheets of paper. Each sheet of paper is considered as being a two-dimensional array of dots, a dot being either white or black. Typically, there might be about 80 dots per centimetre of paper. A naïve representation of a sheet might be to use a collection of bits, one bit per dot. However, far more economical methods are possible, and the international standards for fax transmision utilize all three of the techniques just described above.

The International Telecommunications Union (ITU-T) T.4 standard for Group Three fax apparatus uses compression on a row by row basis. That is, each row of dots is given a compressed representation. The code used involves both run length encoding and Huffman coding. First of all, a row is treated as being a sequences of alternating runs of black and white dots. Thus, it can be represented by a sequence of integers giving the alternating black/white run lengths. However, for most sheets of paper, the pattern of black on white is not random, since it is composed from words and pictures. Thus, the different possible black and white run lengths occur with differing probabilities. Because of this, a Huffman coding of the run lengths was devised, with the probabilities estimated on the basis of an analysis of typical documents. As a result, the actual representation of each row is a sequence of binary codes for the run lengths of alternating black and white areas.

The ITU-T T.6 standard for Group Four fax apparatus (which is also an optional standard for Group Three apparatus) goes further, by including compression across rows. It uses the above scheme, but also has a difference coding scheme, where the runs on one line can be defined in terms of differences in length from runs on the previous line. In particular, if corresponding runs on successive lines differ by at most three in length, a special code is used. Otherwise, the actual length of the run is still coded directly, using the Huffman code. This technique suffers from the general problem for difference coding — if one row is damaged, then subsequent ones may suffer too. It is used in this case because T.6 was designed for transmission over very reliable channels, whereas T.4 was not.

Example: MPEG video compression

The Moving Picture Experts Group (MPEG) has designed various international standards for the compression of video data. There are relations between the work of this group and work on compression for video telephony in the telecommunications community. There are also connections with the work of the Joint Photographic Experts Group (JPEG) on still photograph compression. MPEG is mentioned again in Chapter 11, which contains a case study concerned with the communication of real-time video information. The details of MPEG compression are complex, and require fairly specialized knowledge, and so only an extremely brief summary is given here, in order to show the general compression techniques that are used.

To achieve the required amount of compression, MPEG involves both inter-frame compression using difference encoding between successive video frames, and intra-frame compression on individual frames. The former poses some problems, because one goal of the MPEG standard was to allow random access to individual frames without the need to uncompress the entire video sequence. Therefore, periodically, complete frames are included in the video stream. In between, there are difference coded frames. One type is the **predicted frame**, which encodes motion from a previous reference frame. Another, and more accurate, type is the **interpolated frame**, which encodes motion between a past reference frame and a future reference frame.

The compression techniques used for individual frames, or for motion between frames, are relatively complex. In summary, the video signal is first compressed by the application of a discrete cosine transform (details of interest in the world of mathematics, but not here). The resulting coefficients are then quantized and compressed using both run length encoding and Huffman coding. Thus, as for fax compression, MPEG compression involves all three of difference coding, run length encoding and Huffman coding. There are three variants of MPEG, which attempt different amounts of compression. These are summarized on page 393 in Chapter 11.

Example: PPP message compression

The Internet Point-to-Point Protocol (PPP) is a collection of agreements that are used to provide a means of communicating Internet information over slow channels, such as telephone lines. Referring back to the classic message formats mentioned earlier in Section 2.2.2, PPP involves communicating HDLC-style messages which contain encapsulated IP messages (PPP can also carry encapsulated messages belonging to other protocols). The IP messages may, in fact, be combined TCP/IP messages.

Given that the communication channel is slow, there is a strong motivation to compress the information transmitted, and PPP includes two mechanisms that can be used to achieve this. An example of PPP message compression in a practical context occurs in Chapter 9, which contains a case study concerned with World Wide Web page access from a home computer.

The first compression mechanism involves the simple fact that certain components of the HDLC-style messages never change value over the course of communicating a sequence of frames. Therefore, the communicating computers can agree never to bother sending these fields. This is a fairly rare type of compression, where some information is completely redundant.

The second mechanism involves reducing the amount of administrative information carried by TCP/IP messages, using a compression method devised by Jacobson. This involves a mixture of removing some redundant fields, but also using difference coding to communicate changes from previous TCP/IP messages transmitted. The details of the method used require an understanding of the components of TCP/IP messages, which is not necessary at this point in the book. Jacobson compression can reduce the quantity of administrative information from 40 bytes to five bytes, or sometimes even to only three bytes, which is a huge improvement.

Redundancy for error detection

For most types of computer communications, information is transformed to allow *error detection* rather than *error correction*. That is, to enable a receiver to recognize that information has been damaged, and then usually discard it. If correction is necessary, this is achieved by transmitting the information again, hoping for better luck, as described in Section 3.3.3. The reason for this is that error correction codes involve adding significantly more redundancy to information transmitted. They are also distinctly more complex, and hence more computationally demanding. For moderate error rates, occasional retransmission is far more efficient.

The basic detection and correction capabilities of codes can be quantified a little more precisely. An (n, k) code is one in which k bits of information have redundancy added to give n $(n > k)$ bits. Then, of the 2^n possible bit sequences that might be received, only 2^k correspond to valid transmitted bit sequences. The remaining $2^n - 2^k$ sequences can only result from damage occurring during communication.

An important feature of a code is the **Hamming distance**. This is the minimum number of bits by which two valid n-bit sequences differ. If the Hamming distance is d then errors affecting up to $d - 1$ bits can be detected, and errors affecting up to $d/2$ bits can be corrected. The Hamming distance of a code depends on the redundancy introduced: the larger the difference between k and n, the larger the Hamming distance possible with an appropriate choice of code.

Parity codes

A very simple example of an $(8, 7)$ code is a **parity bit** appended to a seven-bit sequence. For **odd parity,** the extra bit is chosen so that there is an odd number of **1** bits in the resulting eight-bit sequence. **Even parity** is defined analogously. This code has Hamming distance two, because changing only one bit will result in the eight-bit sequence having the wrong (even or odd, respectively) parity. Thus, one-bit errors can be detected. However, they cannot be corrected.

More generally, **Hamming codes** can be used. These involve the use of a collection of parity bits. The details of how these are computed will be omitted here. However, by way of contrast with the above simple parity bit, it is worth noting that a $(11, 7)$ Hamming code is the simplest member of the family that has Hamming distance of three. Thus, it allows correction of one-bit errors and detection of two-bit errors. As well as the additional computation necessary, three extra bits must be communicated for each seven information bits, to give the appropriate redundancy.

Hamming codes are not frequently used on communication channels between computers. This is for two main reasons. First, because their error-correction capabilities are not required. Second, because errors on most communication channels tend to damage arbitrary size groups of bits, rather than randomly affect single bits within fixed (e.g., seven-bit) groups. One alternative is to use parity bits computed over an entire transmitted bit sequence. The next two sections describe two other commonly used techniques for computing a code over all of the bits in a sequence.

Cyclic redundancy codes

Cyclic redundancy codes (CRCs) involve regarding a sequence of bits to be communicated as the set of coefficients of a polynomial. This polynomial is defined in a number system where there are only two values: 0 and 1. For example, the 10-bit sequence:

$$1\ 1\ 0\ 1\ 0\ 1\ 1\ 0\ 1\ 1$$

would represent the polynomial

$$
\begin{aligned}
&1.x^9 + 1.x^8 + 0.x^7 + 1.x^6 + 0.x^5 + 1.x^4 + 1.x^3 + 0.x^2 + 1.x + 1.1 \\
=\ &x^9 + x^8 + x^6 + x^4 + x^3 + x + 1
\end{aligned}
$$

where x ranges over the set $\{0, 1\}$. Each particular code has a fixed polynomial associated with it. This is called the generator polynomial.

There are various standard generator polynomials used in practice. In order of increasing degree, four common examples are the CRC-8 polynomial:

$$x^8 + x^2 + x + 1,$$

the CRC-10 polynomial:

$$x^{10} + x^9 + x^5 + x^4 + x + 1,$$

the CRC-CCITT polynomial:

$$x^{16} + x^{12} + x^5 + 1$$

and the CRC-32 polynomial:

$$x^{32} + x^{26} + x^{23} + x^{16} + x^{12} + x^{11} + x^{10} + x^8 + x^7 + x^5 + x^4 + x^2 + x + 1.$$

If $T(x)$ is the polynomial that is represented by a bit sequence to be coded, and $G(x)$ is the generator polynomial in use, and $G(x)$ has degree k (i.e., the highest power of x in $G(x)$ is x^k), then the bit sequence actually transmitted is that which represents the polynomial:

$$x^k T(x) - R(x)$$

where $R(x)$ is the remainder polynomial when $x^k T(x)$ is divided by $G(x)$. In fact, this sequence is just the original bit sequence with k bits that are the coefficients of $R(x)$ appended on the end.

Note that $x^k T(x) - R(x)$ is exactly divisible by $G(x)$ — this underpins the error-checking mechanism. The receiver divides the polynomial represented by the bit sequence actually received by $G(x)$. If the remainder is non-zero, then it, knows that an error has occurred. If the remainder is zero, it assumes that no error occurred. This assumption may be incorrect if damage has somehow transformed the bit sequence into a different one that represents a polynomial also divisible by $G(x)$.

At first sight, the use of CRCs appears odd. First, there seems little point in complicating a simple bit sequence by regarding it as the representation of a polynomial. Second, the coding process involves the division of polynomials, which does not seem easy. The explanation is that the interpretation in terms of polynomials allows humans to understand that CRCs have appropriate error-detection capabilities, but the actual coding computations can be done efficiently using very simple bit operations.

A consideration of the error-detection capabilities must focus on the cases where the code fails to detect errors. Suppose that the bit sequence representing the polynomial $C(x)$ is transmitted but that the bit sequence representing the polynomial $C(x) + E(x)$ is received. $E(x)$ is the polynomial represented by damaged bits. As an example, suppose that the bit sequence

$$0\,1\,0\,0\,1\,1\,1\,0\,1\,0$$

is transmitted, but that the bit sequence

$$1\,1\,0\,0\,1\,0\,0\,0\,1\,1$$

is received. Reading from right to left, the first, fourth, fifth and tenth bits have been damaged. Then

$$
\begin{aligned}
C(x) &= x^8 + x^5 + x^4 + x^3 + x \\
E(x) &= x^9 + x^4 + x^3 + 1
\end{aligned}
$$

giving

$$C(x)+E(x) = x^9 + x^8 + x^5 + (1+1)x^4 + (1+1)x^3 + x + 1 = x^9 + x^8 + x^5 + x + 1$$

which is the polynomial represented by the bit string received.

Since $R(x)$ is divisible by $G(x)$, the problem occurs if $E(x)$ is also divisible by $G(x)$. If $G(x)$ is chosen carefully (as CRC-8, CRC-10, CRC-CCITT and CRC-32 are), then the possibility of this happening can be minimized. To take one example, **burst errors** are often the most frequent problem on communication channels. A burst error is when a contiguous sequence of bits is damaged. If the

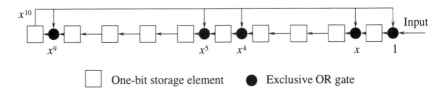

Figure 2.4 Circuit that computes CRC using the CRC-10 generator

length of a burst is r, and it occurs beginning at the i-th bit from the right, then $E(x)$ is a polynomial of the form $x^{i-1}E'(x)$, where $E'(x)$ has degree $r - 1$. If the degree of the generator polynomial $G(x)$ is at least r, then $E(x)$ will not be divisible by $G(x)$, because $E'(x)$ is not divisible by any polynomial of higher degree. Thus, CRC-8, CRC-10, CRC-CCITT and CRC-32 will correctly detect all burst errors of length up to 8, 10, 16 and 32 bits, respectively.

The computation of CRCs appears difficult, because it involves the division of one polynomial by another, to obtain the remainder. However, it is made straightforward by the fact that arithmetic is very simple in the 0-1 number system. Addition and subtraction are the same operation, and are in fact just the bitwise exclusive-or operation (i.e., there are no carries). Thus, the computation can be performed by a carry-free long division algorithm. This is particularly suitable for hardware implementation. Figure 2.4 shows a circuit that can compute a CRC when the CRC-10 generator is used.

Essentially, it is a 10-bit register that allows bits to be shifted through it. The only enhancement is the insertion of exclusive OR gates between certain one-bit storage elements, to introduce feedback. These gates are inserted at points that correspond to one coefficients in the CRC-10 polynomial. The bit sequence that requires a CRC is fed in, one bit at a time, at the right-hand end. When this has been done, the register contains the 10-bit CRC. This computation is a little less efficient for software implementation, because single-bit operations are needed, and these are not normally directly available.

Summation codes

Summation codes are more suitable for software computation than CRCs. These are usually known as **checksums**, but care is necessary since the word 'checksum' is sometimes used to refer to any kind of error-detection code. In essence, summation codes involve regarding a bit sequence as a sequence of binary-coded integers of some fixed size. For example, if the size was eight bits, then there would be a sequence of bytes, treated as a sequence of integers in the range $0, 1, \ldots, 255$. A code is computed by performing some sort of summation of the integers, and it is then transmitted with the information. By recomputing the code, a receiver can check whether any damage has occurred, since hopefully the sum will be different.

An example of a simple summation code is that used within the Internet protocols IP and TCP. The communicated information is treated as a sequence of 16-bit words. These are summed, modulo 65 535. The resulting 16-bit sum is complemented (i.e., each bit in the sum is inverted), and this 16-bit sequence is inserted at an agreed point in the transmitted bit sequence. The effect of this is that, if the complete transmitted bit sequence is summed, the result is equal to zero. This gives an error-detection test for the receiver. The choice of 16-bit arithmetic was made because at the time the protocols were designed most of the computers involved had 16-bit architectures.

The error-detection capabilities of this code are somewhat less than those provided by a CRC, but the code was designed to cope with the most common types of error than can afflict communication channels. In principle, if the communicated information is composed of zero bits and one bits chosen at random, there is a 1 in $2^{16} = 65\ 536$ chance of the code failing, assuming a random error pattern. However, it has been shown by Partridge, Hughes and Stone that, when the binary information is not random, for example, when it is representing characters from English text or when it is binary data containing many zeros, there is a much higher chance of the code not working. In one test, one-thousandth of the possible summation code values appeared nearly one-fifth of the time. This means that the chances of a code value being correct for damaged information is greatly increased.

A more complex example of a summation code is that used for the ISO standard inter-networking and transport protocols . The communicated information is treated as a sequence of bytes. Two zero bytes are inserted into the byte sequence, giving the sequence $M_1, M_2, \ldots, M_n$ say. Then, two modulo 255 sums are computed:

$$C_0 = \sum_{i=1}^{n} M_i \bmod 255 \qquad C_1 = \sum_{i=1}^{n} (n - i + 1) M_i \bmod 255$$

In the byte sequence that is actually transmitted, the two inserted zero bytes, M_k and M_{k+1} say, are changed from zero to:

$$
\begin{aligned}
M_k &= [(n - k)C_0 - C_1] \bmod 255 \\
M_{k+1} &= [C_1 - (n - k + 1)C_0] \bmod 255
\end{aligned}
$$

The effect of this is that, if the transmitted bytes are summed as above, the values of C_0 and C_1 are equal to zero. This gives an error-detection test for the receiver. The error-detection capabilities of this code have conventionally been seen as broadly compatible with those of CRCs; however, the work of Partridge *et al.* has shown that, as for the Internet checksum, certain information types lead to an increased failure rate. Unfortunately, arithmetic modulo 255 is rather more difficult than arithmetic modulo 256 (which just involves ignoring carries from arithmetic operations on bytes), but ISO believed that the choice of 255 made the code more effective. In fact, detailed studies of the respective error-detection capabilities show that this opinion was questionable.

Apart from being easier to compute in software, summation codes have two other merits. First, the extra check data may be included at any point within the transmitted message, rather than always at the end as is the case with CRCs in order to simplify hardware computation. Second, if a small part of a message is altered then it is not necessary to recompute the entire code, as would be the case with CRCs. This is because the original sum of the altered part can be subtracted from the overall sum, and then the new sum of the altered part added in instead.

Encryption

Encryption is intended to deal with both eavesdroppers and imposters. Eavesdroppers are undesirable elements that obtain information by spying on communications. Imposters are undesirable elements that introduce spurious information by pretending to be a trusted communication partner. The idea is to scramble communicated information in such a way that it cannot be understood by eavesdroppers or reproduced by imposters. Note that this process is deliberately called 'encryption' rather than 'encoding', since the latter term covers other types of transformation, such as compression or error-detection coding. In the language of **cryptography**, the science of encryption, eavesdroppers and imposters are unpleasant types of **cryptanalyst**.

There are various levels of attack that might be available to the cryptanalyst. The weakest, already mentioned in the discussion of information theory, is **cyphertext only** where the cryptanalyst only has access to examples of encrypted material. Stronger than this is **known plaintext** where the cryptanalyst has access to matching examples of original and encrypted material. The strongest is **chosen plaintext** where the cryptanalyst has access to the encryption procedure, and so can obtain the encrypted form of any original material. Nowadays, any worthwhile encryption method is expected to withstand a chosen plaintext attack. The encryption process has to be sufficiently complex that even possession of a limited number of chosen encrypted examples does not help.

Most encryption procedures involve use of a fixed algorithm. An appropriate inverse form of this algorithm is used for decryption. Usually, the details of the algorithm are publicly available. Security is introduced by the use of **keys**. These are extra inputs to the encryption and decryption algorithms when they are used to transform information. Traditionally, the encryption key was kept secret by the information transmitter and the decryption key was kept secret by the receiver. In fact, both keys were usually the same, and so were a shared secret between the communicating parties. This immediately raises the question of how the key information is securely communicated. The answer is that some additional, extra secure, communication mechanism is needed. For example, a person might travel with the key information carried in a secure bag. Then, the keys can be used over a computer communication channel.

An alternative solution to the key distribution problem is the use of **public key** encryption. As its name suggests, less secrecy surrounds the keys. In particular, the key necessary for encryption is made publicly available. Only the decryption key is kept secret. This relies on the fact that having full knowledge of the encryption procedure does not allow a cryptanalyst to reverse it in order to obtain a usable decryption procedure. There are no known public key procedures that can be proved secure. However, there are some for which there is strong supporting evidence of security. This is also a problem for secret key procedures. The only provably secure procedure is the **one-time pad**, where a different key is used every time that the encryption algorithm is used. This scheme involves a large amount of secure key distribution.

Example: Data Encryption Standard

The Data Encryption Standard (DES) was defined by the US National Bureau of Standards in 1975, and is a widely used secret key encryption method. Its security was doubted from the very beginning, in that it seemed feasible for a cryptanalyst with huge amounts of computing resource (e.g., some government agencies) to break the method. This suspicion was confirmed in the published literature in 1991. Given this, and the fact that the DES algorithm is freely available around the world, it is unfortunate that the US Government prohibits the exports of products with security features if these involve the use of DES encryption — encryption techniques are classified as being 'munitions'. The International Data Encryption Algorithm (IDEA) is a more recent secret key method that is designed to be far stronger than DES — in fact, there are no known computationally tractible methods to break it. Exercise 2.29 invites the reader to study IDEA, and compare it with DES.

The DES algorithm is applied to fixed-size blocks of 64 bits, and a 56-bit key is used. The same key is used for both encryption and decryption. The encryption algorithm consists of an intricate series of steps, which include both permutation (where a group of bits is reordered) and substitution (where a group of bits is replaced by a different group of the same size). The idea of permutation is familiar from simple puzzles, where the letters of a word are scrambled, and one is invited to identify the word. The idea of substitution is familiar from simple encryption methods where each letter of the alphabet is mapped to a fixed different letter (e.g., the letter three ahead in the alphabet). The decryption algorithm involves reversing the steps of the encryption algorithm.

More precisely, the first step of DES encryption involves a fixed permutation of the 64 bits. This is followed by 16 substitution steps. Finally, there is a simple permutation involving a swop of the most and least significant 32 bits, followed by the same fixed permutation as in the first step. In fact, the permutations do not add any security to the method. This comes from the substitutions, which are dependent on the secret key used. Each substitution step uses a different 48-bit subset of the bits of the key as an input. The overall substitution is deliberately intricate but, for any particular key, it is effectively just a fixed mapping of 64-bit values onto 64-bit values.

When DES is used for independently encrypting 64-bit blocks, it is said to be working in **Electronic Code Book** (ECB) mode. This gives security on a per-block basis, but means that it is possible for an imposter to insert apparently valid encrypted blocks into a stream of encrypted blocks, or to remove encrypted blocks, without detection. To deal with these problems, **Cipher Block Chaining** (CBC) mode can be used instead. For this, before encryption, each 64-bit block is exclusive-ORed with the previous encrypted block. The very first block is exclusive-ORed with a random 64-bit initial value, since there is no previous value. ECB mode and CBC mode are illustrated in Figure 2.5(a) and (b) respectively.

An alternative method is to use **Cipher Feedback Mode** (CFM) mode, which is shown in Figure 2.5(c). This takes into account the fact that it is often more convenient to process byte-size units, rather than 64-bit units. Each byte, and its seven predecessors, are encrypted as a 64-bit unit. The least significant eight bits of the result are then exclusive-ORed with the original byte to yield the encrypted byte. Note that this method involves an encryption computation for every byte, rather than for every 64 bits, so eight times the computational effort is needed. However, the byte orientation is very convenient for hardware implementations, which can perform the required bitwise operations speedily.

Example: Rivest Shamir Adleman cryptosystem

The eponymous Rivest Shamir Adleman (RSA) cryptosystem was first published in 1978, and is a widely used public key encryption method. As well as giving a method for encryption that is still believed to be secure, the method also offers a way of providing authentication as a protection against imposters. In particular, RSA has been used to underpin the facilities of **Pretty Good Privacy** (PGP), a package designed to allow people to exchange information without fear of interference from eavesdroppers and imposters. The widespread free distribution of PGP has caused alarm to various governments accustomed to monitoring their citizens' communications.

As with the DES system, fixed-size blocks of information are encrypted. The larger the size chosen, the more secure the method. For example, 256-bit blocks give moderate security, while 512-bit blocks give good security. Each block is treated as being the two's complement representation of a long integer value, and encryption involves transforming this value. The bits representing the final value are the encrypted form.

For each recipient of encrypted material, a set of three integer values, d, e and n, determine the keys used for encryption and decryption. Both e and n are made public, and (e, n) is the encryption key. Only d is kept secret, and (d, n) is the decryption key.

If b is the block to be encrypted, regarded as an integer, then the transformed value is equal to $b^e \bmod n$. If c is a block to be decrypted, regarded as an integer, then the transformed value is equal to $c^d \bmod n$. Thus, both encryption and decryption involve the exponentiation of a long integer, modulo the long integer

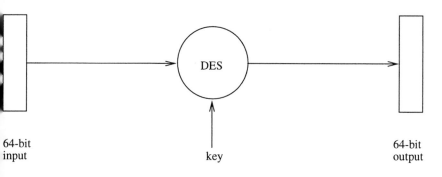

64-bit
input

key

64-bit
output

(a) electronic code book (ECB) mode

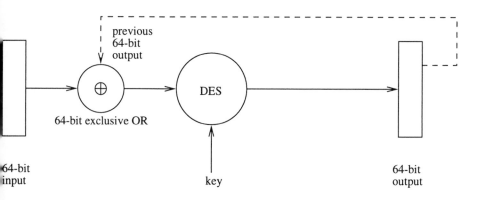

previous
64-bit
output

64-bit exclusive OR

64-bit
input

key

64-bit
output

(b) cipher block chaining (CBC) mode

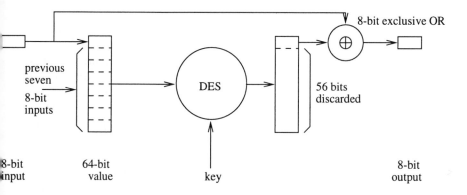

8-bit exclusive OR

previous
seven
8-bit
inputs

DES

56 bits
discarded

8-bit
input

64-bit
value

key

8-bit
output

(c) cipher feedback mode (CFM)

Figure 2.5 DES encryption modes

n. The length of each block is the same as the length of the integer n. The exponentiation operation is fairly computationally demanding, especially when performed by software, and this is one of the drawbacks of the RSA system. However, integrated circuit implementations are now available, which makes the method more suitable for high speed computer communications.

It is essential that decryption reverses encryption, so it must be the case that

$$(b^e \bmod n)^d \bmod n = b$$

for any b. To ensure this property, d, e and n have to be carefully chosen. First, two large random prime numbers, p and q, are selected and then n is set equal to pq. Second, a normal-length random integer e that is relatively prime to $(p-1)(q-1)$ is chosen (i.e., the greatest common denominator of e and $(p-1)(q-1)$ is one). Finally, d is computed as the unique positive integer such that $de \bmod (p-1)(q-1) = 1$. This recipe guarantees that d, e and n work.

Since e and n are made public, it can be seen that d could be easily computed by a cryptanalyst if p and q are known. That is, the most obvious line of attack would be to factorize n to get p and q. The main claim to security of the RSA method hinges on the fact that there is no known fast method for factorizing large integers. If a 512-bit block size (and hence size for n) is used, then the best-known factorization algorithms would take many thousands of years.

Symmetric properties of d and e mean that it is also true that

$$(b^d \bmod n)^e \bmod n = b$$

for all b. Thus, 'decryption' followed by 'encryption' returns the original value. This can be used as the basis for an authentication mechanism that allows a receiver to check that a message actually came from a particular sender, and was not injected by an imposter. A **digital signature** can be communicated by the sender taking a distinctive bit sequence, and applying his or her secret decryption procedure to it. The recipient of the signature can apply the public encryption procedure of the sender, and should obtain the distinctive bit sequence. Imposters are unable to forge the digital signature because they do not know the secret key.

One approach to deriving a 'distinctive bit sequence' is to construct a short digest of a long piece of information. The MD5 algorithm, described in Internet RFC 1321, is an example of an appropriate method. It takes an arbitrary-length bit string, and produces a 128-bit digest from it. Using such an approach, a digital signature can include a 'short summary' of the information bits that are being authenticated by the signature.

It is conjectured that, given an MD5 digest, it is computationally infeasible to construct a bit sequence that would be mapped to it. An imposter might capture digests contained in digital signatures, given that signatures can be 'decrypted' using the public key. However, it is not possible for the imposter to construct alternative bit strings that map to the captured digests, in order to send bogus information accompanied by an apparently valid signature.

The use of digests does not prevent an imposter from replaying signed messages that have been captured earlier. This might be as unpleasant as injecting new messages. For example, a captured message might authorize a withdrawal from a bank account, and so replaying the message could cause the withdrawal to be repeated. To deal with this problem, further precautions are needed, for example, including the time of day or a serial number within the encrypted signature.

2.3.2 Representation of information

In some cases above, the transformation of information also involved a change in representation. For example, Huffman coding involves taking values from some arbitrary data type and mapping them to binary strings. This yields data compression which can be readily measured in information theoretic terms: fewer bits of information. There is also the happy side-effect that bits are likely to be a more easily communicated data type than the original. This is not always the case. For example, run length encoding takes a sequence of values from one data type and maps it to a pair: an integer length and a value. Conceptually at least, this information has a more complex structure.

In general, when all appropriate information transformations have been done, there might be no direct physical implementation for communicating the information data type. To achieve implementation, it is necessary to change the representation into something that is more communicable. The information itself is not altered by the change in representation. This is in contrast to the transformations of the previous section, which are designed to improve the quality of the implementation.

Just as many of the quality-improving transformations are also applicable to the storage of information in single computer systems, so the use of simpler representations for information is also familiar. Users of computer systems, and also computer programmers, are accustomed to working in terms of high-level data types. These have to be implemented in terms of the lower-level facilities available inside computer systems: bits and bytes. As illustrations of implementation using changes in representation, the examples that appear below cover most of the types of information that were discussed earlier in Section 2.2.2, in the context of agreement between computers.

Computer communications examples

Bits

First, there will be a little splitting of the atom. Bits are the fundamental atomic data type throughout the book. However, to avoid accusations of negligence, it is wise to give an outline explanation of how bits are represented in terms of physical media of the type discussed in Section 1.3.1. This section discussed the difference between digital and analogue signals being used on a communication medium.

With digital signalling, the transmitted signal has only a few discrete levels, usually just two levels. For example, for an electrical signal, there might be two different voltage or current levels. For an optical signal, the absence and presence of a light pulse would be the two levels. The two levels then give a natural representation of the binary digits **0** and **1**. A sequence of bits can be communicated by appropriately varying the communicated signal over time. This sort of transmission is usually referred to as **baseband transmission**.

With analogue signalling, the transmitted signal is a continuous waveform within some particular frequency range. This range is termed the **bandwidth** of the signal. Since it is the case that, the larger the bandwidth, the higher the information rate possible, the term 'bandwidth' is often used loosely in computer communications world to mean 'information carrying capacity'. Broadband communication media allow several independent signals to be transmitted simultaneously, each occupying a different frequency range. However, remember that the word 'broadband' is increasingly used loosely in the computer communications world just to mean 'high speed'. The term **carrierband** is used when an analogue medium only carries a single signal.

In order to represent a sequence of bits, an analogue signal must be modulated over time. More specifically, there is a basic **carrier signal** which is a sine wave, and its characteristics are changed over time. With frequency modulation, two different wave frequencies are used to represent the bit values **0** and **1**. With amplitude modulation, two different wave amplitudes are used to represent the two values. With phase modulation, two different phases (wave starting points) are used. Whatever form of modulation is used, devices are necessary to convert the digital information to and from analogue form. These devices are called **modulator-demodulators**, or **modems** for short. They are most familiar to the typical computer user as the devices used to transmit computer information over analogue telephone lines. Modern modems, so-called 'intelligent' or 'smart' modems, perform other functions, such as compression or error detection, as well as the basic analogue representation function.

Characters

Most definitions of character sets, IA5 (ASCII) and EBCDIC being no exception, do not just define which characters are present. They also provide a standard binary coding for each character. In the case of IA5 (ASCII), there are 128 characters in the set, and these are mapped to seven-bit representations. The binary representation could just be regarded as a binary coding of a serial number in the range 0 to 127. However, there is a little more structure to the codes, there to be exploited or abused (depending on viewpoint) by computer programmers. For example, the 32 control characters use all of the 32 codes beginning with **00,** the 10 digits have consecutive codes from '0' to '9', and both the upper-case and lower-case letters have consecutive codes from 'a' to 'z'. Further, the codes for corresponding upper-case and lower-case letters differ only by one bit. Given that the byte is often a natural unit of information to work with, it is common to find the seven-bit IA5 representation being combined with a single parity bit, to allow modest error detection.

The ISO 10646-1 (Unicode) character set has a 16-bit representation. The 16-bit values are organized in blocks, for example, the range 0020–007E (in hexadecimal) contains a basic Latin alphabet, which is just the printable ASCII characters, 0400–04FF contains the Cyrillic alphabet (used in Russian, and other languages), 25A0–25FF contains geometric shapes, and 4E00–9FFF contains unified ideographs for Chinese, Japanese and Korean.

Data types

Just as the XDR range of data types is extremely close to that of the C language, so is the representation of the data types in terms of bytes. Values of the basic types have direct binary encodings, for example, signed integers, enumerations and booleans are all encoded as 32-bit two's complement numbers. Arrays, structures and unions are encoded by concatenating together the binary encodings of their components. Thus, if the binary coding of an XDR data type is to be interpreted to obtain its value, the interpreting computer must know what the XDR format of the data type is. There are no clues embedded in the binary coding.

The representation used by ASN.1 is very different from this. Just as the defined data types are meant to be an abstraction of the range found in high-level computer programming languages, so the binary representation is more abstract than the type of representation that might be used within a computer system. In the Basic Encoding Rules (BER) for ASN.1, the encoding of a basic type or a constructor has three main components:

1. a tag field;
2. a length field; and
3. a data field.

The tag field indicates the basic type or constructor being used. The length field indicates the length of the value in bytes. Finally, the data field contains the actual value. An optional fourth component is an end-of-data flag, which can be used in circumstances where it is not possible to include the length field. The use of a length field means that it is possible to have different-size representations for some of the basic data types. For example, integers can be represented using exactly the number of bytes necessary.

Thus, the ASN.1 representation carries structural information as well. The binary coding can be interpreted without any prior knowledge of what it represents. This is rather more flexible. Also, the use of explicit length fields means that the representation is far less tied to a particular computer system's conventions. A disadvantage of the Basic Encoding Rules for ASN.1 is that their binary representations are rather larger than those of XDR. Also, constructing and interpreting the representations requires more computation by communicating computers. Thus, there is a trade-off between generality and efficiency. Other sets of encoding rules have been defined to address these drawbacks, in particular the Packed Encoding Rules (PER) which produce smaller binary representations.

Programs

The representation of a Java program is very straightforward. A program just consists of lines of text, that is, it is just a sequence of characters. The subtlety lies in the way the sequence of characters is interpreted, but this is a matter for the computers involved, not the communication mechanism.

Communication information

The messages used to implement computer communications obviously have to be represented in some communicable manner. Standard formats for more modern and/or higher level messages are now often expressed as complex ASN.1 data types. These then have a direct binary representation associated. Older standard formats describe the contents of messages, and include binary encodings in a fairly *ad hoc* way. Often, the encodings are very special case to the exact format of messages used. This is the case for all of the classic message formats mentioned earlier.

Here, these formats are presented in terms of their representation by bits. The actual fields of each message format are described as high-level data types (e.g., integer, boolean, etc.), stopping short of saying what the purpose of each field is. This is a topic for later chapters. The messages are illustrated diagramatically in a way that shows the size of each field in terms of the number of bits required to represent the field. In later chapters, this proportionate size is dropped, since it is the type of the field that matters, not its representation. Thus, the main purpose of introducing message formats here is to give examples of representations, and to convey a feeling for the actual appearance and size of messages when they are represented in binary form.

The ATM message consists of five bytes of administrative information, followed by exactly 48 bytes of message contents. There is a one-byte CRC covering the adminstrative information, and it uses the CRC-8 generator polynomial $x^8 + x^2 + x + 1$. The format of the ATM message is illustrated in Figure 2.6. Figure 7.5 shows a logical representation of the ATM message format (in fact, two formats), which puts more flesh on the physical representation shown in the figure here.

The HDLC message consists of two bytes of administrative information, followed by the message contents, followed by a two-byte (16 bit) CRC for the whole frame using the CRC-CCITT generator polynomial $x^{16}+x^{12}+x^5+1$. There is an extended version, sometimes used, that has three bytes of administrative information at the beginning. The format of the HDLC message is illustrated in Figure 2.7. Logical representations of various different HDLC-style message formats appear in later chapters — one example is Figure 9.9, which shows the format used by the PPP protocol.

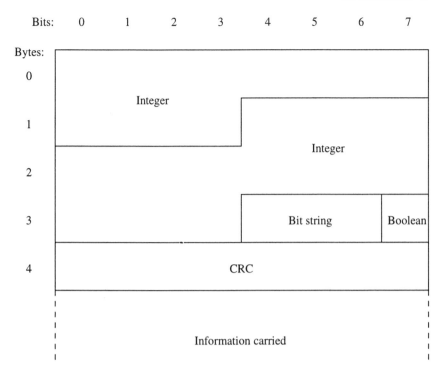

Figure 2.6 Format of ATM 'cell'

The traditional IP message consists of 20 bytes of administrative information (optionally, there may be more), followed by the contents of the message. There is a summation code over the administrative information, which is the standard Internet modulo 65 535 checksum described earlier in Section 2.3.1. The format of the traditional IP message is illustrated in Figure 2.8. As will be explained later in Chapter 8, there is a newer version of the IP message, which will eventually supplant the traditional version. The format of the new IP message is illustrated in Figure 2.9. Figures 8.3 and 8.4 show logical representations of the traditional and new IP message formats respectively.

The combined TCP/IP message consists of the 20 bytes of IP administrative information, followed by 20 bytes (optionally, more) of TCP administrative information, followed by the contents of the message. There is a summation code over both the TCP administrative information and the contents of the message. This is the standard Internet modulo 65 535 checksum, as used in IP messages. The format of the TCP component is illustrated in Figure 2.10. Figure 9.6 shows a logical representation of the TCP message format.

Note that all of the classic messages have space for carrying information. This can be considered as an arbitrary bit string, which is not interpreted by ATM, HDLC, IP or TCP. In general, this is a characteristic of any sort of information-

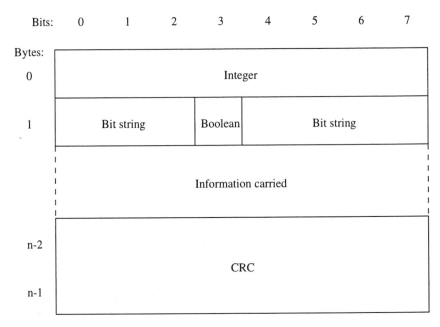

Figure 2.7 Format of HDLC 'frame'

carrying message, which means that a standard implementation technique called
encapsulation can be used. Suppose that there is one sort of message that is
communicable, but another that is not. Then, the idea is to encapsulate messages
of the second type inside messages of the first type. This means that there has to
be a representation of the second message type in terms of the information type
carried by the first message type.

TCP/IP is an example of this: TCP messages are encapsulated in IP messages.
The representation of TCP messages in terms of bit strings allows them to be information bit strings carried by IP. Thus, TCP messages are made communicable
by putting them inside communicable IP messages. The process can be repeated
several times. For example, IP messages might be encapsulated inside ATM messages or inside HDLC messages. An example of this process is illustrated in
Figure 2.11.

Analogue information

Having begun this section with an outline description of how digital bits can
be represented by analogue signals, the section is concluded with the opposite
direction: how analogue signals can be represented by digital bits. The contrast
illustrates the historical development of computer communications, as discussed
in Chapter 1. Once, the main problem was to map digital computer information
into an analogue human communication system. Now, the main problem is to

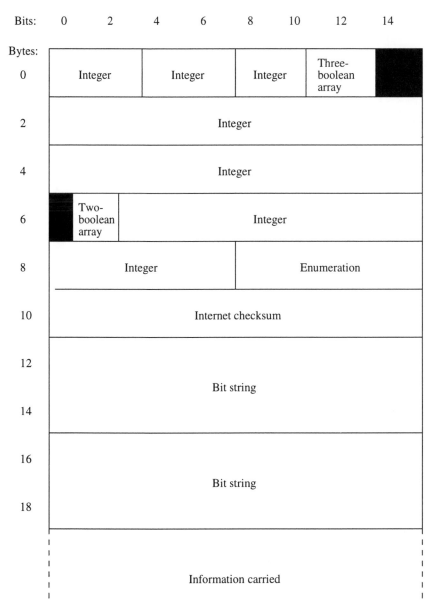

Figure 2.8 Format of traditional IP 'datagram'

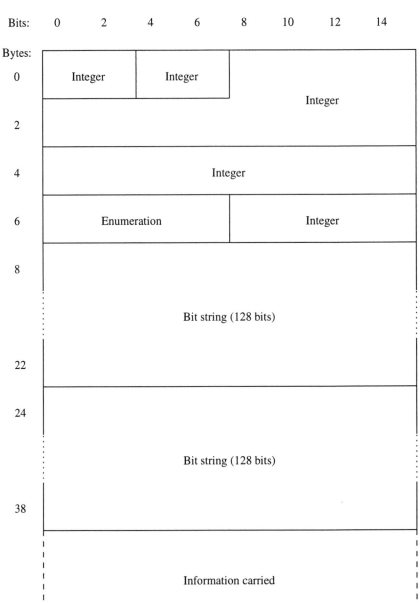

Figure 2.9 Format of new IP 'datagram'

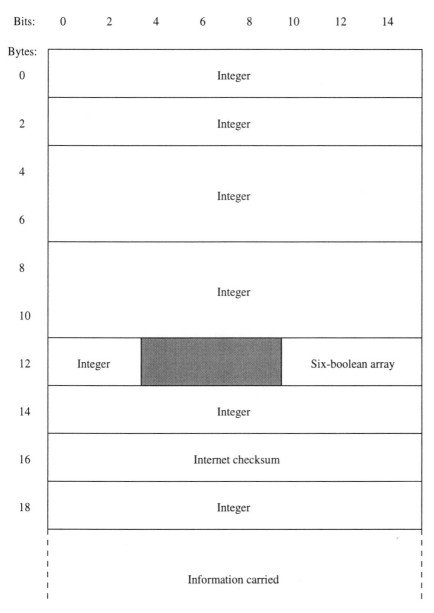

Figure 2.10 Format of TCP component of TCP/IP 'segment'

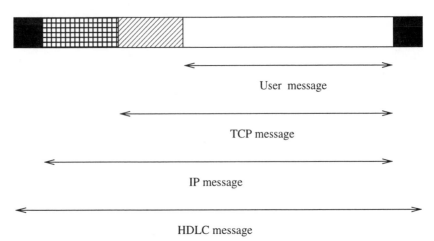

Figure 2.11 Example of encapsulation of messages

map analogue human information into a digital computer communication system. Normally, the main analogue information of interest is derived from video or audio or both together. A device used for converting analogue to digital, and back again, is the opposite form of a modem, and is called a **coder-decoder,** or **codec** for short.

Pulse Code Modulation (PCM) is a basic technique that involves **quantization**. The analogue signal is sampled periodically, and a digital measurement of its amplitude is taken as an encoding of the signal at each time instant. A fundamental result due to Nyquist states that, if a signal has bandwidth b, then $2b$ samples per second are sufficient to allow the original signal to be completely reconstructed from the digital encoding.

For example, the signal corresponding to a human voice over the telephone has a 3000 hertz bandwidth, so 6000 samples per second are sufficient. As mentioned in Section 1.3.2, quantization standards for the telephone system use 8000 samples, and either seven or eight bits per sample, i.e., a range of 128 or 256 quantization levels. Thus, 56 000 or 64 000 bits are required to represent one second of speech. However, the use of compression techniques on the digital information can reduce this substantially, to around 5000 bits per second.

2.4 CHAPTER SUMMARY

Before information can be shared, there has to be agreement between the communicating computers on what the type of the information is. Usually, this type will be strongly influenced by the types of information that the computers are designed to process. However, the computers are likely to differ in terms of detail, or perhaps more fundamentally. Therefore, various standard type definitions exist to form the basis for agreements on bridging differences between computers. Agreements may be static, and pre-arranged before communications. Alternatively, they may be imposed or negotiated when communication begins.

Complex information types are unlikely to have a direct implementation by physical phenomena. The quality of communication can be improved by transformations of information: compression, redundancy for error detection and encryption. Some fundamental limits on the quality improvement possible can be obtained from the results of information theory. In order to implement communication at all, simplifying changes in the representation of information are needed. Ultimately, this is likely to give a representation in terms of bits, which are communicable over physical media connecting computers.

2.5 EXERCISES AND FURTHER READING

2.1 Consider any communications that you carry out in order to share information with someone else, and decide whether these involve unidirectional transferring of information, independent bidirectional sharing of information, or complex sharing of information.

2.2 The author of this book has agreed that its information will be communicated using English. Explain how this absolute agreement with the reader is extended by relative agreements to ensure that the book is fully comprehensible.

2.3 A byte has the bit pattern **10011010**. Assuming that the byte contains the binary representation of an integer, give the value of the integer if the bit sex is (a) 'big endian' — the most significant bit comes first (at the left), and (b) 'little endian' — the most significant bit comes last.

2.4 Obtain lists of the ASCII and EBCDIC character sets. What are the main differences?

2.5 What character set(s) are supported by any computer systems that you use? What range of languages can be supported using the characters available in each character set?

2.6 Find out more about the work of the Unicode Consortium by consulting its WWW page (URL http://www.unicode.org).

2.7 If you are familiar with any high-level programming languages, write down a list of the data types and constructors supported. How well do the listed items map onto (a) XDR items, and (b) ASN.1 items?

2.8 Study the details of XDR by reading Internet RFC 1014.

2.9 If you have experience of computer programming, obtain details of the Java language, and compare its features with those of the programming languages that you are most familiar with.

2.10 Have a look at Internet RFC 1902, which includes a description of a use of a subset of ASN.1. Do not attempt to understand any details, but note what ASN.1 descriptions look like.

2.11 Write down the current time of day in UTC format, in two ways: one without an offset, the other with an offset.

2.12 If your computer is able to make use of specialized services of other computers, try to discover what mechanisms underpin this sharing.

2.13 If you have access to the Internet FTP service, write down a list of the services that are offered. Would you like any other services that are not present?

2.14 Suggest reasons why WIMP systems often allow users to issue commands either using a mouse movement or using a keyboard press.

2.15 Compare the different computer printers that you have used. What common features can you identify in order to develop your own virtual printer model? Does this model allow the printers to be used to their full potential?

2.16 Suppose that a piece of information may have four different values. If the probabilities of the different values being communicated are 0.1, 0.5, 0.2 and 0.2 respectively, work out the entropy of a communicated value. Comment on the difference between this entropy and the entropy if all of the values occur with probability 0.25.

2.17 Suppose that letters of the Roman alphabet are communicated randomly. What is the entropy of each value communicated?

2.18 Combine the results of the source coding theorem and the channel coding theorem to obtain a combined statement concerning the number of transmitted values that are needed to communicate a value of some data type. Your statement should involve both the channel capacity and the entropy of the data type as variables.

2.19 Obtain the typical occurrence probabilities of each letter of the alphabet used in your native language (either by finding a list or by writing a computer program to analyse a large text). Construct a Huffman coding for the alphabet using these probabilities.

2.20 Suppose that a piece of information consists of the temparatures measured by a thermometer every minute over a one-week period. In what cases might (a) run length encoding, and (b) difference encoding, be appropriate for compressing this information?

2.21 If you are interested in fax transmission, obtain the details of the T.4 and/or T.6 compression specifications. By experimentation, check what the compression is for typical pages.

2.22 If you are interested in video compression, obtain further information on the MPEG compression algorithms. What are the differences between MPEG-1, MPEG-2 and MPEG-4? How does MPEG compare with other video compression mechanisms?

2.23 If you are interested in Internet message compression, and have a basic understanding of the Internet protocols IP and TCP already, read Internet RFC 1144, which describes Jacobson compression. Why is this compression not standard for all Internet messages?

2.24 Why does a code with Hamming distance d allow the detection of errors affecting up to $d - 1$ bits, and correction of errors affecting up to $d/2$ bits?

2.25 Consider the bit string **1000010111100101**. Simulate the circuit shown in Figure 2.4, in order to find the CRC for this bit string when the CRC-10 generator is used. To do this, you need to assume that the storage elements all contain zero initially. If the CRC is $c_1c_2c_3c_4c_5c_6c_7c_8c_9c_{10}$, feed the bit string **1000010111100101**$c_1c_2c_3c_4c_5c_6c_7c_8c_9c_{10}$ through the circuit to check that the result is zero. Now, feed through this bit string with one bit flipped, and check that the result is non-zero.

2.26 Explain why the ISO summation code has the required effect, i.e., that with the computed bytes M_k and M_{k+1} inserted, the values of C_0 and C_1 will be equal to zero. Check this by trying a small example using 10 bytes (i.e., with $n = 10$) and with $k = 6$.

2.27 Investigate the history of the DES encryption algorithm, since its introduction in 1975. Would you trust it to protect your secrets?

2.28 Obtain a software implementation of the DES algorithm. Can you follow what is going on? How quickly can this implementation encrypt a single 64-bit block?

2.29 Investigate the International Data Encryption Algorithm (IDEA), which is a more recent alternative to DES, and is used within PGP. How does IDEA compare with DES?

2.30 If you have access to Pretty Good Privacy (PGP), find out what its main features are, and comment on their usefulness to you.

2.31 Construct your own keys for an RSA cryptosystem by choosing suitable values of d, e and n. To make life simpler, let n be between 35 000 and 65 000. Check that your encryption key (e, n) and decryption key (d, n) work by encrypting, then decrypting, the bit string **1000010111100101**.

2.32 Read Internet RFC 1321, which describes the MD5 message digest algorithm. For a very challenging read, have a look at RFC 1521, which describes how both MD5 and DES feature in a particular authentication service: Kerberos. However, note that this requires an understanding of various concepts discussed later in the book.

2.33 Draw diagrams to illustrate the difference between frequency modulation, amplitude modulation and phase modulation, as ways of representing binary values using an analogue signal.

2.34 Find out how high speed (9600 bits per second, or more) modems actually represent bit streams using a modulated analogue signal.

2.35 For any high-level programming language implementation with which you are familiar, find out how each high-level data type is represented in terms of bytes in computer memory.

2.36 What are the differences between the Basic Encoding Rules (BER) for ASN.1 and the Packed Encoding Rules (PER)? What other standard encoding rules are defined for ASN.1?

2.37 Give examples of information encapsulation in communications between humans.

2.38 The bandwidth of a normal analogue video signal is about 4 Mhertz. What sampling rate would be required to represent the signal faithfully in digital form? If each sample contains 10 bits of information, how many bits are needed per second of video?

2.39 Consider again any communications that you carry out in order to share information with someone else. Discuss the information agreements that are necessary, and also the ways in which information is transformed and/or represented in order to implement communication.

Further reading

Many of the topics covered in this chapter have general relevance to information processing and storage, not just to communications. Thus, it is useful background to be familiar with the general capabilities of computer systems. This knowledge can be gleaned from a variety of sources, for example, general textbooks or computer system documentation. A more internal view, relevant to the technical details of communications, can be obtained from familiarity with computer programming. The components of any high-level programming language can be used as a guide to typical information data types and the operations that can be performed on such types.

Information theory supplies a fundamental understanding of the nature of information, albeit without contributing a great deal of practical benefit to real-life communication. Shannon's pioneering paper on the subject appeared in 1949, and can be found in Shannon and Weaver, *The Mathematical Theory of Communication* (University of Illinois Press 1949). A more computer-oriented viewpoint can be found in *Information Theory for Information Technologists* by Usher (Macmillan 1984).

Information transformation is an area with a wealth of literature, in terms of both books and research papers. There is a range of specialist textbooks on information compression, usually referred to as 'data compression'. Some specialize on particular applications, for example, video compression. Error correction codes, as used in everyday computer communications, are very special cases of codes covered in the extensive literature devoted to coding theory. The results of Partridge, Hughes and Stone, mentioned in the section on summation codes, can be found in their paper "Performance of checksums and CRCs over real data", in the proceedings of SIGCOMM-95. Cryptography is also an area with an extensive specialist literature. In some cases, the algorithms used for information transformation are of sufficient general interest, they appear in textbooks on algorithms. For example, both Huffman coding and the RSA encryption method appear in *Introduction to Algorithms* by Cormen, Leiserson and Rivest (MIT Press 1990).

THREE

TIME

> *The main topics in this chapter about time are:*
>
> - sharing of information over a time period
> - synchronous and asynchronous agreement on time between computers
> - segmentation and concatenation of communications
> - flow control to harmonize communication time periods
> - acknowledgement and error handling
> - timeout periods and expiry periods

3.1 INTRODUCTION

In Chapter 2, a simple model of how information might be shared by two computers via a channel is used. However, it has some significant limitations. In particular, two assumptions are:

- both computers, and the channel, are always ready and willing to communicate information; and
- both computers, and the channel, operate at the same speed.

Neither of these restrictions is realistic from a practical point of view.

First, rather than being always ready and willing, computers normally have other work to do as well as communicating. Also, a channel might not be available at all times for use by the pair of communicating computers. Second, there can be significant differences in the speeds of computers, and even bigger differences between the speeds of computers and the speeds of channels. In the past, channels were very much slower than computers, and so had to be managed carefully. However, with the development of new technologies, channels have caught up with computers and, in some cases, now raise new problems by being faster.

In order to deal with the above issues, it is important to describe how information sharing proceeds over time. The matters that arise are very similar to those seen in Chapter 2 when explaining how information itself can be shared. First, it is necessary to have agreement on when information sharing takes place. Second, it is necessary to have implementation that allows required behaviour over time to take place using communications with different — usually simpler — behaviour over time. The implementation work may not just be concerned with achieving communication, but also with achieving it at some particular level of quality.

Human experience

To make the discussion on time clearer, consider the following examples drawn from everyday human experience:

- a one-hour presentation by a speaker
- a day's activity at a tourist information desk
- a telephone call.

These examples will be used throughout the chapter to illustrate various time-related points. Some features of their associated communications are worth noting.

The talk is likely to have an advertised starting time, and is known to have a one-hour duration. During the talk, there will be a fairly continuous information flow. The information desk is likely to have advertised opening hours over part of the day. During the day, there will be sharing of information each time a tourist makes an enquiry. In contrast, telephone calls do not normally take place at fixed times. They begin and end as convenient for the callers. During the call, there will be a fairly continuous sharing of information.

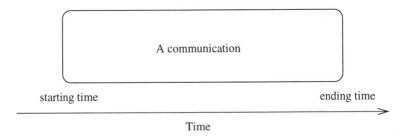

Figure 3.1 Time period of a communication

Computer communications principles

Throughout this chapter and the next, the term 'a communication' will be used to mean an act of information sharing occurring within some time period. This adds more detail to the previous model, where communication was regarded as a continuous and time-unbounded activity. For any particular communication, its time period is given by:

- its starting time; and
- its finishing time

as illustrated in Figure 3.1. This definition seems rather obvious. However, it is subtle because the word 'time' is used, but it is not made clear how time is measured. This has to be agreed between the communicating parties, so that both have a consistent notion of starting time and finishing time.

Computer communications examples

Throughout this chapter, the following examples will be used as illustrations:

- sharing a single bit
- sharing a single character
- sharing a string of bits
- a database transaction
- a file transfer.

The first three are fairly simple in terms of the information communicated. The database transaction involves a control relationship, where one computer is requesting the other computer to perform database operations. The file transfer involves the sharing of a potentially complex piece of information: the contents and structure of a file.

3.2 AGREEMENT ON TIME PERIODS

3.2.1 Absolute and relative time measurement

There are two different ways to measure time periods, and well-known terms from the world of computing and communications — 'synchronous' and 'asynchronous' — can be used for each. Be warned that these terms are often used as shorthand terms for some specific styles of communication, as will be seen later. These uses should not be confused with the very general usage given below.

One English language dictionary defines 'synchronous' as 'keeping time together'. This gives the basic idea here, which is that a clock is available to allow absolute time measurement. The units of measurement will vary depending on the type of communication — days, seconds or microseconds are all possibilities. The starting and finishing times are measured in terms of the absolute units. Relating this to Figure 3.1, the communication would be placed according to two measured points on the time axis.

A typical dictionary definition of 'asynchronous' is 'not synchronous'. This is appropriate here, in the sense that it means there is no clock to supply absolute time measurement. Instead, the starting time will just be defined as 'the time at which communication is seen to begin' and the finishing time will just be defined as 'the time at which communication is seen to end'. That is, communications themselves become the measuring stick for time. Of course, to do this, communications must have easily observable beginnings and ends. Relating this to Figure 3.1, two points on the time axis would mark the position of the communication.

The basic ideas of synchronous and asynchronous communication, as described above, are simple. However, these two simple concepts form an essential basis for numerous communication mechanisms, both human and computer. Indeed, together they form one of the most important principles of computer communications. In order to illustrate this, the remainder of this section on time period agreement is devoted to a variety of examples, first from human experience, and then from computer communications. As well as confirming that the general underpinning notions apply, these examples also illustrate a wide range of common practical mechanisms.

Human experience

At a first attempt, one might classify the three human experiences as being synchronous, synchronous and asynchronous, respectively. The presentation by the speaker is seen as synchronous because of a known starting time and a known duration. The day's activity at an information desk is seen as synchronous because of known opening hours. However, the third example is seen as asynchronous because the act of making the call is what determines the starting and finishing times.

This is described as a first attempt at classification. Although it is natural to think of advertised times as a basis for synchronous measurement, life is not so straightforward. Unfortunately, it is one thing to advertise an absolute time, but quite another to ensure that everybody agrees on when that time actually comes round. The examples will now be considered in more detail.

One-hour presentation by a speaker

The global clock approach might be used for a one-hour presentation by a speaker. Officially, there is a worldwide agreement on absolute time: **Coordinated Universal Time** (UTC) (formerly known as Greenwich Mean Time). However, most people do not have easy access to the official time source. Instead, they rely on approximately correct versions given by clocks and watches. In this case, it might make sense to define a rule that a particular clock at the presentation venue is used to define an agreed absolute time. The time period of the presentation can then be measured by this clock.

In practice though, most presentations do not work this way. For example, the speaker may be guided by his or her own watch while travelling to the venue. If this watch is slightly slower than the chosen clock, the presentation might start a little late. This is not a problem for most audiences, showing that this communication has an asynchronous flavour. There is still a loose synchronous nature — the starting time is not meant to drift by hours or days — but the real starting time is determined by the moment at which information sharing begins.

Tourist information desk

A similar type of argument applies to a day's activity at a tourist information desk. The actual time period of operation is measured asynchronously, since it is determined exactly by the times when the desk is opened and closed by its human operators. However, the agreement on opening hours is approximately synchronous. Of course, in this case, further complications are introduced if tourists from far-flung places are trying to get information on the correct time adjustment needed for the local time zone.

Telephone call

The starting time of a telephone call is sometimes roughly synchronous, for example, where one person has agreed to telephone another at a specific time. However, it is more often asynchronous, and is just measured as the moment at which dialling starts. In almost all cases, the finishing time is asynchronous, measured as the moment at which the telephone is hung up. The actual time at which the caller and the called observe the beginning and end of the call will be slightly different. This is not a problem, as long as the time period is being regarded as asynchronous.

Computer communications principles

Before a communication can take place, the two computers and the channel must agree on rules that give a common view of the time period. For a synchronous approach, the easiest rule is for all to agree that some global clock is used by everyone. However, this is somewhat dangerous because it is then necessary to have sharing of timing information between the clock and the communicators. Such sharing then introduces its own extra timing problems, and these must be solved if accurate agreement is needed.

For computer communications, the situation is a bit different from that experienced by humans. This is because agreement on approximate absolute times is often possible to some extent. For example, the Internet Network Time Protocol (NTP) allows computers to maintain accurate real-time clocks to within one-second accuracy and, in favourable circumstances, to within millisecond accuracy. This is done by the careful sharing of information gleaned from radio transmission of national clock signals. Exercise 3.4, at the end of this chapter, invites the reader to investigate the workings of NTP further.

Rough agreement using a global clock is feasible for some purposes, like ensuring that communication takes place every hour, or every day. However, in the absence of a global clock that can be read instantaneously by communicators, a plausible alternative is to have local clocks that stay in step for the duration of a communication. If a high degree of accuracy is required, such clocks can only be kept in step for fairly short periods. The human analogue of this would be to insist that everyone involved in some communication has to keep their clocks and watches synchronized to the same second. It is a little easier to impose such discipline on two communicating computers.

3.2.2 Examples of communication time periods

In summary, the first example — sharing a single bit — may be regarded as synchronous, and the other examples are asynchronous. Most computer communications are asynchronous.

Sharing a single bit

The sharing of a single bit is the atomic act in the world of digital communications. As a basis for this act, assume that the two computers are connected by a channel capable of transmitting a two-state signal from one computer to the other.

'*Bit synchronous*' transmission

For a pure **bit synchronous** communication, the two computers must have synchronized clocks that determine precisely when the transmitted signal represents the bit to be shared. For this to be possible, earlier agreement must be reached to synchronize the clocks. The most likely way for this to be achieved is for an initial asynchronous communication to set up a rule that, immediately thereafter, bits

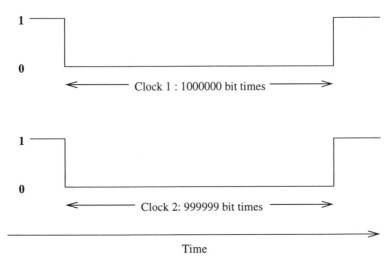

Figure 3.2 Time drift between two independent clocks

will be transmitted at a regular frequency. For example, such a communication might indicate that bits will arrive at a rate of 10 Mbits per second immediately thereafter. In this case, the receiving computer would expect to receive a new bit every 100 nanoseconds. So, the single bit to be shared is now being seen as just one of a sequence of bits. This is because such a synchronization agreement is not made just for one bit alone — this would be pointless. Strictly speaking, therefore, this type of communication should have been deferred until sharing of bit strings is considered in the next two examples.

The main problem with the bit synchronous approach is that it is not possible to keep two independent clocks synchronized for long periods. Suppose that the receiver's clock runs at a fractionally slower rate than the transmitter's clock. Figure 3.2 illustrates the sort of problem that might occur. Here, a long sequence of **0** bit values is transmitted, and this is represented on the channel by a long period in the same state. The receiver might under-count the number of zeros, because it expects each bit to last for fractionally longer than the transmitter intended. If this continues, the bit boundaries gradually drift in time until the difference is longer than one bit time. A similar type of problem occurs if the clocks drift in the other direction.

Clock signals

The normal solution to the above problem essentially makes the communication of each bit asynchronous. However, beware that this solution is often still called 'bit synchronous' by many people. The idea is that a clock signal is transmitted as well as the actual information. This signal lets the transmitter share its timing information with the receiver. The simplest approach is to add a second two-state

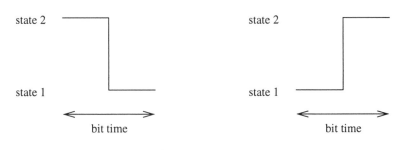

(a) zero bit value representation   *(b)* one bit value representation

Figure 3.3 Manchester encoding of the two bit values

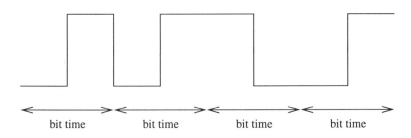

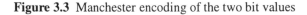

Figure 3.4 Manchester encoding of **1101**

channel, which has its state alternated at each information bit time period. This change of state is an asynchronous indication to the receiver that a new shared bit is beginning (and that the previous one, if any, is ending). The main drawback is that a second channel is required, which may be an unacceptable expense.

Manchester encoding

Another way to share timing information is for it to be carried on the main inform-ation channel. One widespread way of doing this is **Manchester encoding**. To understand this method, it is important to avoid thinking of the two states of a bin-ary channel as literally representing the two possible bit values. Figure 3.3 shows how the **0** and **1** bit values are represented in the most common form of Manchester encoding, and Figure 3.4 shows the bit string **1101** in this representation.

The interesting feature is that each single bit value involves a transition in the state of the channel. This eliminates the problem of long periods of unchanging state, as can be seen from Figure 3.5, which shows a sequence of four **0** bit values. The guaranteed channel transition is an asynchronous indication of the middle of each bit time period. This indication synchronizes a clock that the receiver uses to determine the beginning and end of bit time periods.

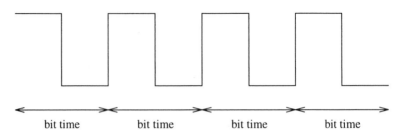

Figure 3.5 Manchester encoding of four **0** bit values

Baud rate

The **baud rate** of a physical channel is the maximum number of state changes that are possible per second. That is, given that a digital channel conveys a signal that has a number of different discrete values, the baud rate measures the maximum number of changes in value that are possible in one second. For a binary digital channel, this is just a measure of the number of changes from one state to the other that are possible.

Manchester encoding gives one illustration of this unit in use: the baud rate of a binary channel carrying Manchester encoded binary information is double the information bit rate. For example, a 200 Mbaud channel can carry Manchester encoded information at a rate of 100 Mbits per second. In this case, the baud rate is higher than the bit rate. In other cases, the baud rate is less than the bit rate. This is because the channel has more than two states, and so several bits can be encoded by one channel state. For example, a modem offering a rate of 33.6 Kbits per second typically uses a baud rate of 1800–2000, with the large number of channel states being represented by different analogue waveforms.

Beware that, very often, the term 'baud' is used casually to mean 'bits per second'. Strictly speaking, knowledge of how information bits are physically represented by channel signals is needed before the two terms can be equated.

Sharing a single character

A single character can usually just be treated as a string of eight bits (that is, a byte). Thus, techniques for sharing characters are really just versions of those used for sequences of single bits, or are special cases of those for sharing arbitrary length bit strings (considered next). However, it is useful to consider characters separately, because a particular asynchronous method is sufficiently common that it steals the term 'asynchronous transmission' when it is applied to binary channels. This method is now usually only used for communication between computers and slower peripheral devices, such as terminals and printers. Its asynchronous nature copes well with serious speed mismatches, such as between a human typing characters and a computer processing the characters.

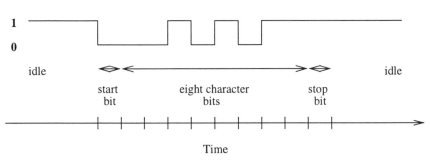

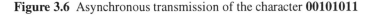

Time

Figure 3.6 Asynchronous transmission of the character **00101011**

'Asynchronous' character transmission

Figure 3.6 illustrates how an eight-bit character is transmitted, in terms of its individual bits. The idea is that, before the string of eight character bits is transmitted, a single 'start bit' is transmitted. Usually, when no communication is in progress, the signal on a binary channel corresponds to the representation of a **1** bit value, and then the start bit has a **0** bit value. The start bit denotes the beginning of a character communication time period. This asynchronous event establishes an agreement that the next eight bits, plus one further bit, will be transmitted synchronously. The rate of arrival is normally fixed for the channel or, if not, is deduced from the rate of arrival of the first few characters transmitted. After the string of eight character bits has been transmitted, a single 'stop bit' is transmitted. Usually, the stop bit has a **1** bit value, and then the channel state remains the same until the next character communication begins. (In earlier times, two stop bits were sometimes used, to add an extra delay while mechanical terminal devices handled the character.) The stop bit denotes the end of a character communication time period.

The disadvantage of this form of transmission is that there is at least a 25% overhead in the number of bits actually transmitted, over the number of information bits. However, this should not be a major problem if the information sharing is genuinely asynchronous at the character level, because there will be gaps between character communication time periods anyway. If characters are more often bunched together, then techniques for handling larger strings of bits are more appropriate.

Sharing a string of bits

Sharing a string of bits is the fundamental operation at the lower levels of computer communication. The earlier description of sharing a single bit mentioned that synchronization on a single-bit basis is not attempted. Instead, an initial asynchronous act establishes an agreement for synchronous transmission of subsequent bits. The example of sharing a single character was a special case of this approach. In general, sharing of a bit string is achieved by first transmitting an agreed distinctive bit pattern to indicate the beginning of a communication time period. This

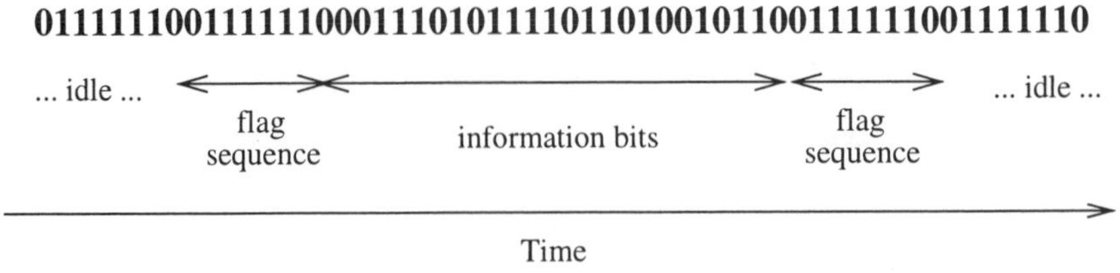

011111100111111000111010111101101001011001111111001111110

Figure 3.7 HDLC framing of **001110101111011010010110**

is followed by the transmission of the information bits. Finally, the end of the communication time period is indicated by the transmission of a distinctive bit pattern. This pattern is often the same as the start pattern, which allows several bit strings to be transmitted one after another, with a single distinctive pattern in between each string.

HDLC framing

HDLC framing is an example of the above approach to sharing a string of bits, and is very common on physical channels between computers. This framing has its name because it is used to surround the classic HDLC message, shown in Chapter 2. The general form of HDLC framing is shown in Figure 3.7. It uses the distinctive eight-bit pattern **01111110** as a 'flag sequence' to mark both the beginning and the end of communication time periods. The information bits shown in between might be the representation of an HDLC message, but this is not an issue of concern here.

On a direct channel between two computers, the flag sequence is normally transmitted continuously when no communication is in progress. This allows the receiver to keep its clock in step with the transmitter's clock all of the time. So, strictly speaking, the beginning of a new communication time period is denoted by the received bit stream changing from repeated **01111110** bit patterns to something different.

Bit stuffing

One problem with the use of special bit patterns to mark the beginning and end of communication time periods is that the same bit pattern might occur within a bit string being communicated. This could lead to the transmitter accidentally appearing to end one time period and then begin another. To prevent this, a technique known as **bit stuffing** is usually used. The idea is that, when the transmitter notices that it may be about to transmit the special bit pattern, it transmits an extra bit that prevents an accident.

For example, the HDLC bit-stuffing rule to protect the flag sequence **01111110** is that, after transmitting five **1** information bit values, the receiver always inserts a **0** bit value. Just as the flag character is pre-agreed between the communicating parties, so too is the associated bit-stuffing rule. Thus, in this case, the receiver

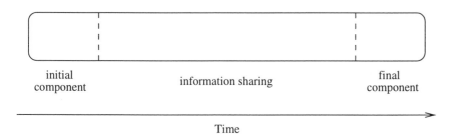

Figure 3.8 Bit stuffing **01011111011** and **10011111101**

| initial component | information sharing | final component |

Time

Figure 3.9 Transaction or file transfer viewed over time

knows that, on receiving the sequence **111110**, the **0** bit value was stuffed and should be discarded. Two examples of HDLC bit stuffing are shown in Figure 3.8.

Database transaction, or file transfer

At this stage, there is little extra to say about the behaviour over time for these two examples. The general idea is similar to that seen for bit strings, except viewed at a higher level of abstraction. It is illustrated in Figure 3.9. There will be some distinctive initial component of the overall communication that denotes the beginning of a new database transaction or file transfer.

For example, if there is some collection of different communicated message types that are used to implement the transaction or transfer, then the initial component will just be one particular type of message. Similarly, there will be some distinctive final component that denotes the end of the transaction or transfer, and this too will be a particular type of message.

This generalized summary captures the spirit of almost all time period agreements in computer communications: easily observable signals precede, and follow, the sharing of information. This forms the basis for an asynchronous style of operation. In fact, the general form of Figure 3.9 captures the forms of Figures 3.3, 3.6 and 3.7 as special cases.

Normally, the extra signals are not just there for timing reasons. As will be seen later, they mark the beginning and end of an enclosing context or environment for the communication. This context or environment affects several different aspects of how the communicators behave during the act of communication.

3.3 IMPLEMENTATION OF REQUIRED TIME PERIODS

Agreement on time periods is only concerned with how the communicators agree on the beginning and end of the time period for a communication. However, it is very often the case that communication is not continuous within the agreed time period. In Section 3.3.1, the effect of pauses during communications is explored. The aim is to explain how communications with complex behaviour over time can be broken down into **sub-communications** with simpler behaviour over time. After this, in Sections 3.3.2 and 3.3.3, it is shown how quality improvements can be obtained by combining the use of sub-communications and the adjustment of the time periods of communications. The implementation techniques for time behaviour are a central part of the communications trade, since time is the dynamic component that has no independent existence away from acts of communication.

3.3.1 Segmentation and concatenation

Human experience

One-hour presentation by a speaker

Consider first the one-hour presentation by a speaker. The audience is likely to be surprised if faced with a continuous barrage of words for 60 minutes. Instead, there should be some identifiable structure to the talk. For example, there may be a number of identifiable sections and sub-sections in the talk. Ultimately, individual sentences might be an appropriate level at which to stop breaking down the structure of the talk over time. Unless the speaker tends to hesitate or to pause for dramatic effect, the words in each sentence will be delivered continuously.

Regardless of the level at which the talk is broken down, each component can be seen as a smaller sub-communication with a time period agreed asynchronously. For example, physical gestures or voice intonation might be used to denote the beginning or end of sections, sub-sections, sentences, etc. within the presentation.

The overall situation is similar to the relationship between this book and its reader. On its own, the book is just a collection of stored information with some structure — chapters, sections, sub-sections, down to sentences. When a reader is involved, communication takes place. The act of reading a particular component fixes a time period for the sharing of information. Importantly, although the author might hope otherwise, there is no reason why the order of reading components of the book must be exactly the same as the order that components appear in the book. For example, the same sharing of information might be achieved by reading chapters out of sequence. However, reading individual sentences in random order might not lead to the same degree of information sharing.

Original communication:

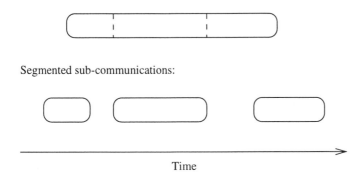

Segmented sub-communications:

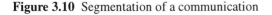

Time

Figure 3.10 Segmentation of a communication

Tourist information desk, and telephone call

Similar observations apply both to the tourist information desk and to a telephone call, although the degree of structuring is likely to be rather simpler in both cases. A day's activities at the information desk can be broken down into individual queries from tourists, occurring at arbitrary points in time. In turn, these might be broken down further into individual sentences. A telephone call will typically have no more structure than just being a collection of sentences separated by pauses.

It may be the case that some sub-communications overlap in time. When a tourist makes an enquiry at the information desk or when two people talk over the telephone, breaking down the communication to the level of separate sentences, for example, will result in non-overlapping sub-communications. However, through accident or rudeness, it is possible that information might be flowing in both directions at the same time. Thus, sub-communications may overlap in time.

Computer communications principles

The obvious simplification to time periods is to use a technique known as **segmentation**. This is illustrated in Figure 3.10, where one communication is segmented into three sub-communications. The actual timing of the sub-communications is not important, just the fact that they are distinct. In this book, the term 'segmentation' is used in a very general sense, to cover any kind of breaking up of communications. In common parlance, 'segmentation' (as its name suggests) usually only applies in a fairly low-level sense, for example, when a longer bit string is communicated by transmitting several shorter strings over a series of time periods.

A less obvious simplification, perhaps, is to use a technique known as **concatenation**. It is illustrated in Figure 3.11, where three communications are concatenated into one **super-communication**. In this book, the term 'concatenation' is used in a very general sense, to cover any kind of joining together of communic-

Original communications:

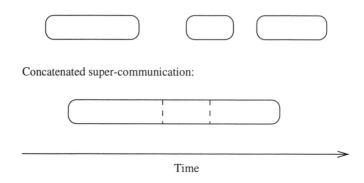

Concatenated super-communication:

Time

Figure 3.11 Concatenation of communications

ations. In common parlance, 'concatenation' (as its name suggests) usually only applies in a fairly low-level sense, for example, when several shorter bit strings are communicated by transmitting a longer string in a single time period.

When a communication consists of a flow of information in one direction only, segmentation and concatenation are opposites of one another. Both involve dividing up communications and putting together communications, but in opposite orders. Dividing up is the easier operation, i.e., the transmitter has the easier task for segmentation, and the receiver has the easier task for concatenation. Putting together is the harder operation. In the case of the receiver for segmentation, it is necessary to ensure that exactly the right sub-communications have occurred, before the original communication is achieved. In the case of the transmitter for concatenation, it is necessary to find an appropriate collection of communications that can be put together to form the actual super-communication. This may involve some prediction to guess whether suitable communications are likely to occur in the near future.

With a unidirectional information flow, segmentation and concatentation have a natural engineering explanation. The sending computer splits up (puts together, respectively) information, and the receiving computer puts together (splits up, respectively) information. When there is an information flow in both directions, the situation is more complicated, since both computers are involved in dividing up, and putting together, the communication. In general, only an external observer can be aware of exactly what segmentation and/or concatenation is carried out for a particular communication. This is because the two computers may observe sub-communications happening in different orders, due to the delaying effect of the channel.

For example, one possible role for segmentation is to divide a bidirectional communication into two sub-communications, each covering one direction of information flow. Such segmentation is very easy for an external observer to see, when it happens. However, it involves a division of information sending between the two computers, rather than one computer dividing up the information it sends. This illustrates that segmentation of communications is more than just the chopping up of information.

In this case, overall control that ensures the two sub-communications implement the required communication must be achieved by having prior agreement between the two computers on how the correct sharing behaviour is to be implemented by putting constraints on the two independent behaviours. Section 3.3.3 contains a simple example of how this might take place, where the constraint is that one computer is responding to information received from the other. In general, such agreements are only practicable when the unidirectional sub-communications are not, in turn, segmented much further.

In some cases, segmentation is used as a means of implicitly representing structure within the information communicated. That is, the information shared by each sub-communication is viewed as one distinct component of an overall structure. If this is the case, then the receiver must ensure that the extra structural information provided by the segmentation is preserved when it concatenates the information received. This is termed **unit-oriented** communication. The simpler alternative, that does not preserve segmentation structure, is termed **stream-oriented** communication. This term reflects the fact that a stream of information is shared, with no added structure.

Computer communications examples

The sharing of a single bit cannot be simplified by segmentation, since bits are viewed as atomic. However, all of the other examples can be simplified this way, using methods that have been hinted at during discussions on how agreement about time periods can be achieved. The sharing of single bits might be simplified by concatenation, but this is just a special case of bit-string concatenation, considered below.

Sharing a single character

The asynchronous method of transmitting eight-bit character representations, preceded by a start bit and followed by a stop bit, cannot be segmented further. This is because each 10-bit string is transmitted synchronously, which implies continuity of transmission without any pauses. However, looking ahead to Chapter 5 and beyond, where multiple communication channels are introduced, an alternative way of transmitting an eight-bit character representation would be to send all eight bits at the same time, over eight independent binary channels.

If this is done, the communication of a character has been broken down into eight separate sub-communications, all of which occur in the same time period. Obviously, agreement rules are necessary to ensure that the receiver does obtain all of the bits 'in the same time period'. An obvious way to achieve this is to make sure of three things: the transmitter sends all eight bits at exactly the same time, all eight channels impose the same delays, and the receiver reads all eight channels at exactly the same time. Such pinpoint precision is unlikely but, in practice, an adequate degree of synchronization can be achieved to make this scheme work.

Sharing a string of bits

For an arbitrary-length bit string, an implementation problem is that limits might be placed on the duration of communications. These appear as restrictions on the lengths of bit strings that it is possible, or desirable, to transmit. The limits may be either lower bounds or upper bounds on length.

If segmentation is used, a longer bit string is communicated by transmitting several shorter strings over a series of time periods. The shorter strings are obtained by dividing up the longer string. The division is based purely on length, since there is not any logical structure within the string. If concatenation is used, several shorter bit strings are communicated by transmitting a longer string in a single time period. The longer string is obtained by concatenating the shorter strings together.

As an example, both segmentation and concatenation can be used to provide the ISO standard transport service, which provides a channel capable of transporting arbitrary-length bit strings between two computers. This service is described in a little more detail on page 133 in Chapter 4, it features in the case study contained in Chapter 10, and its place in the ISO standards family is explained in Section 12.3.2 of the final chapter.

The required service has to be implemented using the services offered by any channels that are available to connect the two computers. Such channel services will carry bit strings, but some may have restrictions on maximum length, for example, whereas others may have better behaviour for longer lengths. Because of this, the implementation of the transport service must use appropriate segmentation and concatenation.

Database transaction

The key feature of a database transaction is its 'all or nothing' property. A transaction consists of a sequence of database operations and either *all* of the operations are performed, or *none* of them are performed. Such behaviour fits naturally with the idea of segmenting communications. The transaction is the higher-level communication, and is divided into several simpler sub-communications carried out over time. This decomposition is likely to occur as a feature of the database service implementation or as a feature of the database user's behaviour, rather than as a necessity for implementing information sharing between the database and its user.

The ISO standard Commitment, Concurrency and Recovery (CCR) service is an example of a general mechanism for segmenting a transaction's time period. An outline of this service is deferred to Chapter 4, since its features also embrace other principles yet to be discussed.

File transfer

The transfer of files between computers is an example of a self-contained piece of computer communication occurring during some time period. Depending on the type of file involved, there may be scope for simplifying the time period. For example, a file with some overall logical structure might be transmitted as a sequence of logical components, separated by pauses. Alternatively, or in addition, a file might be transmitted as a sequence of physical components. The most natural example of the latter is when a file is transmitted as a sequence of blocks, reflecting the way it is stored physically on a computer system.

There is also scope for higher-level communications associated with file transfers. For example, the transfer of a collection of files, perhaps the contents of a directory, might be viewed as one communication. This mass transfer can then be segmented into individual file transfers over separate time periods.

Other classic applications can be decomposed in a similar style. For example, the communication between a remote terminal and a computer system can be segmented into separate log-in periods for different users, and each such period might be further broken down into separate commands issued by the user.

ISO standard session service

Given that applications often have the sort of time structure that file transfers and terminal accesses have, the ISO standard Session Service offers a general way of segmenting the time period of a communication. The place of this service within the ISO standards family is explained in Section 12.3.2. In practice, the service has been distinctly under-used, since most applications just implement their own mechanisms. However, it is still a reasonable illustration of segmenting communications in a general-purpose way. The structural components are shown in Figure 3.12.

The highest level of time period is the **session**, which represents the period during which communication takes place. The service user can then segment the session into logical units called **activities**. The division is determined by the user. For example, a session might be a collection of file transfers, with one activity for each separate file transfer.

Within activities, service users can insert **major synchronization points** to separate logically significant pieces of work in the activity. Finally, users can insert **minor synchronization points** between major synchronization points. Thus, the service offers a four-level scheme for simplifying time periods, which is available to be used to best effect by communicating applications.

3.3.2 Flow control

The foregoing discussion of agreement and implementation issues for time shows that most sharing of information by communication has two key properties. First, the communication is *asynchronous*, that is, the starting time is not fixed relative

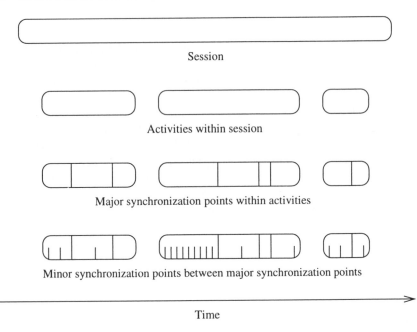

Figure 3.12 Structure of time periods in the ISO session service

to some global clock. Second, the communication can be *segmented* into sub-communications, that is, communication time periods can be expressed in terms of other, simpler, communication time periods. Both of these properties may be required for communication to take place at all. However, very often when these properties hold, they can be exploited to improve the quality of communication for one or more of the communicators. This section explores ways of achieving mutually agreed starting times, rates and durations of communication.

Human experience

One-hour presentation by a speaker

It is unlikely that the speaker's material demands that the presentation must take place at some specific date and time. Instead, prior negotiation will take place to ensure that the chosen time is mutually convenient for the speaker, most of the intended audience and the availability of any communication media required. Such negotiation may be dominated by any of the interested parties, or may be equally weighted towards all parties. Given an advertised starting time, the actual starting time might be further tuned, for example, to suit the speaker's convenience or the audience's readiness.

The rate at which the presentation is delivered also has an element of flexibility. There will be a general consensus on what range of rates is appropriate, beyond which the presentation will seem 'too slow' or 'too fast'. However, within this range, the speaker may choose a rate he or she feels to be appropriate for the material. The audience may also affect the rate of delivery by means of signals conveyed back to the speaker — for example, audible yawns to indicate slowness or interruptions asking the speaker to slow down. Similar considerations apply to the duration of the presentation. Although it is advertised as having a one-hour duration, the actual duration may be fine-tuned to suit the participants.

Tourist information desk, and telephone call

Similar considerations apply to the examples of the tourist information desk and the telephone call. There is considerable scope for varying the starting time of communications to suit the participants. The rate of communication must be adjusted to suit all parties. The duration of communications can be flexible to suit the information being shared.

Computer communications principles

In summary, all three examples of human experience show that people have well-developed mechanisms for achieving quality of communication by careful use of timing. Not surprisingly, similar mechanisms are also used extensively for computer communications. The use of asynchronous features to allow mutually agreeable time periods and rates is normally named **flow control**, although other terms are used in certain specialized circumstances.

To simplify the discussion here, it will be assumed that the information sharing requires a flow of information in one direction only. Thus it is acceptable to refer to an 'information transmitter' and an 'information receiver'. This is a realistic simplification to make since, when the mechanisms described here are applied to genuine bidirectional sharing, they usually apply independently to the different directions of information flow. That is, they are applied after bidirectional communication has been segmented into two unidirectional sub-communications. Practical examples in Chapter 4 will illustrate this point.

The essence of flow control is to ensure that information sharing begins at an agreed time, continues at an agreed rate, and lasts for an agreed period. It has been assumed so far that the communication channel is a passive entity, while the communicating computers are active. This assumption will be continued in the discussion of flow control here, so that the channel does not contribute to negotiations on communication time periods. However, when rather more complex types of channel are introduced later in the book (for example, in Chapter 6), it will be seen that channels also can be full partners in a flow control mechanism.

Domination by the information transmitter

The first flow control mechanisms to be described will involve a dominant information transmitter. Two unsubtle possibilities involving a dominant transmitter are:

- the transmitter communicates at its convenience, and a slavish receiver is assumed to cope;
- the transmitter communicates at its convenience, but the receiver is allowed to ignore the communication if it is inconvenient.

Such rules need to be agreed in advance by both parties. The first approach was common in the early days of computer communications, where a central large computer was a master transmitter, and a smaller computer or terminal was a slave receiver. The second approach is in a similar vein, but gives the receiver a crude form of flow control.

At the lowest levels of computer communication, where hardware transmitters and receivers are attached to physical channels, both approaches are common, since the hardware can be dedicated to communication. However, if the full powers of the receiving computer are required for the communication, then it is usually given a more active role in flow control.

A variant on this scheme is to have a 'benevolent dictator' type of transmitter. That is, as part of the original agreement to communicate, the transmitter gives a specification of the **traffic flow**. This means that it describes how the communication will proceed over time. The receiver only joins in the communication if it is prepared to handle the described traffic flow. The transmitter must respect the agreement so if, during the communication, the actual traffic flow is going to differ from that agreed, the transmitter must employ **traffic shaping** to ensure that its transmissions match the agreed profile. This type of scheme is of much interest in modern high-speed computer communications, where predictability of information flows is important. The mechanism is also desirable, because it allows computers to transmit promptly, without being delayed while polite enquiries are made about the readiness of the receiver.

Domination by the information receiver

At the other extreme to a dominant transmitter, there may be instead a dominant receiver, two possibilities being:

- the receiver demands communication at its convenience, and a slavish transmitter is assumed to cope;
- the receiver demands communication at its convenience, but the transmitter is allowed not to communicate if it is inconvenient.

Similar symmetric comments can be made as were made about the dominant transmitter case. In fact, combinations of dominant receiver and dominant transmitter approaches are not unknown. Historically, a central large computer would be the dominant transmitter *and* receiver, and smaller computers or terminals would be the slave receivers and transmitters.

The technique used by a dominant receiver to issue its demands is usually an example of a general mechanism called **polling**. The receiver sends a request for input to the transmitter, which is then obliged to respond immediately (or, at least, as soon as practicable). The transmitter's response to a poll might be the communication of information, if the transmitter has any, otherwise it will just be an indication that there is no information to communicate.

Ticket style flow control

For most types of present-day communication, there is a more balanced relationship between the communicators, with no domination by one or the other. Agreement on flow control rules may be either static or dynamic, or a mixture of both. The above master-slave relationships are extreme examples of static agreements. More common are agreements that give a transmitter flexibility in choosing the beginning of communication time periods, but within constraints imposed by the receiver.

The arrangements are usually akin to the use of open tickets for public transport, where the possession of one or more tickets allows a traveller to use the transport at some future time. This is in contrast to polling, where the ticket is of a type that only allows travel at a particular time. Here, the word 'ticket' is used to describe a special information type communicated for flow control, but note that the alternative word 'token' is often used in practical applications. Two schemes for the issue of flow control tickets are:

- the receiver issues tickets automatically;
- the transmitter requests tickets, and the receiver responds.

Normally, possession of one or more tickets allows a transmitter to begin a communication at a future time of its choosing. This flexibility may be problematic for the receiver. The second scheme should allow the extent of the future to be constrained, assuming that tickets are only requested when a transmitter is about to communicate. A further possibility, not common in practice, is to build an expiry time into the ticket, so that it can only be used for a limited time into the future.

Apart from allowing a time period to begin, flow control tickets place restrictions on the permitted duration of communication. These are normally expressed indirectly, in the sense that tickets allow a maximum amount of information sharing during a communication time period. The time duration is then a function of the rate of communication. Where a communication is synchronous, the rate will be fixed, and already agreed by both parties. Thus, the maximum duration of a communication can be easily established. Where a communication is asynchronous, as is usually the case at all but the lowest levels, the maximum duration might be estimated (if this needs to be done) on the basis of previous experience of communication rates.

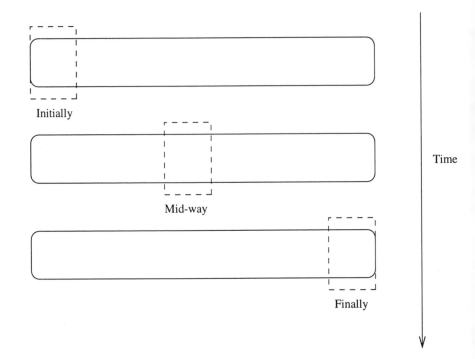

Initially

Mid-way

Finally

Time

Figure 3.13 Movement of sliding window over information

Sliding windows

In Section 4.5.2, examples of connection-oriented protocols are examined. This sort of protocol usually includes a flow control mechanism that employs a variant of a particular implementation of the ticket style of flow control: a **sliding window** mechanism. For this to work, it is necessary that a communication can be segmented into asynchronous sub-communications.

The basic idea is that the communicating computers have a notion of a 'window' through which a portion of the information being shared by a communication is visible. This indicates that, under the flow control agreement, a sub-communication involving the sharing of the visible information may begin at the present time, or in the future. Attempts to share invisible information are not allowed. Initially, the window is over an initial subsequence of the information and then, as communication progresses, it is slid along the information until it is over a final subsequence of the information. This is illustrated in Figure 3.13 (the size of the window stays the same in the figure, but this is not a necessary feature).

One extreme case of this general mechanism is when the size of the window is the same as the size of the information. This means that there are no restrictions on how the communication should proceed over time. At the other extreme, when the size of the window is zero, no communication is allowed.

The units of size for the sliding window are pre-agreed by the communicating parties. The choice reflects the degree of segmentation possible, with one unit typically corresponding to the smallest unit of information that can be transmitted as a sub-communication. When the window size is one, only a single sub-communication may be in progress at any time. Normally, however, the window size is greater, which allows more than one sub-communication to take place simultaneously. A desirable aim is for sufficient sub-communications to be in progress simultaneously to ensure that the channel is always full of information in transit.

The reponsibility for deciding the size of the sliding window at any time, and for moving the sliding window, rests with the receiver. When it makes changes, these have to be communicated to the transmitter. Conceptually at least, this can be seen as an additional, separate, act of information sharing. Obviously, the new communication only exists in order to manage the main information communication, but it can be regarded as independent.

The reason for using a sliding window mechanism is to exploit the fact that the main information sharing communication and the sliding window information communication can be carried out simultaneously. Specifically, the two communications involve information travelling in opposite directions. Thus, in normal operation, while the main information transmitter is using up a supply of tickets granted by the position and size of the sliding window, the ticket supply is being replenished by changes to the sliding window. The aim is to ensure that subsequences of the main information can be communicated whenever it is convenient for the transmitter. This is an improvement on a simple **stop and wait** system where, after each sub-communication, the transmitter must wait until a new ticket is obtained from the receiver.

3.3.3 Acknowledgement and error handling

Human experience

One-hour presentation by a speaker

As the discussion on flow control in Section 3.3.2 shows, there are several ways in which the timing of a presentation can be adjusted to maximize the quality of information sharing. There are also ways in which the progress of the communication can monitored.

If the presentation is assumed to be of high calibre, and the audience is assumed to be exceptionally attentive, then the speaker could perform uninterrupted for one hour. Then, it would be safe to assume that all parties share the same level of information at the end. In practice, such perfection is unlikely. Instead, the speaker is likely to watch the audience for signs of incomprehension, or the members of the audience are likely to interrupt with requests for clarification. In the event of problems, the speaker will probably repeat some earlier part(s) of the presentation, perhaps in a modified form. This type of behaviour is possible because the presentation does not rely on a continuous flow of information over time.

Reviewing this, the overall communication can be segmented into a series of smaller sub-communications, which are themselves asynchronous. In terms of the communication channel model, each sub-communication is concerned with sharing some information between speaker and audience. It is possible to further break up each sub-communication into two parts. The first involves the speaker passing information to the audience. The second involves the speaker finding out whether the information passing is succeeding. The final aim of the overall communication is to ensure that the audience know all of the information, and that the speaker knows that they know it.

Tourist information desk, and telephone call

Similar considerations apply as in the case of the speaker's presentation, since there is also a need to ensure that the required information sharing is achieved by the communication. This is implementable by segmenting the communication time period appropriately.

Computer communications principles

The use of decomposition to give effective information sharing is normally named **acknowledgement**, and it is closely associated with **error handling** in most cases. With a flow control mechanism that ensures communication takes place at a time convenient for all parties, and with a perfect channel, there is no need for any acknowledgement or error-correction mechanism. This is because the information transmitter knows that information sharing will be achieved by the transmission. In less ideal circumstances, there will be some doubt at the information transmitter. Thus, true information sharing cannot be said to have been achieved. To deal with this problem, it is necessary for the receiver to communicate extra information confirming that it has successfully received the information sent by the transmitter. When this has been done, information sharing has been achieved.

At this point, it is reasonable to ask the contorted question, 'but how does the receiver know that the transmitter knows that the receiver successfully received the original information?' In fact, an infinite sequence of increasingly complex questions might be posed. To avoid such problems, it is normal for the receiver to assume that full information sharing has been achieved, once it has notified the transmitter. If this assumption is not correct, and the transmitter has a different view, then the onus is put on the transmitter to resolve the confusion. This is done using mechanisms described below.

Acknowledgement

To summarize, when information sharing involves an information flow in one direction only, then the communication normally can be broken up into a pair of sub-communications. The first one is the actual information sharing. Then, later in time, the second one shares information (in the opposite direction) about the success of the first one. The second sub-communication is usually termed an **acknowledgement**. If the information receiver signals success, the sub-communication is termed a **positive acknowledgement**; if it signals failure, the sub-communication is termed **negative acknowledgement**.

There are various reasons why a receiver might signal failure. One possibility is that the flow control arrangements were inadequate and, although the receiver has noted a communication attempt, it is unable to participate. The negative acknowledgement is then essentially a *post facto* flow control feature. A more common possibility is that an error-detection mechanism is in use, and the receiver has spotted that a communication attempt was faulty. In either case, the effect of the negative acknowledgement is that both communicating parties have the same view of the state of information sharing. Typically, the transmitter will then attempt to do the information sharing operation again.

Error correction

After being made aware that a communication attempt has failed, the transmitter is likely to make another attempt. This is an error-correction mechanism. However, an essential prerequisite is that the transmitter is made aware of failures. Unfortunately, this is not always possible. First, if a communication attempt is completely lost on the channel, the receiver will be unaware that it has failed to happen, unless there is a synchronous flow control agreement that fixes a certain communication time period. Second, the acknowledgement information communication is usually subject to the same sources of error as the main information communication, and so it is possible that acknowledgement information is lost.

To deal with these matters, there is normally an absolute time bound placed on the duration of an act of information sharing, that is, on the time required to communicate both the main information and then some acknowledgement information. If this time is exceeded, and the transmitter still has doubts about the success or failure of a communication, it assumes the worst, and will take action. Very often, it just behaves as though a negative acknowledgement has been received.

Retransmission

If a transmitter tries a communication again, after suspecting that a previous attempt failed, there are various possible states of belief at the receiver when the retransmission happens:
- the communication was successful;
- the communication was faulty or could not be dealt with; or

- there had been no communication attempt.

It is therefore important to ensure that any repeated attempt cannot cause confusion at the receiver. Confusion is most likely in the first case above, since the receiver will be faced with a second successful communication of the same information.

If it is harmless for information to be communicated more than once, such confusion at the receiver does not matter. Otherwise, however, each unit of information transmitted must be packaged in such a way that the receiver can recognize, and discard, duplicates. Where a communication has been segmented into subcommunications, it is common for each information subsequence to be tagged with an index indicating its position within the total information. This gives the necessary uniqueness. Such tagging also assists the receiver in concatenating subcommunications completely, and in the correct order. An alternative to ensuring that communicated information has an 'idempotent' property is to use additional control information communication to alert the receiver to an imminent repeat attempt.

Timeout periods

The maximum duration allowed for a communication, before unilateral action is taken, is called a **timeout period**. Since most communication is asynchronous, the selection of appropriate timeout periods is fraught with danger. If the chosen period is too short, then it is possible that a repeat communication attempt might overlap in time with an earlier attempt. This could cause extra confusion at the transmitter. For example, if acknowledgement information is received, it might refer to either communication attempt. If the chosen period is too long, communication will be slowed down. The transmitter will wait an undue length of time before making a repeat attempt, when there is no possibility of further developments from the original communication.

For computers with predictable behaviour, connected by a physical communication channel operated at an agreed rate, accurate timeout periods can be estimated. However, for any situation more complex, the estimation is far harder, and usually involves dynamic calculation based on recent past experience of communication times.

Expiry periods

In some cases, transmitters can enforce timeout periods by including an **expiry time** with information communicated. The intention is that, after transmission, the information only has a fixed time to live. If a communication channel or a receiver discovers that it is handling information that has passed its sell-by date, the communication attempt is abandoned and treated as an act that did not occur.

Although this approach can assist in making timeout periods work, it is more often used to prevent things like excessively long retention of information by channels, or because the information itself is time-critical. Thus, while closely related to acknowledgement timeout periods, expiry times usually exist for other quality improvement purposes.

Flow control combined with acknowledgement

The two previous sections have described flow control and acknowledgement as independent activities. This is true conceptually: acknowledgement forms an integral part of an information sharing operation, whereas flow control is an additional management communication. In practice however, the two activities are often combined so that receipt of a positive acknowledgement also implies permission to start a further communication in the future. Viewed from the other angle, for example, the movement of a sliding window might also imply positive acknowledgement of a sub-communication newly removed from the scope of the sliding window. Such combination of features is a special case of two general principles — multiplexing and splitting — to be introduced in Chapter 5. At this point, a principled explanation will be deferred. Instead, the next chapter, Chapter 4 shows how all of the ideas introduced in this chapter can be combined in practical circumstances.

3.4 CHAPTER SUMMARY

The sharing of information must be placed in a time-related context. A communication is defined as being the sharing of information within a time period with measurable starting and ending times. In a few cases, normally at the lower levels of communication, the measurement is synchronous, i.e., with respect to a clock. In most cases, the measurement is asynchronous, i.e., with respect to observation of the information sharing act. This demands that there are easily observable starting and ending indications.

Communicating parties must agree on the time periods for communications. It is not usually possible or desirable to use a global clock, or closely synchronized local clocks, to achieve synchronous agreement. Instead, distinctive initial and final components of the shared information are agreed upon, in order to achieve asynchronous agreement.

Except at the lowest levels, communications do not usually proceed as a continuous activity over time. To implement communications with complex time behaviours in terms of communications with simpler time behaviour, decomposition is necessary. Most often, this is some form of segmentation: one communication is divided into several sub-communications. Another opposite possibility is concatenation: several communications combined into one super-communication.

The combination of asynchronous agreement and segmented implementation may also be used to improve the quality of communication. The first major improvement is through flow control, where the timing of communication is adjusted to the mutual benefit of the communicating parties. The second major improvement is through acknowledgement, where the directionless notion of a communication as an act of information sharing can be broken up into sending and acknowledgement sub-communications. Acknowledgement can act as a basis for error-handling mechanisms involving retransmission or expiry.

3.5 EXERCISES AND FURTHER READING

3.1 Consider any communications that you carry out in order to share information with someone else, and compare the relative speeds of you, your partner, and the channel used, in each case.

3.2 For the communications considered in the previous problem, how is the communication time period agreed? Is the agreement synchronous or asynchronous?

3.3 How accurate is the time-of-day clock in any computer system that you use? How accurate is the internal clock used to drive the computer?

3.4 Have a look at RFC 1305, which describes the Internet standard Network Time Protocol (NTP). Do not try to understand all of the details, but note how complex it is. Also have a look at RFC 1769, which describes the Simple Network Time Protocol (SNTP), a simplified version of NTP that can be used when less accurate absolute times are required.

3.5 Suppose that a collection of geostationary satellites covering the whole world broadcasts a clock pulse every microsecond. Discuss the usefulness of such a facility for aiding absolute (synchronous) time agreements in computer communications.

3.6 Using the Manchester encodings for **0** and **1** given in Figure 3.3, draw pictures of the signals transmitted over time to represent (a) **11010**, and (b) **01010**.

3.7 An encoding scheme represents four bits of information using a pattern of five state changes on a binary digital channel. What channel baud rate is needed to achieve a 100 Mbit per second information bit rate?

3.8 For 'asynchronous' character transmission, it is usually possible to select 'one and a half' stop bits, as well as one or two. What does this mean from a time point of view?

3.9 Suppose that the eight-bit pattern **10101010** is used as a flag sequence to mark the beginning and end of time periods for the communication of bit sequences. Devise a bit-stuffing rule that will avoid confusion if this pattern occurs within a communicated bit sequence.

3.10 Explain how the HDLC bit-stuffing rule could be simplified, so that stuffing is less frequent. Can you discover why this simplification was not used?

3.11 Next time you make a telephone call, note how the overall communication proceeds over time. Write down all the different reasons why communication is not continuous in both directions.

3.12 Find examples of concatenation of communications between people. Why is concatenation useful in each case?

3.13 Give an example that shows why the segmentation of a communication that involves a bidirectional information flow into two *independent* unidirectional sub-communications is invalid in general. [Hint: recall that most communication is asynchronous.]

3.14 Suppose that you switch on a personal computer, use it for several pieces of work, and then switch it off again. Identify structuring, over time, of the overall communication between you and the computer.

3.15 What flow control methods are used during a conversation between two people who have met in the street?

3.16 Suppose that two computers that are 200 km apart are connected by a physical channel which can transmit information at two-thirds the speed of light. The two computers use a sliding window mechanism for flow control purposes. If the channel can transmit 10 Mbits per second, what choice of sliding window size might be appropriate?

3.17 Why are much larger sliding window sizes used when channels are implemented using satellites, rather than terrestrial cables?

3.18 During a conversation with another person, note how you and your partner acknowledge each other's information.

3.19 Consider communications between a bank customer using an automatic teller machine and a bank computer. What problems might be introduced if retransmission is used as an error-correction method?

3.20 In what ways are expiry times introduced into human communications?

Further reading

This chapter is intended to be self-contained — drawing on existing human experience of time issues in communication, in order to introduce the main ideas impacting on computer communication. The next chapter, Chapter 4, expands the coverage of how temporal matters are handled in practice. Later on, Chapters 6 to 11 include material concerned with the role of time in particular areas of computer communications.

FOUR

TIME PACKAGES

> *The main topics in this chapter about time packages are:*
>
> - Temporal behaviour patterns in communication services and protocols
> - Unsegmented communications, and connectionless services
> - Simple handshake packages
> - Multi-stage handshake packages
> - Connection-oriented packages
> - Implementation of connection-oriented services using connectionless services

---

4.1 INTRODUCTION

Chapter 3 explains the important issues affecting how two communicating computers behave over a time period. This chapter examines ways in which the behaviour over time is organized. A time package constitutes an agreement on when a communication begins and ends, and may also include an agreement on how the communication must be implemented using simpler sub-communications.

There are not very many fundamentally different forms of behaviour over time found in computer communication. It is true that there are an infinite number of ways in which two computers might organize a dialogue over time, but the more complex types of behaviour are built up by implementation in terms of a small range of simple behaviours. This general idea is familiar from the world of computer programming, where the ability to obey a range of simple instructions in different orders over time can be used to achieve complex computational effects. Here, simple communication patterns can be combined in different orders over time to achieve complex communication effects.

The implementation method of segmenting a communication into several simpler sub-communications is central to the packages discussed in this chapter. Frequently, segmentation is constrained by the fact that the sub-communications must occur in a particular order over time. That is, the sub-communications have to be *sequential*. This may reflect the fact that the total information communicated is often not fixed before the communication starts. It evolves as communication progresses, and so the full effect of a communication can only be quantified by an observer after its completion. Thus, preserving the ordering over time is important when implementing.

This sort of behaviour is familiar from the world of human experience. Most conversations between two people do not begin with precise goals for the information to be shared. There is a dialogue between the two speakers, with each reacting to what the other says. After the conversation, the speakers know what information (if any) has been shared.

Four classes of time package are described in this chapter. First, there is the most simple type: a single communication with no segmentation. After this, there are two classes based on 'handshakes' between two computers. A handshake is a precisely segmented dialogue, involving a short sequential exchange of information. Finally, there is the most complex type of package, which is analogous to behaviour used during a human telephone call. This forms an enclosing context for implementation of arbitrarily complex dialogues.

Examples of the different time packages described in this chapter can be found within the specifications of *communication services*. Essentially, a communication service is something that makes available channels with particular characteristics, so that these can be used for communications which have matching characteristics. In the context of this chapter, the characteristic of interest is the time behaviours that are allowed by the channels made available. Service specifications are prior (and binding) agreements on how users of the service must communicate over time. There is a close correspondence between the packages described here and the generic types of communication service offered in practice. However, this is not the whole story.

Communication protocols are used to implement required communication services using different communication services that are actually available. The agreements specified in protocols include a description of how the communicating parties behave over time. Therefore, examples of the different kinds of time package described in this chapter also appear within communication protocols. Of course, all aspects of the temporal behaviour of a protocol are constrained by what is allowed by the communication service that it uses.

The examples included in the descriptions below are drawn from a range of practical service and protocol specifications. For the purposes of this chapter, only aspects of each service or protocol that impact on time behaviour are included. Therefore, there is no full explanation of the overall intentions behind, and mechanisms involved, in each service or protocol. Such details are deferred to later points in the book, after a full account of basic communications phenomena has been given. The point of including the examples here is to illuminate the time packages, not the particular services and protocols.

4.2 UNSEGMENTED TIME PACKAGE

An unsegmented time package just describes the simplest form of asynchronous unidirectional communication: a communication involving the transfer of information from one computer to another, with the act of communication being continuous over time. That is, there is no segmentation, either for general-purpose implementation or for specific matters, such as acknowledgement or error correction. Two possible time behaviours with this package are illustrated in Figure 4.1. From the transmitter's point of view, the communication begins when it starts sending information. From the receiver's point of view, the communication begins when it starts receiving information. The ending of the communication is defined similarly, with respect to the act of transmission. Thus, the timing is asynchronous.

4.2.1 Connectionless services

There is a generic style of communication service that is founded upon the unsegmented time package. Such a service is termed **connectionless**, since it does not involve a 'connection'. The significance of this negative property is explained in Section 4.5.1, where connection-oriented services are introduced.

The term 'datagram service' is often used synonymously with 'connectionless service', particularly when referring to the service offered by message switching networks, the subject of Chapter 7. This is because the word 'datagram' conveys a more natural explanation of the service, stressing the similarity with the basic service offered by the human postal system. That is, a single piece of information is transmitted from one point to another. A datagram is the computer communication equivalent of a letter and the envelope that it is transmitted in. The letter contains the actual information and the envelope contains extra information to assist in the delivery of the letter.

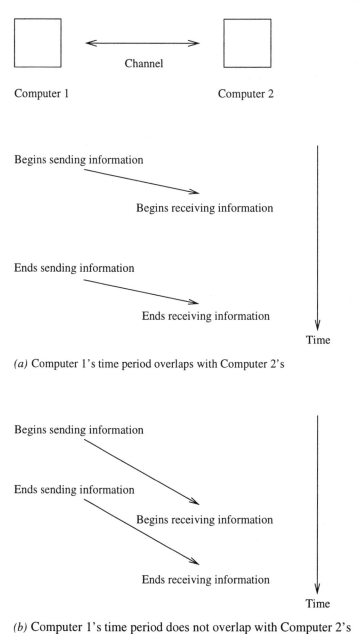

(a) Computer 1's time period overlaps with Computer 2's

(b) Computer 1's time period does not overlap with Computer 2's

Figure 4.1 Communication involving an unsegmented time package

In practice, there is usually a significant quality issue affecting connection-less services, which results from the very simple time package used. They do not normally guarantee delivery, unless they are explicitly described as *reliable* connectionless services. There is some probability that the communication will not be successful. There are three different ways in which failure can occur, although one of them might be termed over-success rather than failure. Figure 4.2 shows the problems.

First, as shown in Figure 4.2(b), the communication might fail altogether, in the sense that the intended receiver is unaware that any communication attempt took place. Second, as shown in Figure 4.2(c), the wrong communication might take place, such as the receiver only obtaining part of the information or obtaining a garbled form of the information. Third, as shown in Figure 4.2(d), the communication might take place more than once, so the same information sharing is repeated. The types of failure possible vary, depending on how the connectionless service is implemented. A further quality issue is the communication latency, that is, the delay between the transmitter starting to send, and the receiver starting to receive.

The examples of unsegmented time packages below focus entirely on connectionless services, and their properties. Such a package may form a component of a protocol as well. However, this would just correspond to an agreement that one computer may sometimes send a message to the other, independently of other protocol components. Thus, it is not particularly interesting to look at specific examples.

4.2.2 Examples of unsegmented time packages

Physical transmission service

A connectionless service is normally the basic offering from a physical transmission system. A message, regarded as a sequence of bits, is transmitted, usually bit synchronously, over a physical medium. Errors are caused by problems affecting the medium, and can result in the loss of messages or damage to messages; duplication is unlikely to be possible. The communication latency is determined by the communication speed possible over the medium, and is fairly predictable for any particular technology.

IEEE 802.2 LLC1 service

The IEEE 802 series of standards is concerned with **local area networks**, and its technical details are covered extensively in Chapter 6, with further comments on the standardization process itself in Chapter 12. The various IEEE 802.2 service standards cover communication between computers, making use of an underlying physical transmission service supplied by some type of local area network. The family of services offered is called **LLC** for Logical Link Control, and has three members. The simplest member, LLC1, offers a connectionless service. This is

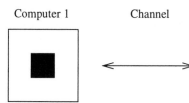

(a) Information known before transmission attempt

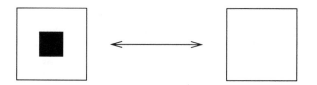

(b) Information known after transmission fails completely

(c) Information known after transmission is faulty

(d) Information known after transmission causes duplication

Figure 4.2 Possible failures for a connectionless service

a best-effort service, and gives no guarantee of delivery. Messages might be lost or damaged but, given the properties of most local area networks, messages are not duplicated. The communication latency is determined by the characteristics of the local area network used, and might not be easily predicted. This matter is discussed further in Chapter 6.

Internet IP service

The Internet Protocol (IP) underpins the operation of the Internet, and is described in detail in Section 8.3.2. It provides a connectionless service between two computers — in fact, this can accurately be described as a datagram service. The service might be implemented using absolutely any type of communication channel, and the distance between the two computers might be anything between one metre and the circumference of the world. Because of the scope of the service, the success guarantees are fairly weak — messages might be lost, corrupted or duplicated, and the communication latency may be large. The exact nature of the service depends on exactly how the two computers are connected together. Where it is useful for communicating computers to assess the quality of the service, this normally has to be done by measuring past examples of communication.

ISO connectionless services

Various ISO standards define connectionless services for different types of computer communications. The overall ISO philosophy on services, and a summary of the main standardized services, can be found in Section 12.3.2. As examples, there are standardized connectionless services that correspond closely to the three previous examples above: physical transmission, IEEE 802.2 LLC1 and Internet IP. The internationally neutral term 'data unit', as yet another name for a message, arises from the context of ISO's work on connectionless services. It is the unit of information transmitted by such a service. In most of the ISO service standards, a connectionless offering was a later addition. The original standards were strongly connection-oriented, and were only extended to add connectionless ideas a number of years later.

4.3 SIMPLE HANDSHAKE TIME PACKAGE

The unsegmented time package illustrated the simplest type of unidirectional communication. This section looks at the simplest type of segmented bidirectional communication: the simple handshake. For this, the communication has two sequential unidirectional stages, as shown in Figure 4.3. First, one computer transmits information to the other. Second, the other computer transmits information back. There are examples of the simple handshake in Chapter 3. One was the use of polling, where a request for input from a dominant receiver was followed by a response from the transmitter. Another was the use of acknowledgements, where

Overall bidirectional communication:

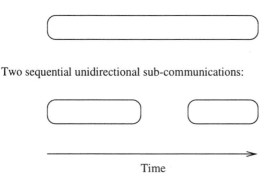

Two sequential unidirectional sub-communications:

Time

Figure 4.3 Communication involving a simple handshake time package

the sending of information was followed by the return of information indicating whether the first information had been received.

The use of simple handshakes is common within communication protocols, when implementing communication services. They may also be offered as the temporal basis of a communication service, with the handshake having some useful significance to the user. For example, when a connectionless service is enriched to involve a simple handshake time package, with the handshake used for acknowledgement, an **acknowledged connectionless** service is obtained.

Two other examples are **command-response** and **request-reply** services. The names used give a hint to the reason for carrying out the handshake: there is a control relationship between the two computers, not just an information-passing relationship. For these services, it is desirable for the service to be reliable, in the sense that a computer initiating a handshake always gets some sort of response back. However, it may be the case that the other computer is never aware of the communication, in which case the response must convey the fact that the communication was unsuccessful. There are examples of such handshakes in the applications examined in the case studies of Chapters 9 and 10.

Implementing such a reliable handshake-style service in terms of an underlying unreliable connectionless service is a case of obtaining a quality improvement. The handshake communication is segmented into two unidirectional sub-communications. The first stage involves sending information from Computer 1 to Computer 2. If this sub-communication is successful, then the second stage involves sending information from Computer 2 to Computer 1. If this second sub-communication is successful, then the overall communication has been implemented correctly. However, either sub-communication may fail, in which case Computer 1 will not receive any second-stage information.

In this case, the appropriate actions are of the type described for error correction in Section 3.3.3. That is, after an absolute time period has elapsed without any response, Computer 1 may either give up the communication attempt or try the first stage of the handshake again. However, it cannot be sure whether or not Computer 2 received the original information, and this may be problematic whichever action is taken after the timeout period.

Implementing a connectionless service using an underlying handshake-style service is rather straightforward. The sender of information just starts a handshake, but ignores any response received back. The recipient of the information may either not complete the handshake by not responding, or complete the handshake with an information-free response. The latter action is probably better, since it will short-cut any timeout mechanisms that may underpin the handshake service.

4.3.1 Examples of simple handshake time packages

IEEE 802.2 LLC3 service

The LLC3 service is the intermediate member of the ISO 802.2 LLC (Logic Link Control) family (the '3' is because it was the third, and last, member to be added). It is an acknowledged connectionless service for local area networks. It is usually built on top of an underlying (unreliable) connectionless service. However, as can be seen in Chapter 6, some types of local area network (for example, rings) supply an acknowledged connectionless service as a natural primitive. If this is the case, the LLC3 service can be implemented straightforwardly.

ISO Reliable Transfer Service

The ISO Reliable Transfer Service (RTS) is a handshake-style service intended for use by computer applications that require communication facilities. Its place within the spectrum of ISO service standards is indicated in Section 12.3.2. As its name suggests, reliability is its central feature. One example is electronic mail, where each mail posting is treated as a message that is sent reliably. This is roughly analogous to the registered post service, where the receipt of a letter can be confirmed to its sender.

The service is designed to be implemented using the features of the ISO standard session service, introduced on page 97 in Chapter 3. During the course of a session, there can be a succession of acknowledged message transfers, that is, a succession of simple handshakes. Each handshake is regarded as an activity, and reliability is achieved by using features of the session service that cause activities to be restarted if failures occur. In cases where the messages are lengthy, the activity can be sub-divided using major synchronization points. These also act as boundary points if activities are restarted, in that any communication successfully completed before a major synchronization point is not repeated.

ISO Remote Operations Service

The ISO standard Remote Operations Service (ROS) is also a handshake-style service intended for use by computer applications. However, it is specialized as a request-response service. The first part of the handshake involves sending a message that indicates an operation to be performed, together with parameters for the operation. The second part of the handshake is a response to the request, either containing the results of the operation or reporting that the operation could not be performed or completed. This service is a simple example of a client-server relationship. The client computer requests a (specialized) service from the server computer, and the server computer returns the results of carrying out the service.

Remote Procedure Call

The idea of remote procedures calls (RPCs) is to present the client-server model in a way very familiar to computer programmers: as a procedure call. The client is the caller of a procedure, and the server executes the body of the procedure. The difference in RPCs is that the procedure calling sequence and the procedure body are executed by different computers. This introduces some new problems that do not afflict normal procedure calls. First, errors that occur in communication, or at the remote server, must be handled in some way by the client. Second, global variables are not available as a way of sharing data between the caller and the procedure. This means that any required data must be passed to the procedure as parameters or from the procedure as results. Third, the time taken by RPCs is likely to be rather greater because of the communication time required.

Of course, the problems listed are inevitable for any client-server system, however it is packaged up. For remote procedure call, the goal is to hide the problems as much as possible, so that the end user can just use procedures without having to worry about whether they are executed locally or remotely. The normal implementation method to achieve this involves the use of **stubs**. A stub is a small procedure that runs on the client computer as a proxy for the remote procedure. The client program calls the stub, which then is responsible for conducting a request-response handshake with the server computer. In particular, the stub can hide any communication-related problems as best it can. Note that it is not necessary for the server to be organized as a collection of procedures. The server only needs to present an appropriate request-response interface to the outside world.

An example of a standard RPC service, also mentioned in Chapter 2, is the one devised by Sun Microsystems Inc, which is now widespread on the Internet. It is used to underpin the Network File System (NFS), for example. There is also an ISO standard RPC service. The protocol used to implement the Sun standard RPC service specifies the format of messages used for calling remote procedures, and messages used for returning results from remote procedures. There is no reliability guarantee — the quality of the RPC service depends on how the handshake-style service is implemented. If it is just implemented using an exchange of messages using an unreliable connectionless service, then RPCs might fail. In such a case, either a stub or the procedure caller must take responsibility for dealing with failures.

Overall duplex communication:

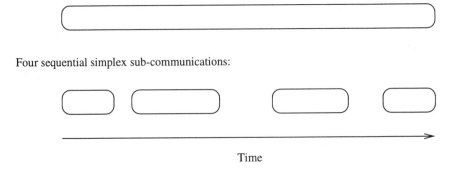

Four sequential simplex sub-communications:

Time

Figure 4.4 Communication involving a four-stage handshake time package

4.4 MULTI-STAGE HANDSHAKE TIME PACKAGE

As a basic time package, the simple handshake is very useful and widespread. However, it has its limitations. One obvious limitation is in its use for control purposes between two computers. A command-response or request-reply service may not be rich enough for some control purposes, and a three-stage or four-stage handshake may be more appropriate. An example of a four-stage handshake is shown in Figure 4.4. When pointing out a small number of primitive time packages that can form the temporal basis for control services, all possible control relationships cannot be catered for. However, there is some scope for identifying appropriate control paradigms that occur in several or many contexts, and then mapping these to suitable time packages. In general, this task is really the province of the distributed systems expert, rather than the communications expert.

Within communication protocols themselves, one need for multi-stage handshakes arises when the two computers need to negotiate dynamic agreements on how a communication is to take place. A dialogue continues until both computers are in agreement. Another reason for using multi-stage handshakes is that, because of the characteristics of the computers or the channel, it is impossible to use a simple handshake protocol to implement a service of the required quality. A more complex handshake mechanism is then needed.

Although it is not possible to characterize handshakes in general, usually a feature of the final stage of a handshake is that the sending computer has some sort of passive role. For example, in simple handshakes, these roles included acknowledging, responding and replying. This feature accounts for the fact that the handshake is ending, since there is no need for the other computer to react.

4.4.1 Examples of multi-stage handshake services

ISO Commitment, Concurrency and Recovery service

The ISO standard Commitment, Concurrency and Recovery (CCR) service is designed to ensure coordination between two (or more) computers, in the sense that a communication either succeeds completely or has no effect at all. This property holds, regardless of whether computer or channel failures occur. The service involves communication between a master computer and a slave computer with a four-stage handshake mechanism:

1. Master sends a set of requests to slave (these may be broken down into separate sub-communications).
2. Slave prepares to carry out the requests. It sends a reply to master, saying whether or not they can be carried out.
3. If slave is able and master wishes to proceed, master sends command to carry out the requests; if not, master sends command telling slave to ignore the requests.
4. If master tells slave to act, slave carries out the requests. Slave sends an acknowledgement to master.

This is based on the two-phase commit mechanism, first invented as a means of ensuring the integrity of transactions on distributed databases. Each phase involves a simple two-stage handshake. The first phase makes the two computers prepare for the communication. The second phase makes the two computers either complete or abandon the communication. The CCR service is meant to be implemented using a reliable transfer service, which can deal with communication problems. Thus, the four-stage handshake is provided to deal with the problem of either computer failing in the middle of the overall communication.

Negotiation in the PPP Link Control Protocol

The Link Control Protocol (LCP) of the Point-to-Point Protocol (PPP), is used for transmission of IP messages over telephone lines among other things. Section 9.4 contains a fairly detailed description of the use of PPP and LCP in practical circumstances. The LCP protocol contains a non-trivial example of negotiation using a (potentially open-ended) multi-stage handshake time package. In LCP, various options are negotiated at the beginning of a communication, including maximum information size, authentication protocol, quality protocol and data compression method. The actual meaning of these options, and their possible values, are not important here. It is their negotiation that is of interest. An example of a fairly protracted, but ultimately successful, handshake is shown in Figure 4.5.

The computer initiating the communication must send a configuration-request message to the other computer. The request includes a list of zero or more negotiable options and suggested values for them. This is the first stage of the handshake. The other computer then has a choice of three different types of response.

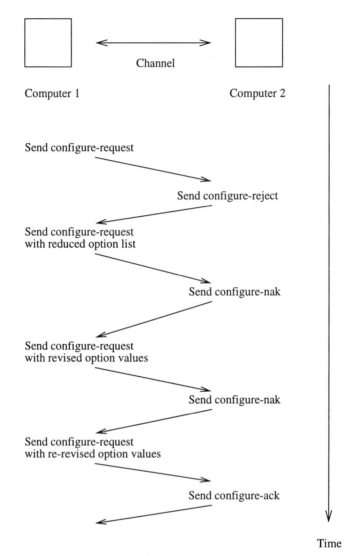

Figure 4.5 Example of Link Control Protocol negotiation

If all of the options and suggested values in a configuration-request message are acceptable to the other computer, it must return a configuration-ack message. This contains the same list of options and values as the configuration-request, and confirms their acceptance. Then, the negotiation and the handshake are complete. This is illustrated at the end of the dialogue in Figure 4.5.

If any of the options in a configuration-request message are unrecognizable or are not acceptable for negotiation by the other computer, it must return a configure-reject message. This contains those members of the list of options and values that are unacceptable. On receiving a configuration-reject, the first computer continues the handshake with a new configuration-request that does not contain any of the unacceptable options. This is illustrated at the beginning of the dialogue in the example of Figure 4.5.

If all of the options are acceptable by the other computer, but some of the sugested values are not, it must return a configuration-nak message. This contains the options that had unacceptable values, together with alternative acceptable values. Options and suggested values that were not in the original list may also be added, to prompt negotiation of these. On receiving a configuration-nak, the first computer continues the handshake with a new configuration-request message that is modified to reflect the values suggested by the other computer. It is not obliged to exactly satisfy all of the suggestions. This is illustrated twice by the middle of the dialogue in the example of Figure 4.5.

It is possible that two negotiating computers have conflicting policies, and so the negotiation can never converge on an agreement. It is also possible for computers to have negotiation policies that lead to convergence, but take many iterations to do so. To prevent excessively long, or infinite, handshakes, it is recommended that a counter of configure-nak messages sent is used. If the counter reaches a chosen value (five is the default), a configure-reject is sent instead to abandon the negotiation attempt.

Initialization in the Transmission Control Protocol

The Internet Transmission Control Protocol (TCP) is considered in rather more detail in Section 4.5.2 below. Here, one component of the protocol is useful as an example, since it illustrates the use of a three-stage handshake time package to overcome implementation problems. The handshake is used at the beginning of a TCP communication, and the problem is caused by a feature of implementation using an unreliable connectionless service. The reason for the handshake is to agree one integer value before the communication starts (the significance of this value is explained in Section 4.5.2). No negotiation is required because the first computer gets to choose the value and the other computer just needs to be informed of it.

When TCP is used over a connectionless service, the first stage of the handshake is for one computer to send a message containing the chosen integral value. If a simple handshake was used, and the other computer just sent back an acknow-

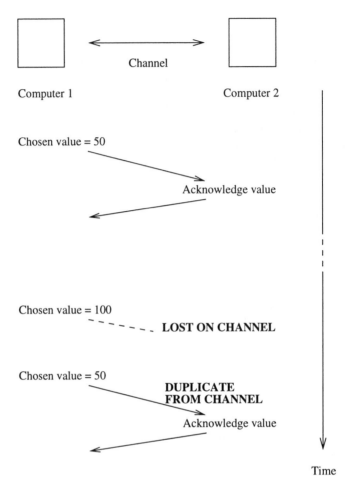

Figure 4.6 Problem at beginning of TCP communication

ledgement of receiving a value, a problem might arise. An example of this problem is illustrated in Figure 4.6. In the example, there is first a successful handshake, the agreed integer value being 50. Then, later in time, there is another attempted handshake, this time the integer value being 100. Unfortunately, this handshake is afflicted by the problem. First, the message containing the chosen integer is lost. Further, after being delayed, a duplicate copy of the earlier message containing its chosen integer arrives at the other computer instead. Such a scenario is possible when an unreliable connectionless service is used. The effect is that the other computer receives the wrong integral value.

If the other computer just sends back an acknowledgement of acceptance, as shown in Figure 4.6, the agreement protocol fails. To deal with this, the second stage of the three-stage handshake actually used in TCP is for the other computer to send back a message containing the integral value it received. If the first computer receives back the value it sent, the third stage of the handshake is for it to send back a message as an acknowledgement. However, if it receives back the wrong value, the third stage of the handshake is for it to send back a message commanding the other computer to abandon the communication attempt.

Authentication protocols

Multi-stage handshakes are common in authentication protocols, which are used to allow computers to prove their identities to other computers. One interesting time-related feature of such protocols is that absolute time values usually feature in some of the messages exchanged, as mentioned on page 64 in Chapter 2. This is to prevent imposters replaying handshakes that they have observed earlier.

One well-known authentication protocol is the Kerberos protocol. This is mentioned in Exercise 2.32, where readers are invited to read the relevant Internet RFC — a task that is made a little easier by an understanding of the material in this section. Essentially, the Kerberos multi-stage handshake involves two simple handshakes, one after the other. A computer that wishes to identify itself to another computer first conducts a simple handshake with a computer that offers specialized Kerberos services. Using information gained from the first handshake, it then conducts a simple handshake with the computer it is interested in. The effect achieved is as follows.

Before the simple handshakes, both computers have their own secret keys that have been previously issued by the Kerberos computer. After the handshakes, both computers have a shared secret key that allows them to communicate with each other using encrypted messages. In addition, both computers are convinced that their partner is genuine. In fact, the shared key is used to encrypt both stages of the second simple handshake, as well as being available for later use. Absolute time values are included in the encrypted messages sent in each stage of this handshake, to prove that the handshake is fresh.

4.5 CONNECTION-ORIENTED TIME PACKAGE

The final type of time package to be discussed is one that is **connection oriented**. The progress of a communication with a connection-oriented time package is similar to making a telephone call. Given this, connections are sometimes referred to as **calls**. Other synonyms sometimes used for 'connection' are **association**, **conversation** and **session**.

A connection is a communication with three distinct sequential components over time, as shown in Figure 4.7. The beginning of the communication is marked by a dialogue to set up the connection: this is termed **opening a connection**, **establishing a connection** or just **connecting**. The end of the communication is

Virtual circuit communication:

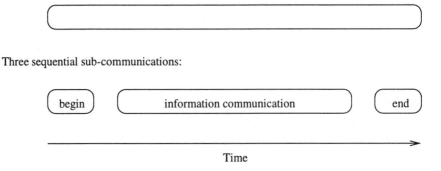

Three sequential sub-communications:

Time

Figure 4.7 Communication involving a connection-oriented time package

marked by a dialogue to shut down the the connection: this is termed **closing a connection**, **tearing down a connection** or just **disconnecting**. Thus, the overall communication resulting from the lifetime of a connection has clearly delineated beginning and ending points, and is a excellent exemplar of the asynchronous style. In essence, Figure 4.7 just shows the natural segmentation of the communication shown in Figure 3.9 into three sub-communications.

4.5.1 Connection-oriented services

The overall connection-oriented time package is what characterizes a connection-oriented communication service. These services can be contrasted with the *connectionless* services of Section 4.2.1, where there is no segmentation to give a connection style package. Strictly speaking, a connection-oriented service has no extra properties. However, in practice, it usually implies some extra time-related properties that improve quality. In the context of message switching networks, the subject of Chapter 7, channels supplied by connection-oriented services are sometimes termed **virtual circuits**. However, this terminology is also used more widely, with 'virtual circuit service' often used synonymously with 'connection-oriented service', just as 'datagram service' is used to mean 'connectionless service'.

Between the setting up and shutting down phases, communication of information takes place. In general, this communication involves information sharing, with information flowing in both directions. For the sake of simplicity, the discussion in the rest of this section on connection-oriented time packages will assume unidirectional communication only: one computer is only a transmitter and the other computer is only a receiver. This allows adequate technical explanation, with discussion of the bidirectional case being deferred to page 163 in Chapter 5, where some additional helpful technical apparatus will be introduced.

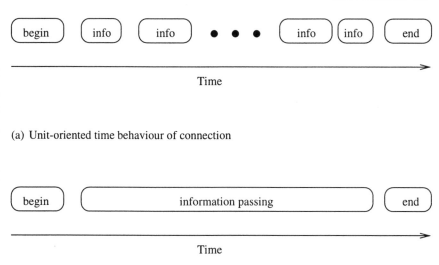

(a) Unit-oriented time behaviour of connection

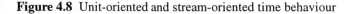

(b) Stream-oriented time behaviour of connection

Figure 4.8 Unit-oriented and stream-oriented time behaviour

On page 95 of Chapter 3, a distinction between unit-oriented and stream-oriented segmentation was described. This leads to two alternative ways in which the time behaviour of a (unidirectional) connection-oriented service can be packaged. The difference is shown in Figure 4.8. In the unit-oriented case, the information-passing phase is segmented into sub-communications for units of information. Thus, the provided time structure might be used by the communicating computers to reflect some natural structure of the information communicated. In the stream-oriented case, the information-passing phase appears as one single communication. Of course, this may be implemented as a number of sub-communications, and indeed this segmentation may be determined by how users of the communication service interact with it, but conceptually such matters are not relevant to users of the service.

For both unit and stream orientation, an extra feature of almost all connection-oriented services is **sequencing**. That is, regardless of implementation method, information is received in the same order that it is transmitted, and so the service behaves like a wire between the two computers. In other words, any segmented sub-communications do not overlap in time, and the receiver sees them beginning in the same order as the transmitter.

As well as sequencing, many connection-oriented services have additional quality features. One is that of reliability, that is, information transmitted is always received. Thus, the service emulates a reliable wire between the two computers. If there is no guarantee of reliability, this is explicitly signalled by describing the ser-

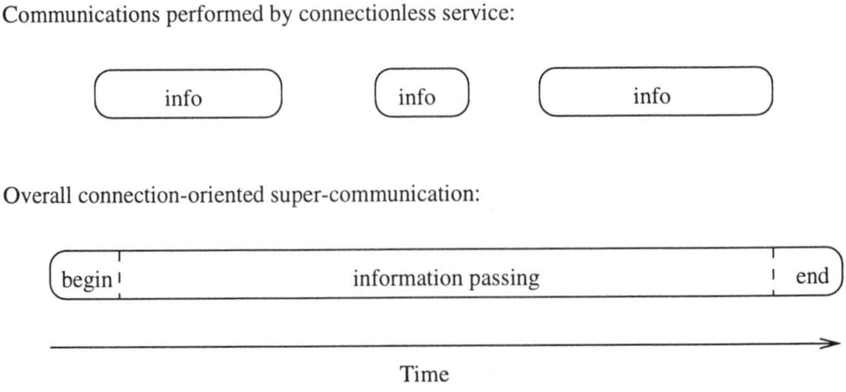

Communications performed by connectionless service:

Overall connection-oriented super-communication:

Time

Figure 4.9 Concatenation of connectionless communications

vice as being unreliable. Another normal feature is flow control, usually involving ticket-style flow control by the receiver using a sliding window mechanism.

Implementing a connectionless service or a handshake-style service in terms of a connection-oriented service is fairly straightforward. A connection allowing a unidirectional information flow is adequate for the former; a bidirectional connection is required for the latter. Any sequencing and flow control facilities provided by the service are not required. However, a reliability feature is desirable if a reliable connectionless or handshake-style service is desired. The main overhead is that the start up and shut down dialogues are necessary as part of the connection-oriented time package.

Thus, to send a message as a connectionless service, a connection would be opened, then a single unit of information would be sent, and then the connection would be closed. If the pattern of usage of the connectionless service is such that several messages are sent over a period of time, then optimization is possible by using concatenation in the implementation. With this optimization, the connection is not closed immediately after a message is sent, but instead is left open for some time period to avoid opening a new connection for the next message. This use of concatenation is illustrated in Figure 4.9.

Conversely, implementing a connection-oriented service in terms of an underlying connectionless service requires a fairly complex protocol. The dialogues for setting up and shutting down connections are just examples of handshakes, simple or otherwise, as discussed in the previous two sections. The initial dialogue may involve the negotiation of features of the connection, for example, flow control arrangements. The actual communication of information is where most work has to be done by the protocol. This may involve implementing sequencing, reliability and flow control, all of which can be done as a combined operation.

The allocation of increasing **serial numbers** (sometimes alternatively called 'sequence numbers') to the information sent in each sequential information-carrying sub-communication is central to this combined operation. Each message transmitted from sender to receiver then contains both a unit of information and its serial number. This allows the receiver to detect missing information, misordered information and duplicated information, caused by unreliability in the underpinning connectionless service. The serial numbers can also be used by the receiver when sending back information for acknowledgement or flow control purposes. Examples of how particular protocols work are included in the next section.

4.5.2 Examples of connection-oriented time packages

Asynchronous Transfer Mode

Asynchronous Transfer Mode (ATM) is the underpinning technology for the Broadband ISDN communication services of the future. There is technical discussion of the underpinning service for ATM in Section 7.4.5, a summary of the main ATM-supported services is in Section 7.5.6 and more details of the connection-oriented ATM service itself feature in the case study of Chapter 11. Here, only the general temporal properties of ATM services are summarized.

The lowest level service of ATM is a connection-oriented service, which allows the unit-oriented transmission of messages (always called 'cells' in the ATM context) between computers. As mentioned in Chapter 2, each ATM message is 53 bytes long, and consists of a five-byte header followed by 48 bytes of information. Because of this, assuming that the bits of each message are transmitted synchronously, the time periods of the sub-communications for each message have the same duration. For example, given a 155 Mbit per second transmission capability, each time period will last around 2.75 microseconds.

The connection-oriented service has a sequencing property, in that messages are received in the same order that they are transmitted. However, the service does not have guaranteed reliability, since some messages may be lost and the information content of some messages may be damaged.

The protocol used to implement this service assumes that its underpinning service preserves sequencing, but that corruption of messages is possible. To improve on the latter feature, the protocol introduces an error-detection capability, and a one-bit error-correction capability, for the five-byte header only. This uses an eight-bit cyclic redundancy code with the CRC-8 generator polynomial $x^8 + x^2 + x + 1$. There is no error correction by retransmission — messages with headers that have more than one bit in error are simply discarded. The overall effect is sufficient to ensure that the service does not deliver messages with damaged headers, except in rare cases when the CRC check fails to spot an error.

When a connection is established, a guaranteed transmission rate is negotiated, and the channel and receiver are expected to cope with this rate. This is the benevolent dictator style of receiver dominated flow control, as mentioned on page 100 in Chapter 3. No extra protocol features to implement flow control are needed, since the underpinning service is assumed able to deal with the agreed flow rate. If the transmitter does exceed this agreed rate, then some of its messages may be lost. This is just a feature of the connection-oriented service offered.

ATM also includes a range of **adaptation layers** in order to make the basic connection-oriented service more usable by communication applications. Essentially, the role of an adaptation layer is to deal with segmentation of communications into 53-byte message sub-communications supported by the basic ATM service. There is a description of the different defined ATM adaptation layers in Section 7.5.6.

There is a reliable connection-oriented ATM adaptation layer, which is supported by the Service Specific Connection Oriented Protocol (SSCOP). This protocol is described in Section 11.3, and is a further example of a connection-oriented time package with reliable transmission of information.

Internet TCP service

The Transmission Control Protocol (TCP) implements the standard connection-oriented service offered on the Internet. This service is stream-oriented, and features sequencing, reliability and flow control. The workings of TCP are described in detail in the case study of Chapter 9. The description here is concerned with the basics of how TCP implements its service. The service underpinning TCP is the connectionless service implemented by the Internet Protocol (IP), which is one of the examples in Section 4.2.2. This means that TCP's information may be lost, scrambled, duplicated or delayed. As in earlier discussions, the relevant aspects of TCP will be described as though it implements a unidirectional communication service. However, the full story is that the service is bidirectional, which allows some optimization, mentioned on page 163 in Chapter 5.

As explained earlier, serial numbers are an important feature of protocols to implement connection-oriented services. In the case of TCP, each consecutive byte of information communicated has a consecutive 32-bit integer associated with it. This does not mean that a 32-bit integer has to be transmitted with each information byte (which would mean a fivefold increase in the amount of data communicated). In general, a sequence of consecutive bytes are transported as a unit in each message transmitted by the IP service, and so each message need only carry the 32-bit serial number of the first byte in the sequence. Note that segmentation to create sub-communications of byte sequences as units has no structural significance, since TCP gives a stream-oriented service.

The numbering of information bytes starts at a value chosen by the sender, and which is notified to the receiver during the initial dialogue. In fact, this value is the integer value that is the subject of the three-stage TCP handshake that was used as an example in the previous section. It is important that different starting values are used each time a TCP connect is established, otherwise late-arriving duplicate messages from earlier communications might arrive, and cause confusion by having serial numbers in the same range as a current communication.

The receiver sends back messages, via the IP service in the other direction, for acknowledgement and flow control purposes. These messages carry a 32-bit serial number and also a 16-bit window size. The serial number indicates the next information byte expected, and so acknowledges all bytes with earlier serial numbers. The window size indicates how many bytes the receiver is prepared to accept from the serial number onwards. So if, for example, the serial number was 3581 and the window size was 294, the receiver is prepared to accept bytes $3581, 3582, \ldots, 3873, 3874$. When it receives an information-carrying message containing one or more bytes starting with its next expected serial number, the receiver moves on its position of the sliding window, and notifies the transmitter of this movement.

The receiver can deal with messages arriving out of order, that is, messages carrying serial numbers higher than expected, either by discarding them or by retaining them until after the expected predecessor(s) arrives. It can also deal with duplicates, that is, messages arriving carrying serial numbers lower than expected, by discarding them. Serial numbers wrap round to 0 if they reach $2^{32} - 1$. Given that this allows 4096 Mbytes to be communicated before repetition, it is assumed that no late-arriving duplicates can cause confusion by having serial numbers that overlap with any repeated use of the same numbers.

To cope with damaged messages, each one carries a standard Internet checksum, of the type described on page 58 in Chapter 2. The receiver recomputes the checksum and, if it is incorrect, the message is discarded — effectively lost in transit. To deal with messages being lost in general, or delayed for an unduly long time, the transmitter uses a timeout and retransmission mechanism. A timer is started each time that a message is sent. The problem is estimating an appropriate timeout period, since the maximum time it may take for a message to be sent and acknowledged normally can vary enormously, depending on the channel between the two computers.

The method used by TCP centres around maintaining an estimate of the **round trip time** — the time between sending an information-carrying message and receiving an acknowledgement message back. This is updated each time an acknowledgement is received, by computing a weighted average of the previous value and the newly observed value. A weighting of (7/8)th previous to (1/8)th new has been found to yield good results in practice. In a modification suggested by Karn, and forming part of **Karn's algorithm**, updates are not performed for messages that have been retransmitted, only for those that are acknowledged first time. This is because it is not clear to which transmission attempt acknowledgements refer.

In early implementations of TCP, the timeout period was chosen as a multiple (greater than one) of the estimated round trip time. The multiplicative factor was a constant, usually equal to two. Now, an estimate of the variance of the round trip time is taken into account as well, to deal with situations where there are wide variations in delays. This estimate is computed in addition to the estimate of the average round trip time. A typical choice of timeout value is the estimated round trip time plus four times the estimated variance.

Karn's algorithm is concerned with the choice of timeouts for repeated retransmissions. Just ignoring round trip times for retransmitted packets is not adequate, since it could lead to under-estimates of the round trip time and so premature retransmission. The approach taken is that, each time a particular message is retransmitted, the timeout period is increased by a multiplicative factor. Typically, this factor is a constant equal to two.

Of course, all the unreliabilities of the IP service can afflict the acknowledgement messages travelling back from receiver to transmitter. However, these also include a checksum, which allows damaged messages to be discarded. Only messages that move the sliding window forward need be taken into account. When the sliding window is moved forward, the relevant timeout clock(s) can be stopped.

HDLC-derived protocols

The High-level Data Link Control (HDLC) family of protocols are internationally standardized, and are used for implementing communication services using physical transmission services. The standard HDLC message format was introduced in Chapter 2. The ITU-T (International Telecommunications Union — Telecommunications sector) standard Link Access Protocol Balanced (LAPB) is a member of the family that is of most interest here. It is designed to offer a similar quality of connection-oriented service as TCP. However, because LAPB is designed for communcations using direct physical channels, it can assume an underlying connectionless service that does not duplicate messages and does not reorder messages. The only assumed flaws are that messages might be lost or damaged. The IEEE 802.2 LLC2 connection-oriented service is the remaining member of the family also containing LLC1 and LLC3, both described earlier. It is implemented using a protocol with timing behaviour the same as that of LAPB.

Rather than fully describe LAPB in the manner that TCP was explained in the previous subsection, only its major differences from TCP are described here. LAPB supports a unit-oriented service, unlike TCP. Sequences of bytes are presented as units to the service at the transmitter, and emerge intact at the receiver after being sent in sub-communications. Serial numbers apply to each unit transmitted, rather than to each byte transmitted. The standard variety of LAPB has three-bit serial numbers, and an extended variety (which is used by LLC2) has seven-bit serial numbers. These sizes are much smaller than those used in TCP, because serial numbers can be reused much more rapidly when there is no danger of duplicate messages arriving.

Again, because there is no duplication, serial numbers begin at zero for each communication. As a result, a simple handshake is used for the initial dialogue. The choice of standard or extended serial numbers is negotiated by this handshake — the sender indicates its choice and the recipient can either accept or reject it, but not change it.

The receiver sends back messages for acknowledgement and flow control purposes. These carry a three-bit (or seven-bit) serial number. Unlike TCP, they do not also carry a window size, since this is fixed permanently for the pair of communicating computers. This fixed size must be smaller than the number of possible serial numbers, i.e., at most seven for three-bit serial numbers or at most 127 for seven-bit serial numbers. Otherwise, confusion can occur. For example, if a transmitter was allowed to send eight messages using three-bit serial numbers, then when a receiver says that serial number 0 is expected next, it could mean either that it has received nothing or that it has received all eight messages. The only (extreme) way in which the window size can be changed is that the receiver is able to send a message that sets it to zero, to prevent any transmissions temporarily.

To cope with damage in transit, each LAPB message carries a 16-bit cyclic redundancy code (CRC), with the CRC-CCITT generator polynomial $x^{16} + x^{12} + x^5 + 1$. In the case of LLC, the message formats used are a little different, and make use of a 32-bit CRC already part of the message format for the underlying service. This CRC uses the CRC-32 generator polynomial. The timeout mechanism uses a timeout period that is fixed permanently for the pair of communicating computers. As with the fixed window size, this reflects an assumption that the time behaviour of the underpinning service is more predictable.

Unlike TCP, there is a negative acknowledgement mechanism that can be used by the receiver. This takes advantage of the fact that messages are delivered in order. If the receiver receives information with a serial number higher than expected, it assumes that the intervening information was lost, and sends back a negative acknowledgement requesting the retransmission of the lost information *and* all information transmitted subsequently. (There are other protocols in the HDLC family that allow selective retransmission of lost information only.) This mechanism is likely to save time, compared with waiting for a timeout to expire and trigger retransmission.

There is also a relic of master-slave flow control arrangements that are central to LAPB's more authoritarian ancestors in the HDLC family. The information sender can demand an acknowledgement from the receiver at any time in order to check the status of the sliding window. In particular, this mechanism is used after a timeout, to ensure that the transmitter gets a speedy indication of what needs to be retransmitted.

ISO transport service

The ISO standard connection-oriented service fills the same niche as the Internet TCP service: it offers sequencing, reliability and flow control. The only significant difference is that it is unit-oriented, rather than stream-oriented. There is a family of five different ISO standard protocols that can be used to implement the service. The choice of protocol depends on the underlying service used.

Transport protocol TP4 is very similar to TCP, which is unsurprising since its design was strongly influenced by TCP. It assumes the worst type of underlying service: unreliable connectionless. The operation of TP4 is covered in detail in the case study of Chapter 10. In contrast to TP4, transport protocols TP0 and TP2 both assume a perfect connection-oriented service. These two protocols are only concerned with initialization and termination of connections, and with segementation or concatenation of communications during the information-carrying phase. The other two protocols, TP1 and TP3, both assume a slightly worse type of service: occasionally, the underlying connection might break down completely. Serial numbers are used in order to assist recovery after such failures. Essentially, the protocol establishes a new connection using the underlying service, and then continues transmitting from the point where the previous connection failed. The main difference between TP0 and TP2, and between TP1 and TP3, is described in context, on page 165 in Chapter 5.

In the TP4 protocol, standard format serial numbers are seven-bit, but it is also possible to use an extended format with 31-bit serial numbers. Serial numbers always begin from zero for a communication. However, all messages also carry a connection reference number, chosen when the connection is established. This allows historic duplicates to be discarded, since the combination of the reference number and the serial number gives a similar effect to TCP's serial number.

Acknowledgement messages carry a serial number indicating the next unit expected from the information transmitter. They also carry a credit field, which is four-bit if seven-bit serial numbers are being used, and 16-bit if 31-bit serial numbers are being used. This field gives the size of sliding window that the receiver is allowing. Optionally, a selective acknowledgement scheme can be used, where acknowledgements indicate which units have actually been received, and which are still outstanding.

All messages — information-bearing and acknowledgement — carry a 16-bit ISO standard checksum of the type that is described on page 58 in Chapter 2. A fixed timeout period is used during the lifetime of the connection, and it is chosen at the time of establishment on the basis of information known about the underlying connectionless service. The transmitter notifies its choice to the receiver during the initial dialogue, which is a three-stage handshake as for TCP. At this stage, also the choice of serial number size is agreed, as is whether selective acknowledgements are to be allowed.

4.6 CHAPTER SUMMARY

There are a few elementary patterns of behaviour over time that characterize most communication between two computers. These patterns are often embodied in the types of service offered by communication systems. They also feature within the protocols that are used to implement communication services in terms of simpler, or at least different, communication services. The most frequent types of behaviour are:

- unsegmented (connectionless) — a single unidirectional communication

- handshake — a multi-stage sequence of communications in alternate directions
- connection-oriented — a bidirectional communication with explicit beginning and ending dialogues.

When presented as services, the norm for connectionless is for there to be an element of unreliability: losses, damage and perhaps duplication. The norm for handshakes and connection-oriented is reliability. However, some connectionless services are reliable, and some handshakes or connection-oriented are unreliable.

Within protocols, the most heavyweight implementation problem is implementing a sequenced, reliable and flow-controlled connection-oriented service using an unreliable connectionless service. From the point of view of time behaviour, the two key features are the use of handshakes for acknowledgement and flow control purposes, and the use of timeout periods to deal with communication losses. The use of serial numbers carried with information transmitted is also central to making such protocols work.

4.7 EXERCISES AND FURTHER READING

4.1 Acquire a script for a scene in a play where two characters conduct a dialogue, and decide what overall information sharing is achieved by the dialogue. Look for patterns of sub-communications within the dialogue, and assess their contributions to the overall effect.

4.2 Write down a list of computer communication applications for which an (unreliable) connectionless service is adequate.

4.3 If you are familiar with computer programming, write down all the actions that are taken when a procedure (or function) is called within a program. For each action, write down the complications that might arise if the procedure is executed remotely rather than locally.

4.4 Have a look at RFC 1831, which describes the Internet RPC protocol, to become familiar with issues that arise when implementing remote procedure calling.

4.5 Consider two people bidding against each other at an auction. The auctioneer acts as a passive channel between the bidders. What informal protocols do the bidders follow when conducting a multi-stage handshake to determine a piece of shared information: the final price paid?

4.6 Consult the database literature, to discover how and why two-phase commit mechanisms are used for database transactions.

4.7 Read Section 4 of RFC 1661, which describes a general option negotiation algorithm, then see its application to the Internet standard PPP LCP in Section 5 and 6 of the same RFC.

4.8 Why does TCP connection initialization need a *three*-stage handshake, i.e., why is the third stage necessary? Consult Section 3.4 of RFC 793, which specifies TCP, if you need more information.

4.9 For a challenging read, have a look at RFC 1521, which describes the Kerberos authentication service, and study the use of handshakes in the protocols used to implement the service.

4.10 List examples of human communications that are connection-oriented. For each example, discuss the importance of each of the following three extra properties: sequencing; reliability; and flow control.

4.11 Suppose that a connectionless service is being used to transmit messages, each one containing a character corresponding to a key just pressed by a person at a keyboard. If the service is implemented using an underlying connection-oriented service, discuss how concatentation might be carried out to avoid using one connection per character.

4.12 Serial numbers are often attached to everyday objects and documents. Consider the purposes that they serve, and comment on whether these include dealing with losses, damage or duplication.

4.13 Why do you think that the protocol implementing the ATM service applies a CRC only to the header of each 53-byte message, rather than to the whole message?

4.14 Discuss the advantages and disadvantages of a flow control mechanism like that used in ATM, where a transmission rate is negotiated when a connection is established.

4.15 TCP uses 32-bit serial numbers and a 16-bit window size. Using information about speeds and latencies for typical physical transmission media from Section 1.3.1, construct realistic scenarios in which: (a) a delayed duplicate message and a fresh message might have the same serial number due to wrap-around; and (b) a transmitter uses up the whole of its sliding window space before the receiver can acknowledge any of the transmitted information.

4.16 The original method for computing TCP retransmission timeouts, described in RFC 793, led to problems. If you have access to SIGCOMM proceedings, read Karn and Partridge's paper "Round Trip Time estimation" in SIGCOMM-87 and Jacobson's paper "Congestion Avoidance and Control" in SIGCOMM-88. The first includes Karn's algorithm, and the second includes the idea of including the round trip time variance in the timeout calculation.

4.17 As a challenge, look at RFC 1644, which describes an experimental Internet protocol: T/TCP. This is an extension to TCP to allow the implementation of efficient request-response services.

4.18 Research the history of the HDLC family of protocols, in particular tracing back to IBM's Synchronous Data Link Control (SDLC) protocol, which was the ancestor of HDLC. How closely related are SDLC and LAPB?

4.19 Why does LAPB give a choice of either three-bit or seven-bit serial numbers?

4.20 TCP uses dynamically chosen timeout periods, whereas TP4 does not. What are the merits and demerits of the different approaches?

Further reading

Like the previous chapter, Chapter 3 on Time, this chapter is intended to be largely self-contained. It takes the ideas of Chapter 3, and shows the ways in which they are most commonly packaged together in communication services and within communication protocols. Any further reading would be optional, and concerned with finding out more details of the examples used during the chapter. In some cases, a far greater level of detail can be found in Chapters 9 to 11, which contain case studies designed to show exactly how practical communications work. More information about Internet standard protocols can be found by studying the appropriate Internet RFC documents.

FIVE

SPACE

The main topics in this chapter about space are:

- sharing of information by computers over a channel
- identifiers for computers and channels
- absolute and relative agreement on space between computers
- broadcast, multicast and unicast channels
- computers as communication filters and switches
- introduction to computer networks and inter-networks
- splitting and multiplexing of communications

5.1 INTRODUCTION

In this chapter, the model of communication is extended into its final form. In earlier chapters, a communication is seen as an act that shares information between two computers via a channel, within some time period. This has a significant limitation: communications take place in a little private world that contains only two fixed computers and one channel between them. Such an assumption is a large restriction in terms of modelling communication between all computers of the world.

In order to deal with this, it is important to describe how information sharing is carried out in **space**. That is, which computers and channels are involved. The issues that arise are similar to those already seen when considering both information and time. First, it is necesssary to have agreement on where information sharing takes place. Second, it is necessary to have implementation that allows required communication spaces to be created using channels that allow different — usually simpler — communication spaces. The implementation work may not just be concerned with achieving communication, but also with achieving it at some particular level of quality.

Human experience

To make the discussion on space clearer, consider the following examples drawn from everyday human experience:

- a television programme
- a tourist sending postcards to friends
- a telephone call

These examples are in a similar vein to those used in Chapter 3, but are slightly varied to illustrate various space-related points effectively.

The space for all of these examples will be regarded as consisting of a collection of communicating people, together with some sort of communication medium. In the first example, the people are the programme presenters at a television station and the viewers of the programme, and the medium is a television broadcast. In the second example, the people are the tourist and his or her friends, and the medium is the postal system. In the third example, the people are the two callers, and the medium is the telephone system. In all three cases, the media involve some kind of technology. However, for other examples such as the presentation by a speaker or the tourist information desk from Chapter 3, the medium would just be direct human speech.

Computer communications principles

For the remainder of this book, the term 'a communication' will be used to mean an act of information sharing among a group of two or more computers via a channel, occurring within some time period. For any particular communication, the computers involved are described by a set containing a unique identifier for each one. The identifiers are unique within the enclosing universe of computers under consideration. This universe is visible in its entirety to an external observer, but not necessarily to the individual computers. The channel also has a unique identifier among all possible channels. A communication is illustrated in Figure 5.1.

The information-passing capabilities of channels must be extended under the new definition of a communication. Rather than just being able to transfer information in one or both directions between two computers, a channel must be able to pass information in whatever ways necessary to achieve the required information sharing among the group of computers. Each communication space involves just *one* channel. With this definition, the channel can be regarded as the component that carries out the communication. Like the computers, the channel can have an existence independent from particular communications and so is, in this sense, a static component of the communication.

Just as the term 'message' is used consistently throughout this book when information is being discussed, so the terms 'computer' and 'channel' are used consistently when space is being discussed. Before proceeding further, it is worth noting some synonyms used in practice for 'computer' and 'channel' when describing practical computer communication systems. Other terms for computer include: **device**, **end system**, **host** and **station**. Later in this chapter (on page 153), a list of synonyms for specialized computers forming integral parts of communication systems is included. Other terms for channel include: **cable**, **circuit**, **interface**, **link**, **medium**, **path**, **pipe**, **tunnel** and **wire**. The word 'connection' may also be used to mean 'channel' in some practical contexts, but it is used consistently throughout this book to mean a particular type of time package, as described in Section 4.5.

5.2 AGREEMENT ON SPACES

For a communication to take place, the computers and channel must agree on rules that give a shared view of the space. There are two main problems. First, there must be an agreed scheme for identifying computers and channels. Second, given an identifier scheme, there must be agreement upon which computers and channel form the space.

5.2.1 Identifier schemes

With time, there is a natural physical measurement system available, and this could give a direct identifier scheme for time periods. For space, it is not so straightforward. A natural physical system would be one based on geographical position, using a suitably accurate positioning system. It is technically feasible to find very accurate positions for objects, for example, using the satellite-based Geographical Positioning System (GPS). However, such a scheme is not particularly appropriate, because physical position is not a very good way of deriving unique identifiers for computers or channels that are usable over a non-trivial time span. This is particularly so these days, with the advent of mobile computing, since it

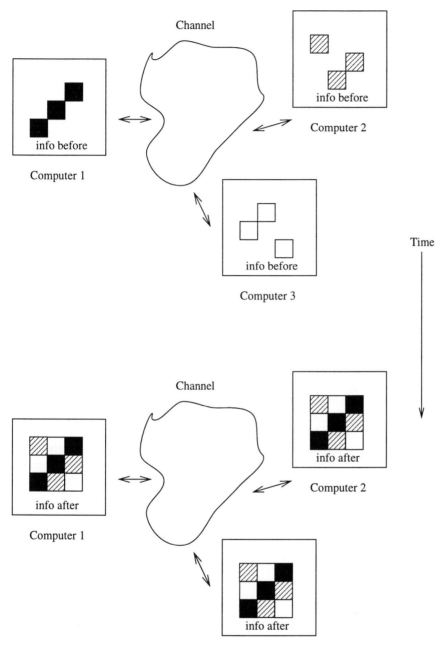

Figure 5.1 A communication

is no longer the case that computers stay in the same place semi-permanently. As an alternative, some type of identifier scheme chosen for human and/or computer convenience is used.

Human experience

Tourist sending postcards to friends

For the postal system, people are identified by their names and addresses. The effect is that individual people have a unique identifier that can be used to direct mail to them. Here, each friend has an identifier, as does the tourist, based on his or her current location. The medium is a part of the worldwide postal system, and a possible unique identifier could be derived from the address of the postbox used by the tourist together with the addresses of the friends. The exact way in which names and addresses appear varies from country to country, and indeed also according to individual taste. However, most addresses consist of several components which have geographical significance. These might include apartment/house name or number, street name, town/city name and country name.

Telephone call

A natural way of identifying telephone callers is just to use their two telephone numbers (assuming that each number is only used for one call at a time). The medium is a part of the worldwide telephone system, and a possible unique identifier could just be derived from the two telephone numbers. Telephone numbers are structured to have several components. In very many cases, this structuring has a geographical significance. Thus, the first component of an international telephone number usually identifies a country. Subsequent components can then be chosen in any way by the particular country, without having to worry about clashes with numbers in other countries. As an example of within-country structuring, in many cities, a second component identifies the city, a third component identifies an area of the city, and finally a fourth component is a unique identifier within that area of the city.

One problem with both the telephone number scheme and the postal address scheme is that, when people move, their telephone number and address usually have to change. That is, their location in the identifier structure changes.

Computer communications principles

Flat or hierarchical

The identifier scheme may be either **flat** or **hierarchical**. In a flat scheme, identifiers have no added structure, and so allocating a new unique identifier involves picking a string that has not been used before. In a hierarchical scheme, identifiers consist of a series of components that position them inside a hierarchy. The hierarchy is chosen for administrative and/or technical convenience. Postal addresses and telephone numbers are both examples of this. For example, a unique 10-level postal address usable by aliens might be:

```
491 Princes Street
Edinburgh
Scotland
United Kingdom
Europe
Earth
Solar System
Milky Way
Local Group
The Universe
```

Allocating a new unique identifier only involves picking a string that has not been used before in the appropriate local part of the hierarchy. This is rather easier in the locally flat scheme than it is within a globally flat scheme, unless the total number of identifiers being used is small.

Names or addresses

When the identifiers have a mainly human significance, for example, organizational, functional or positional, they are usually referred to as **names**. When the identifiers have a mainly computer or communications significance, they are usually referred to as **addresses**. To ensure uniqueness, it is often the case that addresses are fairly long and convoluted. This is not a major problem for computers to handle. However, if humans are involved as users, then some means of allowing more human-friendly ways of identification than just the computer identifiers is required. Therefore, there may be both names and addresses for particular computers or channels. Translation of names to addresses (or vice versa) is usually achieved by a **directory** service. The general idea is familiar from telephone directories where, using a personal name, plus perhaps address or occupation, it is possible to look up the unique telephone number for a person.

Computer communications examples

IEEE 802 identifiers

IEEE 802 identifiers is used in IEEE (Institute of Electrical and Electronics Engineers) standard local area networks, which are discussed in detail in Chapter 6. Identifiers have 48 bits, with the first two bits being **00**, to indicate that the remaining bits are uniquely assigned under a scheme administered by the IEEE. This is shown in Figure 5.2. In principle, the identifier scheme is flat. However, because blocks of identifiers are assigned to different equipment manufacturers by the IEEE, there is a two-level hierarchy in practice. Each network interface has a fixed identifier for its lifetime, with a first component indicating the manufacturer and the second component being a serial number from that manufacturer.

Figure 5.2 IEEE standard 48-bit identifier

Figure 5.3 Basic structure of an ITU-T X.121 standard identifier

ITU-T X.121 identifiers

As explained later in Section 5.3.1, a computer network is a collection of computers and channels, together with agreements on how they inter-communicate. ITU-T X.121 identifiers are used in long distance computer networks operated by state or private telecommunications authorities. This particular type of network is discussed in detail in Section 7.4.3. An X.121 identifier consists of up to 14 decimal digits and, unsurprisingly, is reminiscent of a telephone number in appearance. The basic structure is shown in Figure 5.3. The first three digits are a country code, and the fourth digit is a network number within a country. For countries expected to have more than 10 internationally connected networks, more than one country code is allocated (e.g., the USA has codes 310 to 329). The remaining digits can be allocated in any way by a particular network. As an example, the next few digits might be used for a town or city code, as in telephone numbers. The final two (13th and 14th) digits, if present, are usually used as an identifier for a particular process within the computer identified by the first 12 digits of the identifier.

Internet Protocol identifiers

The Internet Protocol (IP) is mentioned as an example in Chapter 4, and is described in detail in Section 8.3.2. The Internet is an example of an inter-network. The characteristic of an inter-network is that it is a collection of networks of differing types. Inter-networks, in general, are the subject matter of Chapter 8. A particular feature of an inter-network is that the component networks might have different standard identifier schemes for computers. Thus, an inter-networking service must use an over-arching common identifier scheme that allows inter-working between the different networks. IP has such a scheme associated with it.

For IP version 4, which was used from the beginnings of the Internet and only began to be very gradually replaced by IP version 6 in 1996, identifiers are 32 bits (four bytes) long. Each computer in the Internet has a unique identifier, and this has two hierarchical components. The hierarchy reflects the fact that the Internet is composed of a collection of computer networks. Thus, in an Internet identifier for a computer, the first part is a network identifier and the second part is a computer

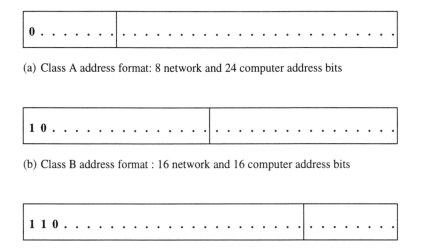

(a) Class A address format: 8 network and 24 computer address bits

(b) Class B address format : 16 network and 16 computer address bits

(c) Class C address format: 24 network and 8 computer address bits

Figure 5.4 IP identifier structure

identifier within the network. There are three basic types of identifier, Class A, Class B and Class C, as illustrated in Figure 5.4. Identifiers beginning with the three bits **111** are used for other purposes, mentioned in Section 8.3.2. The two-level identifier hierarchy allows network numbers (or blocks of network numbers) to be allocated to organizations, which can then allocate unique computer numbers within the network.

The designers of the identifier scheme allowed for up to $2^7 = 128$ large networks with up to $2^{24} = 16\,777\,216$ computers (Class A), up to $2^{14} = 16\,384$ medium networks with up to $2^{16} = 65\,536$ computers (Class B) and up to $2^{21} = 2\,097\,152$ small networks with up to $2^8 = 256$ computers (Class C). At that time, this seemed reasonable, since the world view was one of a relatively small number of large or medium size networks and a large number of small networks.

However, in the Internet, the problem since has been a fast growth in the number of networks containing 300 or more computers. Class B identifiers must be used for such networks, and there are relatively few available. For the next generation Internet Protocol version 6, identifiers are increased to a 16-byte length, and have a richer hierarchical structure. This scheme is designed to satisfy all envisaged demand for identifiers, as well as making allocation easier. This new form of IP identifier is described in Section 8.3.2.

ISO NSAP identifiers

The ISO standard connectionless inter-networking service, which is the ISO analogue of the Internet IP service, is described in Section 8.3.3. ISO NSAP (Network Service Access Point) identifiers for computers are used in this service, and

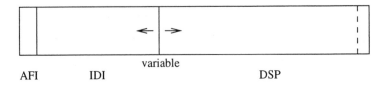

Figure 5.5 Basic structure of an ISO NSAP identifier

the standard protocol that implements it. The scheme used does not begin from scratch, as with IP version 4 identifiers. Instead, it unifies various existing identifier schemes used for both computer networks and telecommunications networks.

A full identifier can be up to 20 bytes or 40 decimal digits long. The basic structure is shown in Figure 5.5. All NSAP identifiers have a hierarchical structure with three components. There may be further structure within the second and/or third components. The first component consists of two decimal digits, which form an **Authority and Format Identifier** (AFI). This indicates which identifier scheme is used for the second component of the identifier, and the authority responsible for allocating identifiers within this scheme. It also indicates whether the third component is expressed in decimal or binary notation. There are seven identifier schemes included, coded by decimal numbers as follows:

Authority and scheme	Decimal	Binary
ITU-T public data network (X.121)	36	37
ISO data country codes (ISO 3166)	38	39
ITU-T telex (F.69)	40	41
ITU-T telephone (E.163)	42	43
ITU-T ISDN (E.164)	44	45
ISO international codes (ISO 6523)	46	47
Local non-standard scheme	48	49

Note that X.121, an earlier example, is one of the schemes included. The ISO data country codes scheme has a code for each country in the world, whereas the international codes are used for projects that span more than one country. The local scheme identifier is included to allow private identifier systems to be used.

The second component of an NSAP identifier is the **Initial Domain Identifier** (IDI). This is an identifier within the scheme specified by the AFI. For example, if the AFI is 36 or 37, then the IDI is a 14 decimal digit X.121 identifier. If the AFI is 38 or 39, the IDI is a three decimal digit country code, e.g., 840 for the USA. Other schemes also have appropriately lengthed IDI components. In the case of AFIs 48 and 49, the IDI component is null.

The third component of an NSAP identifier is the **Domain Specific Part** (DSP). This contains the actual identifier for a computer within an identifier scheme administered by an authority specified by the combination of the AFI and IDI fields. The maximum length of the DSP component depends on the length of the IDI field, since the total length of an NSAP identifier cannot exceed 40 decimal digits (which corresponds to 160 bits, since each decimal digit is equivalent to four bits).

By convention, the final byte (or, equivalently, the final two decimal digits) of the DSP is an identifier for a process within a computer identified by the previous digits of the DSP. Apart from this, the DSP usually has a hierarchical structure, which follows the internal structure of inter-networks and networks. There is more discussion of such hierarchies on page 305 in Section 8.3.3, and also on page 379 in Chapter 10.

5.2.2 Absolute and relative spaces

Given that an identifier scheme is available, the communicators must agree on the space used. In a simple situation, such as the earlier model of communication, the agreement is implicit: there, the space consisted of the two computers and the connecting channel. In other, very static, circumstances, there might be permanent absolute agreements on the set of computers that communicate and on the channel used. However, apart from at the lowest levels of communication, such agreements are unusual. Mobile computing makes permanent agreements even less likely. Instead, the space is usually agreed upon when a communication actually begins, or is about to begin.

Human experience

Television programme

For a programme broadcast by a television station, the channel is the only component of the space that must be known by all of the communicators. The people watching the programme will usually know which television station is transmitting it, since television channels are normally only used by a single station. However, this is not necessary. Instead, viewers may learn the identity of the station from the transmission itself. The television station will not normally know who is actually watching the programme. This is because the information flow is a unidirectional broadcast. In summary, the communication space is relative to the actual communication, and participants have partial knowledge of it. In more specialized cases, such as closed-circuit television, the space might be absolute, and communicators will have exact knowledge of it.

Tourist sending postcards to friends

The tourist is aware of all of the communicators in the space, since he or she is choosing to whom to send postcards. The tourist also selects the channel, by posting the cards in one or more postboxes and addressing each card to identify a mailbox for delivery. Thus, the tourist has a complete view of the space. The space is relative to the actual act of communicating. Each recipient has only a partial view of the space, obtained from the communication. This view shows that the recipient and the tourist are communicators (unless the tourist sends an anonymous postcard, of course). The postcard is unlikely to reveal which other people were recipients of postcards. A recipient also does not have complete knowledge of the channel, only the fact that it included his or her mailbox. Given that the information flow is unidirectional, the postcard recipients do not require any further knowledge of the space.

Telephone call

The communication space for a normal telephone call is relative to the call starting. The caller knows who the called is, by selecting the number dialled. The called's telephone might automatically supply the number of the caller, or the caller is likely to identify himself or herself at the beginning of the call. Given that both parties know each other's telephone number, then both parties have exact knowledge of the telephone channel. Even if the human recipient does not know the caller's number, the telephone equipment does, and so a bidirectional information flow is possible.

Computer communications principles

As in the case of time, there is some subtlety in the simple definition of a communication. Again, this occurs because it is not made clear how the space is measured by the participating computers. There are two possible mechanisms. The first is an absolute scheme, where the communicators are described as a particular collection of computers and a channel. The second is a relative scheme, where the communicators are just described as 'the computers and channel involved in the communication.' In summary, in the first case, the communication depends on the space and, in the second case, the space depends on the communication.

In general, it is not easy to have pre-arranged absolute sets of communicating parties. Until a communication actually starts, it is not easy to decide exactly who the communicators are. Thus, most spaces are relative to the actual communication starting. In the absence of some absolute map of the communication space, each computer will have its own local version. This need not be a perfect record of the actual space. For example, if a computer has a completely passive role in the communication, i.e., it only experiences an information inflow, then it might not need to know exactly where the information is coming from. Further, when the implementation of complex spaces is considered later in this chapter, it will become clear that a computer might only need to know about a subset (perhaps of size one) of the communication space, in order to participate fully. Compared with time agreements, agreement on maps of spaces are normally coarser grain, that is, change far less frequently than agreements on time periods.

5.2.3 Examples of communication spaces

Two computers linked by a physical channel

When two computers are directly linked by a physical channel, there is an absolute communication space. It consists of the two computers and the channel. There is no need for any further knowledge to be derived from actual communications. This was the situation in the model of communication used until the current chapter.

Two computers using a connection-oriented service

In this example, two computers communicate using a connection-oriented communication service, akin to the telephone system, rather than using a permanent direct physical channel. Before the connection is established, that is, before the actual communication begins, there is no absolute map of the space. Knowledge of the space is acquired by the communicators in much the same way as happens with the two parties in a telephone call. It is normal for the computer starting up the connection to send its identifier to the other computer as part of the establishment process. The only significant difference from a telephone call is that the identifier scheme is likely to be richer. This is because several processes within a particular computer are likely to be participating in different communications at one time. Therefore, a unique identifier for the channel must be derived not just from identifiers for the two computers, but also extended identifiers for the particular processes within each computer.

Two computers linked using a connectionless service

This example is similar to the previous one, except that the temporal nature of the communication service is different: it is connectionless rather than connection oriented. In space terms, sending a message using a connectionless service is almost the same as communicating using a connection-oriented service. That is, when a computer sends a message, it is fixing the communication space. When a computer receives a message, it then becomes aware of the space and the other communicator. The communication space ceases to exist as soon as the message has been sent, just as the space of a communication using a connection-oriented service ceases to exist when the connection is closed. The information flow is unidirectional, so there is no particular need for the receiving computer to acquire any knowledge of the communication space. In fact, in this example, there might be multiple recipients of the message, if the connectionless service allows it. However, any particular recipient need not necessarily know whether or not there are others.

5.3 IMPLEMENTATION OF REQUIRED SPACES

Agreement on spaces is only concerned with how the communicators agree on the collection of computers, and the channel, for a communication. However, it is very often the case that there is no direct physical realization of the communication space. One implementation method, which is covered in Section 5.3.1, is to involve computers actively in order to restrict or extend the set of computers that are linked by a channel. A second method, which is covered in Section 5.3.2, is to reduce or increase the number of channels that are used to support channels required for communications. Note that both of these implementation methods refer to spaces, rather than individual communications. This is because space, like information, is a static component of communications that can have an independent existence.

5.3.1 Filtering and switching

First, it is useful to introduce three technical terms used to describe different primitive types of information flow between computers in a space:

1. **broadcast:** information flows from one computer to all of the other computers
2. **multicast:** information flows from one computer to a subset of the computers
3. **unicast** or **point-to-point:** information flows from one computer to another

These are illustrated in Figure 5.6. Of course, broadcast and unicast are just extreme special cases of multicast. These three types of communication are the most common in practice, since they make a useful simplification that aids understanding, implementation and use of communications, and the channels that enable them.

Virtually all physical communication channels fall into one of two categories. One is channels that support broadcasting over a collection of three or more computers. The other is channels that support unicasting over a collection of two computers. This categorization was introduced in Section 1.3, where different physical media were classified as being either broadcast or unicast. To repeat a warning in that section, based on everyday human experience, it is tempting to think of radio or microwave as broadcast and copper or fibre optic cables as unicast. This is how they appear in conventional applications like radio and television, and telephone systems. However, one should also recall mobile telephones and cable television as counter-examples.

A further, related, primitive type of information flow is the **anycast**. This has the information flow of a unicast, but also has some characteristics of a multicast. It involves communication between one computer and *any one* of a subset of the computers. To an external observer, an anycast appears like a unicast. To the sending computer, an anycast appears like a multicast. The idea is that communication takes place with one of a collection of computers, but it does not matter which one, because all perform a similar function. For example, the communication might be a request-response handshake between a client and any one of a collection of equivalent servers.

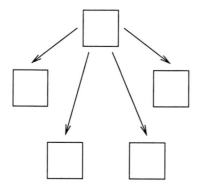

(a) Example of broadcast communication

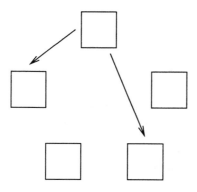

(b) Example of multicast communication

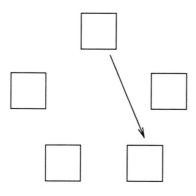

(c) Example of unicast communication

Figure 5.6 Broadcast, multicast and unicast information flows

Communications that involve more complex 'm to n' information flows must typically be implemented using communications with these more primitive types of 'one to n' information flow. This can be done by segmentation of such communications into sub-communications that have simpler information flows.

It is first necessary to look at the problem of implementing communications with one type of primitive information flow, by using one or more communications that involve different primitive information flow(s). Such implementation work is needed if a suitable channel is not available to allow the required communication to take place.

Human experience

Television programme

The idealized communication is of a television station multicasting to all the programme viewers. Unless the television station is operating in a fairly small area, this is unlikely to be the reality. Each viewer tunes into a broadcast transmission. However, this is likely to be from one of several distribution centres. The transmission may come from a UHF transmitter, from a satellite or from a cable centre. The television station transmits the programme to the various distribution centres, again using a variety of technologies. Thus, the implemented multicast communication may involve more than one sub-communication, and these sub-communications might be unicast or broadcast in nature. The whole process is coordinated by the television companies and the television distributors, who may or may not be the same organization.

Tourist sending postcards to friends

Here, the communication involves the tourist multicasting to the group of friends. There is no direct multicasting. Instead, a separate postcard is sent to each friend. Further, each postcard is not directly carried from postbox to mailbox. Instead, it is transferred by various postal staff through various postal sorting offices. Each step can be viewed as a unicast sub-communication. Thus, there are two stages: the multicast is implemented as a set of unicasts; and each unicast is implemented as a sequence of unicasts. The whole process is coordinated by the international postal system, which involves the cooperation of different national organizations.

Telephone call

The telephone communication is of a unicast nature. However, there is not usually a direct physical link between a pair of telephones. Instead, a link is made using a sequence of physical links through various telephone exchanges. Therefore, the required unicast communication is implemented using a sequence of unicast sub-communications. The whole process is coordinated by telephone companies or organizations, one or more depending on the parties involved in the telephone call.

Computer communications principles

Computers as filters or switches

Given that a computer is able to participate in communications using one or more channels, there are two activities it can carry out in order to assist in implementing communications requiring channels that are not directly available. The first is **filtering**, which means selectively ignoring some of the information that is communicated to it. This can be used to reduce the number of computers involved in a communication. The second is **switching**, which means selectively communicating further some of the information that is communicated to it. This can be used to increase the number of computers involved in a communication.

Computers and channels forming networks

For computers to assist in implementing communications, agreements on information, time and space are necessary. These guide the filtering and/or switching actions performed by the computers. The norm is to have agreement packages among a collection of computers, that result in the implementation of services providing channels that allow certain information flows between these computers. Any service that supports unicast, multicast and/or broadcast communications can be expected to allow any computers in the collection to be the source and the destination(s). That is, allowable information flows along channels are symmetrical from any viewpoint among the computers in the space.

A collection of computers and channels that uses general agreements on information, time and space matters among the computers in order to implement idealized channels using the actual channels is called a **computer network**.

The computers involved in a network can have a dual role. They are responsible for implementing the protocols associated with network agreements on information, time and space. However, they can also be users of the network service. The whole point of creating the idealized channels is to allow communication between computers for general purposes. In a single computer system, there may be a natural division between where network duties and general-purpose duties take place, for example, between hardware and software, or between operating system and user process.

Some computers may be solely concerned with implementing the network, and have no other duties. In such cases, they will be termed **filters** or **switches** in this book, depending on role, to reflect their more specialized networking roles. There are various synonyms used for such computers when describing practical communication networks. These include: **bridge**, **gateway**, **intermediate system** (used in contrast to 'end system' — a computer that just uses the network service), **node**, **protocol converter**, **relay**, **repeater**, **router** and **store-and-forward system**.

Note carefully that this implementation process can be repeated many times over. The channels made possible by one or more computer networks can be used, in turn, to construct another computer network that can provide different channels. Because of this, extra terminology creeps in. For example, the term **sub-network** (or just **subnet**) is used to describe a network that is a component, with other other sub-networks, of a larger network. The term **inter-network** (or just **internet**) is used to describe a network that is constructed from a collection of smaller networks.

Computer communications examples

Unicast or multicast channel implemented using a broadcast channel

Suppose that there is a network that enables arbitrary broadcast channels among its computers. This can be used as the basis for implementing arbitrary unicast or multicast channels. To do this, the information to be shared over such channels is transmitted to every computer in the network using the broadcast capability. Computers then filter information received over the broadcast channel. Specifically, there must be an information agreement that all communicated information includes a way of specifying the identifier(s) of the computer(s) that are intended to receive the information. Then, all computers look at the identifier information. Only those with appropriate identifiers accept the rest of the information, and so participate in the communication. The other computers ignore the communication. This style of operation is usually termed 'promiscuous' — all computers can see all communicated information, but have to be trusted not to abuse the privilege.

Unicast channel implemented using different unicast channels

Suppose that there is a collection of computers, with unicast channels between certain pairs of computers. The computers and channels may or may not be organized into networks — it is not important. Given an appropriate arrangement of channels, this can be used as the basis for a network that can implement arbitrary unicast channels. It is necessary to establish information and space agreements so that unicast channels can be implemented between pairs of computers that do not already have direct unicast channels. This is feasible as long as it is possible to find a route formed from direct channels leading from one computer, via intermediate computers, to the other computer. Then, a series of sub-communications can take place to implement the required communication. The original computer, and the intermediate computers, are involved in switching, in order to forward information. To do this correctly, the communicated information must include information about the intended final destination of the information. Intermediate computers filter out the rest of the communicated information. To construct such a network, there must be at least one possible route between every pair of computers in the network. If there is more than one possible route, then there is scope for improving the quality of a communication, by choosing a route that is quicker, cheaper or more reliable, for example.

Broadcast or multicast channel implemented using unicast channels

Broadcast or multicast channels can also be implemented on a network of the type just described. One simple way is just for the information source to conduct a unicast communication with each recipient individually. In the case of broadcasts, the source must have some means of knowing the identity of all computers in the network, so that it can communicate with them. A better way is to enhance the switching behaviour of computers. In this case, the communicated information must include information about the identifiers of the computers that are to receive it. Then, the original computer, and any intermediate computers, conduct more restricted sub-communications over multicast (possibly just unicast) channels with one *or more* direct neighbours that can assist in implementing the full communication. That is, the remaining broadcast or multicast communication is partitioned between channels to neighbours at each stage. The effect is of a tree of channels, rooted at the information source, and with recipient computers as leaves and possibly nodes.

Networks and graphs

The language, both verbal and pictorial, of **graphs** is sometimes convenient for discussing spatial aspects of networks. Graphs have **vertices** (sometimes called 'nodes') and **edges** (sometimes called 'arcs') that connect pairs of vertices. A **directed graph** has **directed edges**, that connect ordered pairs of vertices, with the underlying idea of moving from the first vertex of the pair to the other.

In the case of computer networks, a network can be regarded as a directed graph, with computers represented by vertices and unicast channels represented by directed edges. The direction of edges corresponds to the information flow of channels. There need be no extra representational apparatus for multicast or broadcast channels, since these can be modelled by multiple unicast channels, as seen in the previous example. A graph representation of a small example network is shown in Figure 5.7. Note that all edges are bidirectional, which is normally the case in practice. That is, if it is possible to have a unicast channel in one direction, it is possible to have one in the other direction.

Given directed graph representations, some of the ideas from the examples above fit in well with concepts from the world of graphs. A **path** between vertex s and vertex t in a directed graph is a sequence of directed edges

$$(s, v_1), (v_1, v_2), \ldots, (v_{k-1}, v_k), (v_k, t)$$

for some k. A graph is said to be 'connected' if there is a path between all pairs of vertices. Thus, paths and connectedness are directly relevant to constructing networks supporting arbitrary unicast channels, as in the second example above. Connectedness means that there are routes between all pairs of computers. Figure 5.8 shows an example of a (contorted) path between the top-left and bottom-right vertices of the graph introduced in Figure 5.7.

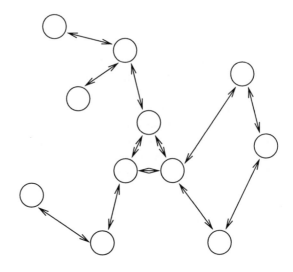

Figure 5.7 Example graph representing network

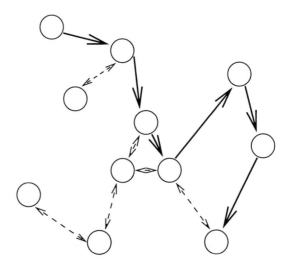

Figure 5.8 Directed path in example graph

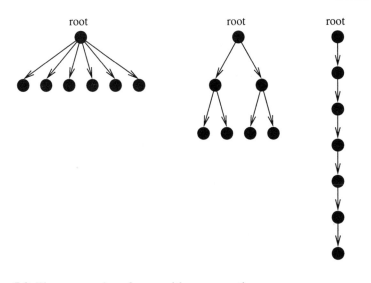

Figure 5.9 Three examples of trees with seven vertices

Trees can also be defined within graphs. A tree in a directed graph is a sub-graph composed of a set of ℓ vertices and $\ell - 1$ directed edges between vertices in this set. Each edge is directed at a different vertex. The single vertex that does not have an edge directed at it is the root of the tree, and there must be a path from the root vertex to every other vertex in the tree. Figure 5.9 shows three examples of trees containing seven vertices.

A **spanning tree** of a graph is a tree that includes all the vertices of the graph. Trees and spanning trees are relevant when constructing networks that support multicasts and broadcasts, as in the third example above. Figure 5.10 shows an example of a spanning tree rooted at the leftmost vertex of the graph introduced in Figure 5.7.

There are two main advantages in using graphs as an abstraction of computer networks. One is as a way of hiding detail. For example, instead of having each vertex representing one computer, it is possible to have each vertex representing a collection of computers, perhaps the members of a component network. Edges then represent channels between computers in one collection and computers in another collection. Thus, graphs can be used as a tool for representing different levels of network implementation.

The other main purpose is to give access to the large collection of algorithms and data structures that the computer science community has made available for solving graph-related problems. There are standard methods not just for finding paths and trees in graphs, but also for finding cheapest paths and trees when cost measures are associated with edges.

Examples include Dijkstra's algorithm and the Bellman-Ford algorithm, both of which, given a vertex, find the shortest paths from that vertex to all other

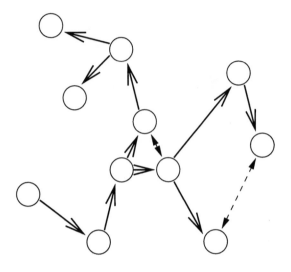

Figure 5.10 Directed spanning tree of example graph

vertices. The Bellman-Ford algorithm is more general than Dijkstra's algorithm, because it allows edges costs to be negative. However, its run time is proportional to the number of vertices multiplied by the number of edges, whereas Dijkstra's algorithm's run time is proportional to the number of vertices squared. These algorithms are useful when designing and operating networks, as will be seen in Chapters 7 and 8. It is not necessary to know the details of how the algorithms work in order to understand how networks work but, for interested readers, the details of Dijkstra's algorithm can be explored through Exercise 5.13.

Chapters 6, 7 and 8 are concerned with the detailed information, time and space agreements that are necessary in order to make unicast, multicast and broadcast networks work. The chapters are differentiated by the complexity of the switching task that has to be undertaken by computers in order to implement the required channels. Filtering also features, but not as a central concern. In essence, switching is a process of receiving a message as a result of one communication, and then selectively forwarding it using another communication.

In the networks of Chapter 6, computers either do no switching, or do decision-less switching in the sense that messages received on one channel are always transmitted on one or more other channels. It is assumed that communications over component channels have similar information, time and space characteristics. The effect of the non-existent or unsubtle switching is that the basic capability of the network is to supply broadcast channels, because no computer in the network introduces selectivity in its switching. Therefore, such networks are entitled **Message Broadcasting Networks**. This style of networking has its roots in physical channels that enable broadcasting.

In the networks of Chapter 7, computers do selective switching. Messages received on one channel are selectively transmitted on one or more other channels. It is assumed that communications over component channels have similar information, time and space characteristics. The effect of the switching is that, depending on the type of selectivity chosen, the network can supply unicast, multicast and/or broadcast channels. Focusing on the method of implementation, such networks are entitled **Message Switching Networks**. This style of networking has its roots in physical channels that enable unicasting.

Finally, in the networks of Chapter 8, computers again do selective switching. However, the switching goes beyond the simple relaying of messages from one channel to another. A switching computer must also deal with an assumption of significant differences in the information, time and space characteristics of communications on its different channels. This situation usually arises because the component channels are, in turn, implemented by networks with different information, time and space agreements. For this reason, and following the normal convention of the computer communications trade, the networks of the chapter are entitled **Inter-networks**.

Before embarking on an in-depth study of networking, however, there is another basic space implementation technique to consider. This has various uses, including the sharing of channels and the introduction of parallel transmission over channels.

5.3.2 Splitting and multiplexing

The second basic implementation technique for space involves the complementary procedures of **splitting** and **multiplexing**. These refer to the ways in which several channels can be used to implement one channel, and in which several channels can be implemented by sharing one channel.

Human experience

Television programme

Continuing from the last section's example (on page 152), suppose that the television programme is a transmission of a musical concert. The quality of sound in television broadcasts is usually inferior to the stereo quality available on radio. Therefore, pictures and sound are often broadcast simultaneously over the two different media. This is a simple example of splitting: one channel is implemented using two independent channels.

The particular television programme is only one of several or many that can be received at any time. However, these are all delivered using a single physical medium, whether it is electro-magnetic radiation or cable. This is a simple example of multiplexing: several independent channels are implemented using a single channel. In this case, **frequency division multiplexing** (FDM) is used: each television programme is broadcast in a different frequency band.

The frequency bands are all contained within the overall frequency range of the medium. Thus, the different communications do not interfere with each other.

Tourist sending postcards to friends

A tourist who wished to save money on postal costs when visiting a far-away destination could make use of multiplexing. He or she could place all of the postcards into an envelope and send it to one friend, then ask the friend to distribute the postcards. This would be a typical example of why multiplexing is used. That is, to share an expensive communication channel between several channels. Of course, multiplexing also happens at a higher level in this example. The tourist's postcards are only a few among very many items in the postal system. At various stages in the journey, postcards will travel with many other items in a single mail container.

Telephone call

Multiplexing is a common feature of the telephone system. At most stages, many calls share the same physical links. For example, one caller might be within an organization with a private switchboard. Then, all of the organization's calls are likely to share the same physical link to the external telephone system. As another example, many national or international telephone calls share physical links between cities or countries at any one time. The use of carrier channels for doing this is described in the discussion of leased lines in Section 1.3.2.

Computer communications principles

Just as segmentation and concatenation are opposites for one another for time period implementation, so splitting and multiplexing are opposites of one another for space implementation.

Splitting

An alternative term for splitting is **downward multiplexing**. The set of communications that involves one channel is partitioned into subsets of communications that involve different channels. The set of computers involved in the space of each communication is not affected. This process is illustrated in Figure 5.11. The benefit of splitting is that it allows the introduction of parallelism — several communications can proceed at the same time if they have independent channels. This allows the overall quality of the communication system to be improved, for example, by making individual communications faster.

Multiplexing

An alternative term, matching the alternative term for splitting, is the longer **upward multiplexing**. The sets of communications that involve several different channels are combined into a set of communications that involves one channel.

Communications over single channel

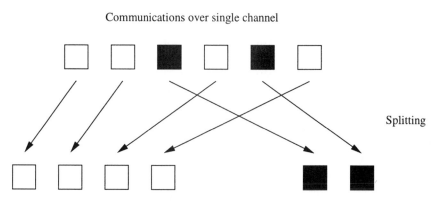

Splitting

Communications over one channel   Communications over another channel

Figure 5.11 Splitting a channel over several channels

Communications over one channel   Communications over another channel

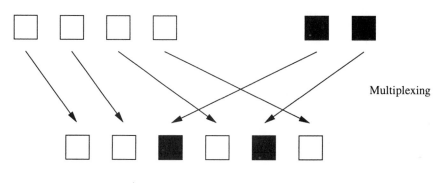

Multiplexing

Communications over single channel

Figure 5.12 Multiplexing several channels on to a single channel

The sets of computers involved in each of the original sets of communications must be the same. This process is illustrated in Figure 5.12.

One frequent restriction on multiplexing is that all of the communications in the resulting set must have disjoint time periods, since the channel only allows one communication at a time. A simple counter-example is a bidirectional channel between two computers: this allows two independent unidirectional communications to take place at the same time, in opposite directions. However, this is not a major exception to the general rule, since the channel is really just a pair of independent unidirectional channels.

If attempted multiplexing does lead to a clash, then **contention** is said to occur. In general, to achieve the disjoint property, each communications that uses a channel being multiplexed must have its time period adjusted appropriately so that it does not overlap with others. This preliminary time implementation step is akin to flow control, but is carried out for the benefit of the channel rather than the computers. It relies on the multiplexed communications being asynchronous.

The advantage of multiplexing is that it allows the sharing of channels — several channels can make use of the same channel. This gives a cost saving, compared with providing multiple parallel channels. There may also be cost savings for the computers themselves, since they only have to deal with one channel at a time, rather than several.

Computer communications examples

Simplex and duplex channels

When a channel is used to connect two computers, there are several different possible communication behaviours that it might allow. In the model of a communication system introduced at the beginning of Chapter 2, a channel allows information to flow in either direction between the computers, at the same time if necessary. This is known as a **full duplex channel**. However, to simplify many of the communications described so far, they have had a unidirectional information flow. A channel that can only support communications in one direction is known as a **simplex channel**. In between these two types of channel, there is a third type: the **half duplex channel**. This type allows information to flow in either direction, but not in both directions at the same time, One use of a half duplex channel is for handshake-style communications.

In space terms, the different types of channel — simplex, half duplex and full duplex — are increasingly powerful. That is, a simplex channel can be implemented easily using either of the others, and a half duplex channel can be implemented easily using a full duplex channel. It may be possible to implement the features of a full duplex channel using a half duplex channel, essentially by multiplexing two unidirectional channels onto the single half duplex channel. However, to do this, it is necessary to segment any bidirectional communications into unidirectional sub-communications that have disjoint time periods. A half duplex or full duplex channel can be implemented easily using two simplex channels, one for each direction of information flow. This is an example of splitting.

Multipeer channels

A **multipeer** channel (alternatively termed a 'multipoint' channel) is a channel connecting n computers, that supports all of the n possible broadcast information flows. Typically, but not necessarily, at most one flow can be in progress at one time. However, the important point is that this channel can support the n different flows in any sequence over time. In Chapter 6, a dominating issue is how n separate broadcast channels, one for each possible source, can be multiplexed onto a single multipeer channel.

In the opposite direction, the example of splitting a duplex channel over two simplex channels can be generalized to give a way of splitting a multipeer channel over a set of broadcast channels. If there are n different sources for the multipeer channel, then n broadcast channels are needed, each providing the information flow needed for a different source.

One example of this situation in practice is the **multicast group,** which is a collection of computers that send multicasts to each other. A multicast group of n computers requires n one to $(n-1)$ multicast channels to implement the complete information flow of the group. Thus, over the n computers involved, the multicast group is just a multipeer channel implemented using broadcast channels within the group. In the context of an enclosing network, each of these broadcast channels is implemented using a multicast channel provided by the network.

Piggybacking

The discussion of connection-oriented time packages in Section 4.5 was restricted to a service offering only a unidirectional information flow from one computer to the other. That is, the service included just a simplex channel. However, to implement the service, the full duplex nature of an underlying channel was needed in order to send back control information. Real connection-oriented services provide a full duplex channel to their users, to allow a bidirectional information flow, i.e., information sharing, between computers.

From the point of view of the communication service, the duplex channel provided to a user is implemented as two simplex channels that are essentially independent. The only dependency between the channels is the fact that they are created and destroyed simultaneously, when the connection begins and ends, respectively. This means that, in principle, two duplex channels are appropriate for implementing the two simplex channels, each used for user information in one direction and control information in the other direction. However, in practice, there is only *one* underlying duplex channel, and so multiplexing is necessary. That is, a user information communication channel in one direction must be multiplexed with a control information communication channel for the other direction, and vice versa.

To improve efficiency, it is not always necessary for the two types of communication to be disjoint in time. Most protocols for implementing connections use information-carrying messages that also have space for control information for the other direction of transfer. This is known as **piggybacking**, to convey the idea of control information hitching a lift along the channel. As an example, all TCP messages carry both the serial number of the information present and also the serial number and window size for the other direction. In fact, there is only one type of message, so all acknowledgements are piggybacked, on zero-size information messages if necessary. The LAPB protocol also allows piggybacking,

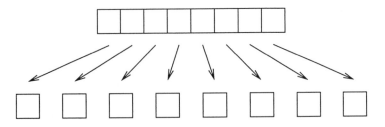

Figure 5.13 Splitting a character communication channel

although there are explicit control information datagrams as well. However, the TP4 protocol does not include piggybacking as a feature.

Sharing a single character

On page 88 in Chapter 3, the asynchronous method of transmitting eight-bit character representations could not be segmented in time further, since the 10 necessary bits are transmitted synchronously. An alternative way of transmitting each character is to use segementation and splitting. If eight separate binary channels are available, then each character communication can be segmented into eight bit communications, and then the eight bits can be transmitted in parallel, with a significant time saving. This is illustrated in Figure 5.13. In general, this approach can be used for larger strings of bits. For example, 32-bit or 64-bit buses are used within computers or between computers and peripheral devices. Parallel channels with synchronized data transfer are not normally used for communication between computers, both because of timing problems and because of expense.

Time division and statistical multiplexing

A normal requirement for multiplexing is to avoid contention. That is, to ensure that multiplexed communications have disjoint time periods. As already mentioned, achieving this is a flow control problem, in which the need for flow control arises from the channel rather than from the communicating computers. In some cases, flow control required by the computers might be adequate to allow multiplexing. For example, if there is a master computer and other slave computers, then the time-disjoint communications enforced by the master will ensure that there is no contention for the channel.

For the classical multiplexing problem of sharing one actual unidirectional channel between several required unidirectional channels, there are two main approaches to flow control — one synchronous, the other asynchronous. With **time division multiplexing** (TDM), the required channels take turns to use the actual channel, each being allowed a fixed period of time. Thus, the communications being multiplexed have synchronous time periods which are disjoint. This arrangement ensures that the multiplexing is fair, but is wasteful if some of the communications are irregular and so waste their guaranteed time slots. TDM has a long history in the telecommunications world, and the methods of sharing a carrier channel described in Section 1.3.2 illustrate this.

An alternative, which is the most common in computer communications, is **statistical multiplexing**. This treats the communications to be multiplexed as being asynchronous. Ideally, communications on the required channels occur during disjoint time periods, in which case there is no problem. However, if there is contention, the multiplexing scheme is able to adjust the time periods to make them disjoint. This is done by delaying communications until a later time when the shared channel is free, using buffering associated with the channel. The effect is that a transmitting computer begins a communication at its choice of time, but that the receiving computer sees the communication beginning at a rather later time. The 'statistical' nature of the multiplexing refers to the fact that the buffering is adequate to cope with almost all cases of contention. In some extreme cases, such as communications taking place on all required channels at the same time, it may not be able to cope and information will be lost.

ISO standard transport service

As mentioned on page 133 in Chapter 4, the ISO standard connection-oriented transport service may be implemented by one of five standard transport protocols: TP0, TP1, TP2, TP3 or TP4. Three of these — TP2, TP3 and TP4 — allow for multiplexing. This capability is the main feature that distinguishes TP0 and TP2, and TP1 and TP3. Several channels provided by the transport service may be multiplexed onto a single channel provided by the underlying service. This allows a possibly expensive channel to be shared and so to be used more productively. The TP4 protocol also allows splitting. A channel between two computers using the transport service might be split between several different channels provided by an underlying service. This allows several communications to proceed in parallel, using separate channels. The parallelism is not synchronous, so the protocol must ensure that correct sequencing of information is preserved.

5.4 CHAPTER SUMMARY

A communication involves information sharing among a collection of computers using a single channel. The computers have to be given unique identifiers. Identifier schemes may be either flat or hierarchical, the latter making allocation of unique identifiers easier. All of the computers involved in a communication have to have some knowledge of the map of the space. This might result from an absolute agreement, where the space is fixed before communication. More usually, the map of the space is relative to the communication. That is, computers acquire knowledge of the space when a communication begins, or as it continues.

The communications that can be carried out using computer communication systems are usually forced to have one of three forms:

- unicast — one computer transmits information to another
- multicast — one computer transmits information to several others
- broadcast — one computer transmits information to all others

and most physical channels support either unicast or broadcast transmission directly. In order to implement any particular pattern of information sharing, an appropriate channel has to be implemented using the types of channel available.

A main way in which this can be done is by using computer networks, which are collections of computers and channels between the computers. Protocols, involving agreement on information, time and space issues, are needed in networks. Using these, networks harness the capabilities of the computers in order to provide desirable channels using the available channels. The two main tasks carried out by computers in networks are filtering and switching. The capabilities of networks can be categorized by the complexity of the switching tasks carried out, giving message broadcasting networks, message switching networks and inter-networks.

To implement channels efficently, it may be necessary use different underlying channels that link the same set of computers. This may involve splitting, where communications over one channel are partitioned over several different channels. Alternatively, it may involve multiplexing, the opposite of splitting, where communications over several different channels share the same single channel.

5.5 EXERCISES AND FURTHER READING

5.1 For everyday communications that you have with other people, or with groups of people, identify the communication spaces, state how much you individually know about the spaces, and explain how you acquire that knowledge.

5.2 Investigate the history of the Global Positioning System (GPS), and find examples of its usage in everyday life.

5.3 For a computer system that you use, find out any addresses or names that it has for communication purposes, and investigate whether these are from a flat or hierarchical identfier system.

5.4 The Internet **Domain Name Service** (DNS) is a distributed mechanism for mapping names to addresses. Look at RFC 1034 and RFC 1035 to discover how it works, but do not attempt to understand all of the details.

5.5 Read RFC 1884, which describes the IP version 6 addressing scheme.

5.6 Find out why the designers of the IP version 6 addressing scheme decided not to just use the existing ISO NSAP identifier scheme.

5.7 Discuss computer communication applications where it is necessary to know the exact composition of the communication space, and others where it is not.

5.8 Devise examples of how multicasts could be made available in a natural way for human communications.

5.9 Suppose that you live in a house shared with others. Give an example of how you might act as a filter, and as a switch, when collecting a mail delivery for the house.

5.10 For each computer network that you have access to, list the number of computers involved, the types of channel that the network implements, and the names of protocols used for agreement within the network.

5.11 The Internet is the most famous example of an inter-network. How many networks are within the Internet at the present time?

5.12 Graphs are a convenient abstract representation of computer networks. Give examples of other real-life systems that could be represented using graphs.

5.13 Look up the details of Dijkstra's shortest path algorithm, which should be found in any standard book on algorithms. For the example in Figure 5.8, use the algorithm to find the shortest path from the top-left vertex to the bottom-right vertex, assuming that every edge has the same cost ($= 1$, say).

5.14 Describe clearly the difference between splitting (described in this chapter) and segmentation (described in Chapter 3).

5.15 Describe clearly the difference between multiplexing (described in this chapter) and concatenation (described in Chapter 3).

5.16 The description of piggybacking involved multiplexing because it forms part of the implementation of a service that, essentially, provides its user with two unidirectional simplex channels. Therefore, the user has to perform splitting of its information-sharing communications to use these channels. Explain why, if the service user could directly make use of the full duplex channel that underlies the service, then piggybacking of acknowledgements would just be an example of concatentation.

5.17 Give examples of practical situations where time division multiplexing would be efficient.

5.18 Investigate the ISO standard Transport Protocol, to discover other differences between TP0–4, not mentioned in this chapter, or in Chapter 4.

Further reading

This chapter is intended to be self-contained — drawing on existing human experience of space issues in communication, in order to introduce the main ideas impacting on computer communication. The next three chapters, Chapters 6–8, expand the coverage of how computer networks are designed and used. Later on, Chapters 9 to 11 include material concerned with the role of space in particular areas of computer communications. More details about graphs, and graph algorithms, appear in textbooks on algorithms, for example, *Introduction to Algorithms* by Cormen, Leiserson and Rivest (MIT Press 1990).

CHAPTER SIX

MESSAGE BROADCASTING NETWORKS

> *The main topics in this chapter about message broadcasting networks are:*
>
> - information, time and space basics of message broadcasting
> - local, and metropolitan, area networks
> - implementing a multipeer channel using simple channels
> - multiplexing broadcasts onto a multipeer channel
> - guided technology networks: ethernet, token ring/FDDI, token bus, DQDB and 100 BASE VG-AnyLAN
> - unguided technology networks using radio and infra-red transmission

6.1 INTRODUCTION

As Chapter 5 explains, there are increasingly complex ways in which computers can contribute switching abilities to assist in supporting a computer network. This chapter concentrates on the simplest cases: no switching is done; or non-selective switching is done. Because of the lack of discrimination, such networks can be regarded as **message broadcasting networks**, which have the basic characteristic that every communication that takes place has a space that contains all of the computers in the network. Therefore, if n computers are involved in the network, then n different one to $(n - 1)$ broadcast channels are made available by the network. In addition, broadcasting networks can supply unicast and multicast channels as well. These are implemented using filtering by computers, as described on page 154 of Chapter 5.

Message broadcasting networks may be implemented using broadcast-style channels, or using unicast-style channels. Reflecting the fact that the computers play a modest networking role in the switching sense, broadcasting networks are normally low-level implementations, in that the channels used are physical channels, rather than channels created by further lower-level network implementations. Therefore, this chapter focuses on such networks. However, in principle, the underlying ideas could be used to implement higher-level broadcasting networks using channels provided by other lower-level networks.

Traditionally, message broadcasting networks have been nearly synonymous with **Local Area Networks** (LANs), which are networks with a physical radius of no more than one or two kilometres. This is because the high communication speeds needed to make broadcasting a realistic possibility were only attainable over short distances. However, this assumption is changing with advances in technology. **Metropolitan Area Networks** (MANs), which are networks with a physical radius that can span a city, are often organized as broadcasting networks. Meanwhile, the techniques used for message switching networks, traditionally associated with longer distance networks and the subject of Chapter 7, are now being used in some LANs and MANs, as well as in experimental Home Area Networks (in the home) and Desk Area Networks (on the desk). Thus, it is now more useful to have a classification based on networking style, rather than based on physical distance spanned.

6.1.1 Information basics

The messages transmitted by broadcasting networks contain at least three basic components:

- source identifier: the identifier of the sending computer
- destination identifier(s): the identifier(s) of the intended recipient computer(s)
- information: the information being communicated by sending the message

Note that, for low-level broadcasting networks, messages are usually called 'packets'. In fact, this terminology is normal for all types of networks based on physical channels, including the types described in Chapter 7. Thus, the terms 'packet broadcasting network' and, particularly, 'packet switching network' are often encountered in practice.

The source identifier is included so that a recipient can tell from where the message originated. This is needed because the network supports n different communication spaces, where n is the number of computers in the network. Including the source identifier in each message acts as a convenient way for computers to deduce the communication space from the actual communication. A relative scheme like this is necessary unless the recipient can differentiate between communication spaces by the network service providing independent access to n different channels.

The destination identifier(s) are included so that recipients can tell whether or not the message is intended for them. This allows unicast or multicast communications to be implemented using the broadcasting network. A recipient just filters out any message that does not carry its own identifier.

The identifiers used for computers have network-wide significance, and are of a fixed size agreed for the particular network. One example would be the 48-bit identifiers defined in the IEEE 802 series standards for LANs, and mentioned on page 143 in Chapter 5. This identifier scheme allows worldwide uniqueness of identifiers. For private LANs, an alternative also allowed by the standard is to use 16-bit identifiers. However, it is not possible to mix the two identifier lengths in the same network. The use of fixed-size identifiers, and also locating the identifiers in a fixed position within each message, simplifies the handling of messages by computers.

The information component of a message is just a sequence of bits (in fact, more likely a sequence of bytes) as far as the network is concerned. In principle, the information component might have an arbitrarily chosen length. Some networks impose a fixed length on this component, so all messages have a uniform length. This, again, simplifies handling of messages. If not, almost all other networks impose a maximum length on this component, and hence on the overall message length. This is because, as Section 6.1.3 explains, the collection of broadcast channels is multiplexed onto a single multipeer channel. A restriction on length ensures a degree of fairness in this sharing of the multipeer channel. Some networks also impose a minimum length restriction on messages. The reasons for this are discussed in detail later but, in summary, it may be because either the broadcast channel has characteristics that may cause it to 'lose' very short messages, or the communication of very short messages is inefficient.

The standard network message format, which may include other components in addition to the above, will have a direct representation in terms of bits. This representation is an absolute agreement, fixed for all computers participating in the network. The representation of bits depends on how the broadcast channels are implemented. If a single underlying broadcast-style channel is used, then the representation will be fixed. However, it may vary if a collection of different unicast-style channels is used.

In summary, from an information point of view, there is absolute agreement on the type of information communicated: the standard network message. There is also absolute agreement on how this is represented in terms of bits.

6.1.2 Time basics

Two absolute time measurements are usually possible for message broadcasting networks based on physical channels. The first is the latency, that is, the delay between the time that a computer begins a transmission and the latest time that any computer in the network begins to receive the transmission. This is affected by the physical technology used to implement the network, as well as the physical distances involved in the network. For example, a network of computers in the same room, connected by fibre optic cables, will have a very low latency. A network of computers distributed around the world, connected by satellite links, will have a high latency.

The second absolute time measurement is the rate at which transmission can take place, measured in bits per second. Again, this is affected by the physical technology used to implement the network. In the case of a genuine broadcast technology, the rate is a measurable global parameter. Where the network is implemented using unicast technology, the rate is determined by the slowest link contributing to the implementation. This is measurable, since the links are part of a tightly-knit collection, and their performance is dictated by the demands of the network.

In combination with the absolute agreements on information, the duration of communication time periods becomes measurable. For a fixed message size, the time between starting to transmit a message and finishing the reception of the message at all computers, is measurable. It is equal, in seconds, to:

$$latency + (message\ size/rate)$$

where the latency is measured in seconds, the message size in bits, and the rate in bits per second. This puts absolute time bounds on the time period for the communication of the information in the message. Where there are minimum and/or maximum size restrictions on message size, lower and/or upper bounds respectively can be put on the communication time period. Note that, although the duration of communication time periods can be absolute, the message communications are asynchronous, since they may start at arbitrary times.

In high speed networking circles, a further measure of interest is the product of the rate and the latency, usually called the **bandwidth*delay product**. This gives a measure, in bits, of the amount of information that can be simultaneously present on the channel. That is, it measures the number of bits that can be transmitted during the time it takes for one bit to travel from the transmitter to the receiver.

In terms of time packages, the service offered by message broadcasting networks is normally an unreliable connectionless service. There is no segmentation, so information is presented to the service in message-sized units, and each unit results in a single communication of a message. The bits of a message are communicated synchronously. A simple handshake time package is an elaboration on the unsegmented time package that is sometimes found. With this, a message is broadcast with a single intended recipient. This is followed by an acknowledgement message being broadcast by the recipient, intended to reach the original sender.

Given that distances are usually short in broadcasting networks, and the quality of the physical media is high, the service offered is usually inherently very reliable. No further quality is added by the network, for example, to provide a totally reliable service. Because of this, message formats include an error-detection code component. This allows recipients of messages to detect when errors occur, the normal action being to discard the damaged message. The network itself takes no responsibility for any actions required after this. The effect is the same as if the network had lost the message completely, which is another possibility. Duplication of messages should not occur in individual broadcasting networks.

In summary, from a time point of view, there is absolute agreement on the duration of communication time periods, or at least for bounds on the duration of time periods. However, the beginning of time periods is relative to communications starting. Message communications are not segmented over time, and the bits comprising a message are communicated synchronously. There is neither time-related error correction provided by the network nor provision for time-related flow control by the communicating computers. However, note that there is time-related flow control for multiplexing channels. This is discussed in detail in Section 6.3.

6.1.3 Space basics

As has already been discussed earlier in this chapter, there are often some absolute physical constraints on the area occupied by a broadcasting network. These are necessary to achieve the required communication speeds using the available technology. The two dominant types are Local Area Networks (LANs) and Metropolitan Area Networks (MANs) (the latter sometimes called Campus Area Networks (CANs) in academic environments). The absolute space constraints do not extend as far as fixing the physical position of the computers in the network, but may impose restrictions on the minimum or maximum distance allowed between computers. These restrictions are related to the physical transmission properties of channels.

The identifier scheme for broadcasting networks is usually flat. The most common such networks are under the control of a single organization, and so it would be possible to allocate unique identifiers relatively easily. However, most networking equipment is bought with an identifier already built in. IEEE standard broadcasting LANs are an example of a globally unique flat identifier scheme across all such networks in existence. Strictly speaking, this is unnecessary so long as the networks are not connected together. Such interconnection will be considered in Chapter 8.

The network map presented to users is a simple absolute one. All communications involve *all computers* in the network. More specifically, any individual computer can transmit information to all of the other computers. If computers require information about the particular computers in the network, then this must be configured absolutely, or must be acquired using the broadcasting service provided. When such information has been obtained, computers can introduce selectivity into communications by appropriately setting up the destination identifier component of messages.

Overall, a broadcasting network offers the ability to conduct n different types of communications, where n is the number of computers in the network. Each of these types involves the same collection of computers: all of the computers in the network. The difference is in the information flows allowed by the channels. Each of the types corresponds to a different computer broadcasting information to all of the other computers. No other types of communication are possible. If required, they must be implemented using the basic broadcast channels of the network.

The implementation of the broadcasting network service consists of two steps. First, a particular type of channel has to be implemented: a multipeer channel of the type described on page 162 in Chapter 5. This channel allows all of the n possible broadcast information flows. For networks based on a physical multipeer channel, the name **medium** is very often used for this channel, reflecting its physical nature.

Second, the n different broadcast channels (i.e., one for each of the n possible broadcast information flows) must be multiplexed onto the single multipeer channel. The term **Medium Access Control** (MAC) is often used to describe this process when a physical multipeer channel is used. The multiplexing is trivial if the multipeer channel allows several broadcasts to take place simultaneously, but this is uncommon.

The next two sections cover these two implementation components, and are followed by examples of real broadcasting networks, to illustrate how all of the ideas fit together in practice.

6.2 MULTIPEER CHANNEL IMPLEMENTATION

It is necessary to implement a channel that allows broadcasting by any one computer to all of the other computers. This means that there must be appropriate physical connectivity between the computers. Such connectivity may be of a guided type or an unguided type, or a mixture of both. The way in which the channel is implemented depends on various matters, including the layout of the computers and the distances between them, and the types of physical communication medium that can be used. Here, three main types of multipeer channel implementation are described. All have the characteristic of a regular structure. That is, they involve physical connectivity between computers that is clearly part of a coordinated aim: to achieve broadcasting. Note that, although couched in

Computers

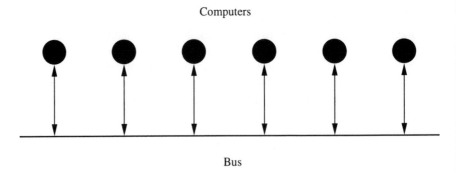

Bus

Figure 6.1 Multipeer channel using common medium

terms of physical connections, the ideas here translate to networks founded on higher-level connections with similar properties.

6.2.1 Common medium

A common medium is the simplest implementation approach. It is just a physical medium that is inherently multipeer in nature. That is, any computer is able to transmit on the medium and the transmission is received by all of the computers in the network. When a guided technology, such as an electrical cable, is used, the channel has a linear arrangement and visits each computer in turn. Such a channel is normally referred to as a **bus,** and is illustrated in Figure 6.1.

Guided technologies are only used over relatively short distances, e.g., a few kilometres at most, due to electrical limitations on the broadcasting of signals. For longer distances or, increasingly, for shorter distances also, unguided technologies, such as radio, can be used. In a recycling of a term that had become old-fashioned in a different context, broadcast networks with unguided channels are nowadays termed **wireless**.

In contrast with the other two types of channel implementation described below, the computers have a largely passive role. They are able to listen to the channel to detect broadcasts, and then to store copies of the broadcast information, without affecting the operation of the channel. The only active role is when initiating a new broadcast. No computer is required to play a part in maintaining or ending the broadcast by performing switching duties. This is a natural property of the medium.

6.2.2 Chain or ring

A chain is an approximation to a bus style of multipeer channel. The computers are arranged in a line (in the logical, rather than physical, sense) with a unicast channel between each pair of neighbouring computers in the line. The only difference for

Computers

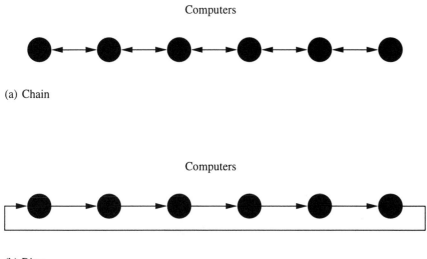

(a) Chain

Computers

(b) Ring

Figure 6.2 Multipeer channels using chain and ring arrangements

a ring is that the two computers at the end of the line are connected by a unicast channel, which turns the line into a circle. It is normal for the unicast channels in a ring to go in one direction only, since bidirectional operation is not required to achieve broadcasting. Figure 6.2 shows a bidirectional chain arrangement, and a unidirectional ring arrangement.

In a chain, when a computer broadcasts information, it must send it to both of its neighbours. The neighbours in turn pass it on to their neighbours. This continues until the information reaches the computers at the ends of the line. The effect is the same as for a bus, except that computers play an active role in propagating the broadcast information by carrying out non-selective switching. As the information passes through the intermediate computers, they can copy it as their own private copy of the broadcast.

In a ring, when a computer broadcasts information, it sends it to one of its neighbours. The neighbour in turn passes it on to its other neighbours. This continues until the information returns to the original broadcaster. Even in bidirectional rings, this simpler approach is preferred to one in which information is passed in both directions round the ring. It avoids some intermediate computer having to detect information arriving from both its neighbours, and then being selective in its switching by not passing the information on any further.

In principle, chain or rings can employ a variety of technologies for the unicast channels between computers. Unlike a genuine broadcast medium, there is no need for uniformity. However, the overall performance of the implementation will be determined by the worst unicast channel in the network. The fact that there is a collection of channels means that the extent of the network is only limited

Computers

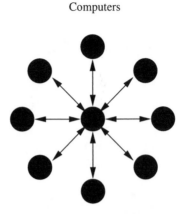

Figure 6.3 Multipeer channel using star arrangement

by a maximum distance between neighbouring pairs of computers, rather than a maximum distance between the two furthest points on the network. Thus, the technology is more suitable for Metropolitan Area Networks than a common medium technology.

The computers in such a network have an active switching role, in forwarding information to neighbours. However, this is normally done by low-level hardware that operates autonomously from the main computer (indeed, it may be separate from the computer), and so forwarding is not affected by loading on the computer or, more extremely, by the computer being switched off.

6.2.3 Star or tree

A star is an approximation to many types of wireless channel. A central computer, or a specialized switching device, acts as the centre of the star, and each computer in the network is connected to it by a duplex channel. A computer can broadcast a message by transmitting it to the centre, which in turn transmits it to all of the computers in the network along the separate channels. Thus, the centre performs non-selective switching to give it a genuine broadcasting capability, whereas the computers perform no switching and only have a single unicasting capability. Figure 6.3 shows a star arrangement.

Clearly, a star network depends completely on the central component. If this fails for any reason, then the network will not operate. The central switch has a reasonable level of complexity, since it must be able to read from a set of n input channels, where n is the number of computers in the network, and be able to write to n output channels, preferably simultaneously. This is in contrast to the modest switching required by an element in a chain or ring, which corresponds to the case $n = 1$ here.

Computers

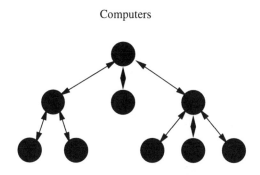

Figure 6.4 Multipeer channel using tree arrangement

A generalization of a star network is to a tree network, which helps to simplify the complexity of the switching required. Groups of nearby computers are formed into mini-stars, the computers being the tree leaves, and the star centres being the nodes parenting the leaves. In turn, these star centres are formed into star networks, with the new star centres being the nodes at the next level up of the tree. Eventually, the top level of star centres is formed into a single star network, with a centre that is the root of the tree. Figure 6.4 shows a rearrangement of the star network of Figure 6.3 into a three-level tree. The advantage is that each component star network only has a small number of attachments: one duplex channel to the centre of its parent star network, and several duplex channels to the centres of its child star networks. Thus, to implement a broadcast, each star centre is only responsible for relaying the message to a small number of channels. In the event that one star centre fails, partial operation of the network is still possible.

A tree network is also advantageous for directly implementing unicast or multicast communications, when the communicating computers are located in a localized part of the tree. This is because it is not necessary to propagate messages throughout the whole network. However, this involves selective switching by the star centres, and so belongs more properly in Chapter 7.

From a physical point of view, many common medium or chain/ring style networks spanning short distances are organized in a star pattern. The common medium, with an interface for each computer, or the chain/ring, with a forwarding interface for each computer, is contained in a single box. Each computer is then connected to its own interface in the box. Thus, visually at least, the computers are connected to a central switching element. However, it is important to note that the box itself performs no switching function between the connections to the computers. It just houses the computer interfaces. Such a box is called a **hub**. Figure 6.5 illustrates hub arrangements for common medium and ring networks.

Computers

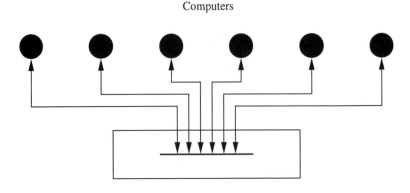

(a) Hub containing common medium

Computers

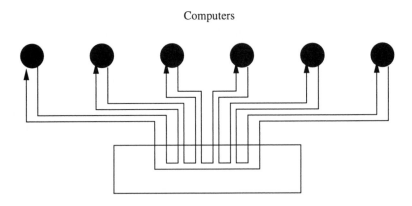

(b) Hub containing ring

Figure 6.5 Hub arrangements for common medium and ring networks

6.3 COMMUNICATION MULTIPLEXING

A message broadcasting network supplies a service that allows communications with different broadcast information flows to take place. However, the multipeer channel implementations described in the previous section do not allow arbitrary numbers of broadcast communications to happen at one time. In fact, the norm is for only one broadcast to happen at a time. Some of the multipeer channel implementations are based on independent unicast channels, and so could allow more than one broadcast at a time, because different messages could be simultaneously transmitted on separate channels. In fact, this is usually avoided in practice, to reduce the complexity of the operational procedures.

Given that only one broadcast can happen at a time, all communications must occur during disjoint time periods, to avoid contention. Furthermore, to ensure fairness of access, most multiplexing schemes include absolute bounds on the length of time period for which one computer is allowed to make use of the multipeer channel.

Regulating access to the multipeer channel is essentially a process of regulating the beginning of communication time periods. Therefore, for the variety of mechanisms that are used in practice, the techniques used are mostly familiar from earlier study of flow control procedures in Section 3.3.2. Here, however, rather than enabling flow control between communicating computers, the mechanisms enable flow control between a transmitting computer and the multipeer channel. In fact, of the mechanisms considered below, only the first one is novel, in that it is not used for end-to-end flow control applications. The novelty comes from the fact that each computer carries out flow control in isolation, guided only by its observations of the channel.

6.3.1 Isolated

Isolated multiplexing of communications involves the individual computers using contention avoidance or detection techniques. These do not guarantee to prevent contention occurring, but seek to minimize it. The most obvious form of contention avoidance is for a computer to listen to the multipeer channel before attempting to try transmitting information. If a communication is already in progress, then it can defer its own communication until the channel is free. However, this does not prevent two or more computers both waiting until the channel is free, and then colliding by transmitting at the same time. One way of reducing the chances of this happening is to introduce a random element into the procedure. Rather than transmitting as soon as the channel is free, computers are obliged to wait for a random short period of time before trying to transmit. This results in a small degree of randomness in the starting time chosen for the repeat communication attempt, which helps in avoiding collisions as soon as the medium becomes free.

The above mechanism allows collisions between information-carrying messages to occur, and this is part of its design. A variation is to use an extra protocol when computers begin transmitting. Rather than just transmitting the message that they wish to send, the computers instead transmit special control messages. These messages have different lengths for the different computers, and so have different communication time periods. The idea is that the computer that transmits for longest wins the competition for the channel, since the others drop out earlier. If the aim is just to eliminate contention, then each computer might have a different length allocated, and so the computer with the higest number always wins. However, if fairness is needed, then computers can use randomly chosen lengths.

This still leaves open the possibility of contention if two computers have chosen the same number.

Although avoidance techniques can reduce the chances of transmission contention occurring, it is not totally prevented. To deal with this, a contention detection method is used. This relies on a transmitting computer being able to detect that its transmission is being interfered with. Detection is done by receiving broadcast information at the same time as transmitting it. If a transmitting computer receives different information to its own, or receives garbled information, then it is alerted to the fact that another computer is transmitting at the same time. Once this abuse of the channel has been detected, the protocol requires that all transmitting computers cease transmitting. This may not be immediate, since there is a requirement for the computers to prolong the broadcast damage so that all computers in the network are made aware that a collision has occurred, and so can throw away any message prefixes already received successfully.

After a collision has been detected, the computers involved have acquired the information that they are in competition for the multipeer channel. Therefore, it is essential to employ a collision avoidance technique, otherwise the problem will be repeated at each future attempt. The standard technique involves waiting for a random period before trying again. Thus, the starting time of a repeated communication attempt is chosen with a random element. Note that this use of randomness is *necessary* after a collision, rather than just desirable as it is before first trying to transmit. This approach is known as the ALOHA method, in honour of its use by the pioneering University of Hawaii radio-based network of 1971, which was the ancestor of all message broadcasting networks. The introduction of randomness into the starting time chosen for communications means that no global control mechanism is needed.

The choice of possible random period lengths is important. If the choice of periods is too small, then further contention is likely. If the choice is too large, then communications might be delayed for unnecessarily long periods of time. There is an algorithm, known as **randomized exponential backoff**, for choosing appropriate random times before trying again. On the i-th occasion that a computer suffers a collision for a particular communication, it chooses a delay in the range

$$0, \ldots, 2^{i-1}$$

'time slots'. Thus, the first time, it chooses either zero or one, the second time, it chooses zero, one, two or three, etc. This continues up to some limit on i ($i = 10$ being a standard value), at which point the range is not increased further, to avoid delays becoming outrageously large.

A problem arises with this scheme if computers are able to transmit messages very quickly in succession. Once one computer has got access to the channel, it may be able to keep it while transmitting many messages one after another. After each of the messages, any competitor will collide with the computer's next message transmission. However, the computer that has just transmitted will choose a backoff period of either zero or one, but its competitors will be choosing a period from an increasingly wide range. Thus, the chances are that the computer in possession will keep its access to the channel. This phenomenon is called **capture**. One solution to this is called the **Binary Logarithmic Arbitration Method (BLAM)**, and involves all computers (not just colliding computers) maintaining the same backoff period range.

The length of a time slot is the maximum time that it takes for a transmitting computer to be sure that it has not suffered from a collision. Given this, and assuming that only one competing computer picks the lowest random number of slots, this computer is guaranteed not to collide with its rivals. The appropriate slot length is equal to the time for a message to propagate to the furthest point on the network, plus the time for any evidence of a collision to propagate back (plus a small safety margin). For a physical multipeer channel, this can be computed as an absolute time period, given knowledge of the distance spanned by the channel, and data transmission rates along it.

Contention avoidance or detection techniques are only really practicable for networks on which there is a fairly low loading and, further, that the load imposed by the computers is fairly randomly distributed over time. If this is not the case, then a lot of time and network capacity will be wasted by collisions, and one of the techniques described below will be more suitable. However, where loading circumstances are favourable, this type of technique has various points in its favour. First, it is completely decentralized, with computers making decisions locally, guided by the raw information that is being received from the channel. Second, it involves a simple protocol, which is always an advantage. Third, it allows computers to transmit information immediately (as long as no contention occurs), rather than be delayed by any arrangements needed under a flow control agreement.

6.3.2 Permission-based

Permission-based multiplexing depends on computers requiring some form of permission before being allowed to broadcast. Only one computer is allowed to have permission at a time, in order to guarantee that multiplexed communications have disjoint time periods. One way to achieve this is to have a master computer that polls the other computers in turn. However, this has the disadvantages of any centralized approach: the operation of the network is only as good as the operation of the master computer.

A less domineering alternative is to use ticket-based flow control, as described on page 101 of Chapter 3, where tickets are issued by the receiver. However, here, the central differences are:

- there is only one ticket in circulation;
- the ticket is issued on behalf of the multipeer channel rather than by the receiver of communicated information; and
- the ticket is immediately passed on to another computer by any computer that receives the ticket and has no immediate use for it.

When a computer receives the ticket, it is allowed to broadcast a message if it needs to and, after it has done so, it must pass the ticket to another computer in the network.

When such a network is first activated, then one computer must be responsible for generating the ticket. This computer might be fixed, under some absolute agreement for the network, or might be chosen dynamically by some form of appointment, competition or election protocol among the computers on the network. Thereafter, there must be a rule to ensure that the ticket is continuously circulated among all of the computers in the network, to give them an opportunity to transmit. For a network which seldom changes, and which contains computers that seldom fail, there may be a fixed ordering of the computers. However, in general, it is necessary to have a protocol that allows computers to be added or removed from the circulation list dynamically. Communication of the ticket, which is just a special form of message, must be by unicast communication between the current ticket holder and the next ticket recipient using the basic broadcast capability of the multipeer channel.

Advantages of a permission-based technique are that contention for the medium is eliminated and that the rights of computers to transmit can be regulated by the chosen ticket passing order. For example, if the ticket is passed around each computer in turn, then transmission rights are fairly shared out; this is a reasonable arrangement for a network which is highly loaded. As an alternative example, if it was arranged that one computer received the ticket every second time that it is passed on, then the effect would be that this computer was given priority for its transmissions.

Disadvantages include the fact that, before transmitting, a computer must wait for the ticket to arrive, although none of the other computers may actually need to transmit at that moment. A further disadvantage is the complexity of the arrangements for the ticket itself. As well as dealing with adding and removing computers, the ticket protocol must deal with computers that fail while holding the ticket. It must also deal with computers that, through error, create multiple tickets. However, these are not insurmountable problems, and such protocols exist and are used in practice, as will be seen in Section 6.4.

6.3.3 Reservation-based

Reservation-based multiplexing is an extension of permission-based multiplexing that is designed to reduce the time that a computer has to wait before being allowed to broadcast a message. A problem with ticket-style schemes, as noted above, is that a computer must wait its turn for the ticket, even if other computers before it have no need of the ticket. A reservation-based scheme is based on the style of flow control where a transmitter requests a ticket from the receiver when it needs to, or is about to need to, transmit.

In this case, any such protocol is complicated by the fact that receivers are not responsible for issuing tickets. Rather, tickets are issued on behalf of the multipeer

channel. One possibility is to designate a particular computer as a master, responsible for issuing tickets, and for the other computers to make reservations with this
master. However, this makes the network completely reliant on the master computer. Ticket-passing, just described above, is an example where the responsibility
for the control mechanism is distributed among all of the computers in the network.
Ideally, a reservation-based scheme should similarly distribute responsibility for
noting reservations and issuing tickets. The basic complication is that, when a
reservation is made, all computers that might conflict with the reservation must
be notified of it. Thus, some kind of broadcasting of each reservation is required.
This is rather different from the unicasting of tickets.

A fatal flaw in this appears to be the fact that broadcasting of reservation
messages should be allowed whenever necessary — which introduces exactly
the same problem that reservation is meant to be solving! One solution is to
generalize what happens with ticket-passing, by allocating short time slots on the
multipeer channel that allow computers to make reservations, between the main
items of broadcasting business. This solution anticipates the technique of time
division multiplexing, discussed in the next section. Another solution is to employ
piggy-backing as a multiplexing technique that combines reservation messages
with existing broadcast messages. Then, computers receive notification of future
reservations along with communications that are happening at the present.

On receiving a reservation message, it is essential that a computer takes it into
account when formulating its own transmission plans. In order for a reservation
scheme to be completely fair, each computer should maintain a count of the
number of pending reservations. When it wishes to transmit, it should broadcast
a reservation message and also note the value of this count. Then it should allow
that number of other computers to transmit first. After this, assuming that all the
other computers have received the same reservation messages, the computer can
proceed with its transmission without fear of collision. Moreover, transmissions
proceed in the same order that reservations were made. This mechanism assumes
that all computers are allowed to have reservations pending. In cruder schemes,
only one computer might be allowed to have a reservation pending, in which case
the use of counters is redundant.

6.3.4 Physical division

The above schemes for multiplexing communications, so that they do not overlap
in time, are very much in the asynchronous spirit. That is, the starting times of
communications are adjusted as appropriate, under the control of the multiplexing
policy. An alternative is to use a method of sharing the medium that is directly
related to its physical properties, rather than to the communications required of it.

A very synchronous approach is **time division multiplexing**, already mentioned in Section 5.3.2. With this, the communication time available is divided
into fixed-size time slots, and then these slots are allocated to the computers on
the network. There will be periodic cycles, during which each computer has at

least one time slot available for its exclusive use. Time division multiplexing removes the need for distributed mechanisms to decide which computer is allowed to transmit next. It also allows guarantees to be made on the absolute time at which communications will takes place. For applications that require real-time information transmission, such guarantees are important.

The main disadvantage of time division multiplexing is its inflexibility. If the communication needs of the computers in the network are fixed and continuous, the scheme works well. However, in general, the scheme is wasteful if the communication needs of the computers vary over time. Time slots will be unused by some computers, while other computers will have to wait. Of course, if computers are frequently added to or removed from the network, there must be some scheme for adjusting the allocation of time slots. However, this would normally be under human control, rather than happening automatically during normal operation.

An improvement is to blend time division multiplexing with the other types of multiplexing. Some time slots are permanently reserved for things like computers with continuous needs, ticket-passing or reservation passing. The remaining slots are allocated using a dynamic multiplexing method.

There are other methods of multiplexing a physical broadcast channel, but these lie beneath the bit frontier, being concerned with how bits are transported over the physical medium. The main method is **frequency division multiplexing**, which can be used where bits are transmitted by the modulation of a waveform. Separate logical channels can be provided over the physical channel by allowing each to transmit within a different frequency range. This allows multiple communications to take place during the same time period. However, this benefit is not gained for free, since the communication rates possible depend on the available bandwidth. Thus, the raw data rate of the physical channel is divided up between the logical channels with, if anything, a reduction in the aggregate data rate.

Sometimes, the allocation of frequency bands to communications varies very dynamically. For example, in radio communication, frequency hopping — changing the frequency range used — may occur as often as for every bit transmitted, in an attempt to avoid interference from radio sources broadcasting on particular frequencies. Embroidering this further, **code division multiplexing** is a multiplexing method that relies on different computers having different pseudo-random frequency hopping sequences, and this allows sharing of the medium without fixed frequency ranges having to be allocated to the computers in the network. There is more discussion of this technique in Section 6.5.

6.4 EXAMPLES OF GUIDED TECHNOLOGY NETWORK IMPLEMENTATIONS

The six examples in this section cover the main message broadcasting networking technologies that are in use for LANs and MANs based on guided media. Two of the six — token ring and FDDI — have roughly similar operational principles, but there are some distinctive differences. These two, and each of the other four, illustrate a different combination of multipeer channel implementation and communication multiplexing technique. Each combination has been designed to suit the practical environment in which the network is expected to operate: the physical medium used and the quality of broadcasting service offered.

6.4.1 Ethernet

Ethernet is a message broadcasting mechanism based on a guided common medium, with contention avoidance and detection used for multiplexing. Here, the term **ethernet** is used as a generic term for such networks. However, the name 'Ethernet' (with a capital 'E') has historic significance, being the name chosen for the pioneering network of this type, first developed by Xerox in the early 1970s. Therefore, strictly speaking, the name only refers to one particular product, although it has come to be used as a generic name for the technology.

A more accurate name is the less catchy **CSMA/CD,** standing for Carrier Sense Multiple Access with Collision Detection. The 'Carrier Sense' part refers to computers checking whether the channel is busy before trying to transmit. The 'Collision Detection' part refers to computers noticing when their transmissions collide with those of other computers, and then withdrawing and retrying gracefully. A further name used for ethernet, encapsulating the type of multiplexing and the type of channel, is the **contention bus**. The major influence on ethernet technology was the pioneering work by Xerox, joined later by DEC and Intel. This led to standardization by the Institute of Electrical and Electronics Engineers (IEEE), specifically the IEEE 802.3 standard for CSMA/CD networks. The following description refers to IEEE 802.3, but the general principles apply to any ethernet style network.

Information

The messages transmitted on an IEEE 802.3 ethernet have the format illustrated in Figure 6.6. The seven preamble bytes are present to allow a recipient to synchronize bit timing with the transmitter, and then the start byte signifies the beginning of the message contents. The destination and source identifiers are standard IEEE 802 identifiers, as described on page 143 in Chapter 5. After them, the two-byte length field gives the length in bytes of the information carried by the message. This cannot just be deduced from the overall length of the message minus the length of the fixed control fields, because the message may be padded to ensure that it is larger than the minimum length allowed for messages. The minimum length restriction is to allow collision detection, and will be discussed a little later. Excluding the preamble and start bytes, the minimum length of a message is 64 bytes, and there is a maximum length of 1518 bytes. After the information bytes and any padding, there is a 32-bit CRC using the CRC-32 generator to allow error detection by a recipient.

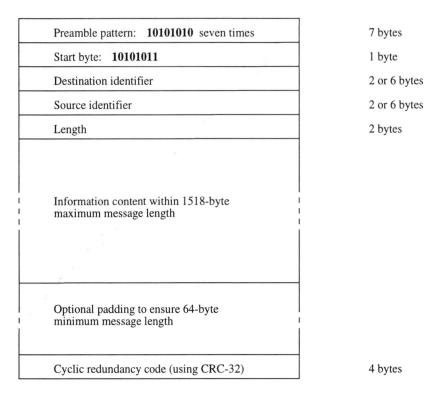

Preamble pattern: **10101010** seven times	7 bytes
Start byte: **10101011**	1 byte
Destination identifier	2 or 6 bytes
Source identifier	2 or 6 bytes
Length	2 bytes
Information content within 1518-byte maximum message length	
Optional padding to ensure 64-byte minimum message length	
Cyclic redundancy code (using CRC-32)	4 bytes

Figure 6.6 Format of IEEE 802.3 ethernet message

Time

The latency of communications and the transmission rate of communications vary between different varieties of 802.3 ethernet. They are determined by the maximum allowed length of the common medium and the bit rate of the medium. These parameters are incorporated in a naming convention for different varieties, for example, the original **10 BASE 5** standard operates at 10 Mbits per second ('10') and has a maximum channel length of 500 metres ('5'). The 'BASE' refers to the fact that baseband transmission is used. The normal rate for ethernets was 10 Mbits per second until the mid-1990s, at which point enhanced standards for a 100 Mbits per second rate began to come into use. The maximum channel length depends on the physical technology used. For example, 10 BASE 5 is for thick (0.5 inch diameter) coaxial cable, whereas **10 BASE 2** is thin (0.25 inch diameter) coaxial cable. The latter is often known as 'cheapernet' since it is cheaper to install than the former. Signals propagate along coaxial cable at approximately 200 metres per microsecond, so the latency of a 10 BASE 2 ethernet would be approximately 1 microsecond. Thus, the maximum communication time period on such a network would be: $10^{-6} + (8 * 1518)/10^7 = 0.0012154$ seconds.

The basic service of ethernets is an unreliable connectionless service. Any communication time period involves the continuous transmission of a message. The bits of each message are communicated synchronously, with Manchester encoding — described on page 87 in Chapter 3 — being used to maintain synchronization.

Space

As the maximum channel lengths quoted above may suggest, ethernets are used as Local Area Networks. Some extension in the distance covered is possible by using devices called **repeaters**. These connect together channels, replaying signals from one to another. For example, with 10 BASE 5, up to three repeaters can be inserted, which allows a distance of up to 2 km to be covered. The identifiers within a LAN are flat, with each computer being programmed with its own identifier. Thus, unicasts or multicasts can be implemented by filtering, each computer only accepting broadcast messages with a destination identifier matching its own. With the IEEE identifer scheme, a multicast has a destination identifier that is the number of a pre-assigned multicast group. There is also a special identifier meaning 'all computers on the network'. Messages broadcast with this identifier as their destination are treated as genuine broadcasts, and are accepted by all of the computers receiving the broadcast message.

Multipeer channel implementation

The ethernet multipeer channel is implemented by a guided common medium that is a bus. As already mentioned, coaxial cable is one possible technology, with the common cable visiting each computer on the network. There are alternative technologies, but these make use of a hub arrangement, rather than having a single medium snaking around the computers. The hub is a central box, with an interface for each computer in the network. The general arrangement of this was shown in Figure 6.5. The computers are connected to their interfaces in the box using a pair of connections: one for receiving and one for transmitting. When any computer transmits to the hub, the hub's function is to echo the transmission on all of the outgoing receiving connections to the computers. Thus, the effect is that the hub behaves like a common medium.

Possible technologies for the receiving and transmitting connections are twisted pair electrical cable or fibre optic cable. The standard for the former is **10 BASE T**, and for the latter is **10 BASE F**. Note the letters 'T' and 'F' replacing the number representing the maximum channel length, which is no longer an issue. The maximum length of the connection between a computer and the hub is 100 metres or 500 metres for the two standards respectively.

There are IEEE 802.3u standards for **fast ethernet**, that is, 100 Mbits per second ethernet. IEEE 802.3u is essentially just a version of 802.3 that has a 10 times faster transmission rate. The possible technologies are all hub-based — there is no version for a genuine physical bus. Therefore, **100 BASE T** is the generic name for the higher speed follow-up to 10 BASE T. The physical technology used comes in three flavours.

The **100 BASE T4** standard involves using four Grade 3 UTP twisted pair cables (hence the 'T4') to achieve the 100 Mbits per second transmission rate. One cable is used just for receiving, one just for transmitting, and the other two for both receiving and transmitting. Each cable has a transmission rate of 33.33 Mbits per second. This gives the required combined rate, since the 100 Mbits per second channel is split between the slower channels. For higher quality Grade 5 UTP cabling capable of faster transmission rates, the **100 BASE TX** standard can be used. It involves only two cables, as for the 10 Mbits per second standard. One is used for receiving, the other for transmitting, both at the 100 Mbits per second rate. Finally, there is the **100 BASE FX** standard, for fibre optic cabling. The major difference from 100 BASE TX is that the distance between a computer and the hub can be up to 2 kilometres, rather than just 100 metres.

100 BASE T is the natural evolution of the hub-style 10 BASE T ethernet. Section 6.4.6 describes a rather different evolution from the 10 BASE T network, to provide a 100 Mbits per second network. This fully exploits the fact that, although logically a bus, the physical connectivity of the network is actually a star or tree. The first 100 BASE T products appeared in 1994, so it is a relative newcomer. However, its successor was not far behind, since the first **gigabit ethernet** products appeared in 1997. These are covered by IEEE 802.3z standards, and are 1000 BASE T ethernets. Fibre optic cable is the normal type of cabling, but four-cable or eight-cable Grade 5 UTP is also an option. As in the move from 10 BASE T to 100 BASE T, so the move from 100 BASE T to 1000 BASE T is largely concerned with the physical transmission of bits rather than new protocols.

Communication multiplexing

The ethernet multiplexing scheme involves contention avoidance and detection, as described in Section 6.3.1. The original Ethernet was the pioneer of this technique in a cable-based network. The use of a minimum message size is crucial to making the scheme work efficiently. If two computers at opposite ends of the ethernet channel both started transmitting very short messages at the same time, then each would have finished transmitting before hearing the other's message, due to the latency of the network. Some computers in between would detect a collision, at the point where the messages collided, but others would be unaware of the collision until after the message from the nearer computer had been successfully received. This is illustrated in Figure 6.7.

The minimum message size is chosen so that the time to transmit it is more than twice the latency of the network. That is, the message will still be being transmitted after the time it takes for the beginning of the message to traverse the whole bus (during which time other computers may start transmitting), plus the time it takes for any competing transmission to reach the first transmitting computer. The choice of a 64 byte minimum size for messages gives a minimum

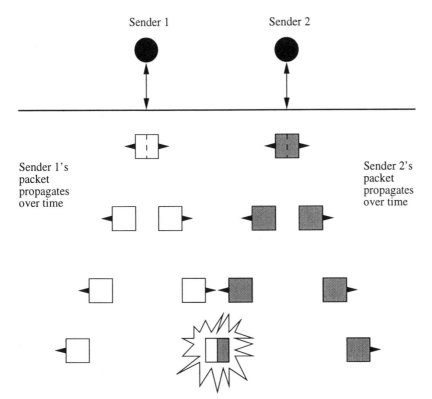

Figure 6.7 Ethernet collision invisible to transmitters

communication time of 51.2 microseconds on a 10 Mbits per second ethernet —
enough for collisions to be detected on a coaxial cable up to 5 km long (if this was
allowed).

Just as the transmission time must be long enough for collisions to be observed
by all of the transmitting computers, it is also necessary to ensure that collisions
are noticed by all of the computers in the network. It is possible that, if two
transmitting computers notice a collision quickly, they will both stop transmitting
before evidence of the collision has propagated through the network. This is
illustrated in Figure 6.8. To ensure that the news of the collision travels, both
stations transmit a short 32-bit 'jam signal' before going quiet. This is recognized
by all computers attached to the bus as a signal to abandon any communications in
progress. Then, using the binary exponential backoff algorithm, each competing
computer waits for a random period of time before trying again.

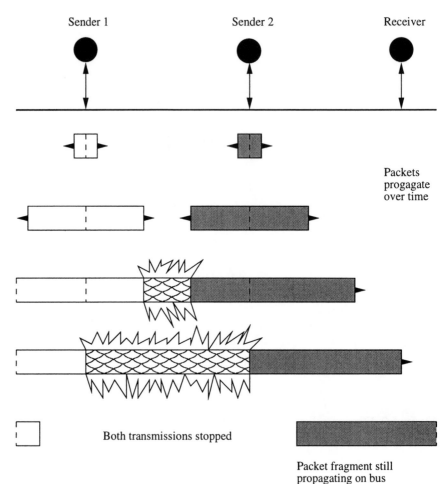

Sender 1   Sender 2   Receiver

Packets progagate over time

Both transmissions stopped

Packet fragment still propagating on bus

Figure 6.8 Ethernet collision invisible to a receiver

6.4.2 Token ring

Token ring is a message broadcasting mechanism based on a ring of simplex channels, with a ticket used for permission-based multiplexing. The term 'token' is used for the ticket used for multiplexing, hence the name given to this type of network. The first network of this type was developed by IBM in the late 1960s. IEEE developed the 802.5 standard for token rings, based strongly on the system developed by IBM. The following description refers to IEEE 802.5, but the general principles apply to any token ring style network. Note that, in 802.5 networks, a reservation-based mechanism is also used to elaborate the basic style of permission-based multiplexing.

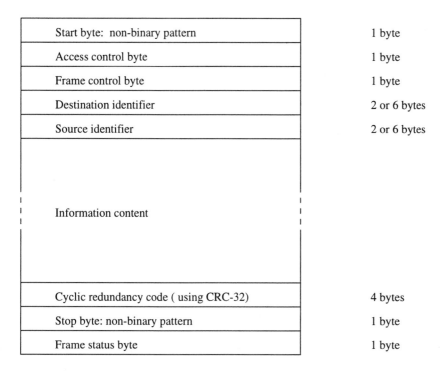

Start byte: non-binary pattern	1 byte
Access control byte	1 byte
Frame control byte	1 byte
Destination identifier	2 or 6 bytes
Source identifier	2 or 6 bytes
Information content	
Cyclic redundancy code (using CRC-32)	4 bytes
Stop byte: non-binary pattern	1 byte
Frame status byte	1 byte

Figure 6.9 Format of IEEE 802.5 token ring message

Information

The messages transmitted on an IEEE 802.5 token ring have the form illustrated in Figure 6.9. The start byte signifies the beginning of the message contents. It contains four bits that do not have valid **0** or **1** values — this is possible because bits are Manchester encoded, and these special 'bits' have signal values that are not valid bit codings. Next, the access control byte contains various fields concerned with the multiplexing mechanism, as discussed below. The frame control byte differentiates between messages that are carrying information and messages that are for ring control purposes. The destination and source identifiers are standard IEEE 802 identifiers, as described earlier.

After the information bytes, there is a 32-bit CRC-32 cyclic redundancy code to allow error detection by a recipient. There is no minimum size restriction on messages. In principle, there need be no maximum size restriction but, in practice, to ensure fairness in sharing the channel, a maximum of 5000 information bytes is typical. After the CRC, there is a stop byte which, like the start byte, contains four special 'invalid bits', as well as two other bits that are discussed below. Finally, there is a frame status byte, which contains acknowledgement information, also discussed below.

Time

The latency of a token ring network depends on a number of things:
- the number of computers in the network;
- the speed at which bits are forwarded by each computer in the ring;
- the distances between computers; and
- the technology used for each unicast channel between computers.

Since it must be possible for a message representing the token to circulate round the rings, with all its bits present at one time, it is important to ensure that the latency of a ring is large enough to accommodate this.

The standard bit rates for an 802.5 token ring are 4 or 16 Mbits per second. For a 4 Mbits per second ring, given that the token message size is 24 bits (only a start byte, an access control byte and a stop byte), a token will be transmitted in 6 microseconds. If the latency is less than this, the token could not circulate independently when the ring is idle. For example, for a typical bit propagation speed of 200 metres per microsecond, the physical latency of a 1 kilometre ring would be only 5 microseconds — too short. To deal with this, it is necessary to introduce artificial delays in the ring interfaces to increase the latency sufficiently. This is only a problem for the special case of the token, since no other messages are allowed to circulate round the ring independently, without intervention by a computer.

The basic service offered is an acknowledged connectionless service. Messages are communicated in a time period with continuous synchronous communication of bits which, as remarked above, are represented using Manchester encoding to allow synchronization. It is possible for an asynchronous sequence of messages broadcast by a computer to be parts of one communication that has been segmented. The stop byte of each message contains an Intermediate (I) bit, which is set in messages that are intermediate in a sequence; the bit is cleared in the final message of a sequence.

Space

IEEE 802.5 token rings are intended as Local Area Networks. The next example (FDDI), in Section 6.4.3, illustrates the techniques extended to Metropolitan Area Network distances. The identifer scheme within a token ring LAN is flat, with each computer being programmed with its own identifier. The message broadcasting service can be used to implement unicasts and multicasts in the same way as for ethernet: computers filter out any messages with destination identifiers that do not match their own.

Multipeer channel implementation

The multipeer channel is implemented using the simplex channels between computers. A transmitting computer sends its message round the ring, in a series of hops between the computers. This allows an acknowledgement service to be provided very easily. The frame status byte at the end of each broadcast message contains an acknowledge (A) bit and a copied (C) bit, which are both zero when the message sets out. In fact, there are two copies of these bits in the frame status byte, to increase reliability given that this byte is not covered by the CRC.

Any computer that recognizes a match of the destination identifier with its own identifier sets the A bits before passing the message on; further, if it makes a copy of the message, it also sets the C bits. Thus, when the message arrives back at its transmitter, this computer can tell (a) whether any computer recognized its identifier, and (b) whether any computer accepted the message. As a further acknowledgement mechanism, the stop byte in each message contains an error (E) bit, which is zero when the message sets out. Any computer that detects an error as the message circulates can set the E bit, and this allows the transmitter to tell whether any errors occurred during the broadcasting of the message.

As for ethernet, the actual ring connections may be located in a single box, which acts as a hub. Computers are then connected to interfaces in the hub by transmitting and receiving connections. Such an arrangement has the added benefit of protecting the ring against computer failures, since the hub can arrange automatic bypassing of any interfaces whose attached computers fail or are switched off.

Because a broadcasting computer is responsible for removing its messages from the ring after circulation, there is the possibility of orphan messages being doomed to roam the ring if their parent computer fails in some way during their progress round the ring. To deal with this problem, and various other administrative problems, the IEEE 802.5 standard includes the fact that one computer is designated as a ring master, responsible for such things as detecting orphan messages. For this purpose, the access control byte of each message contains a monitor bit, initially zero but set by the master computer when the message passes through. If a message arrives at the master computer with this bit already set, then it must be an orphan and is removed from the ring by the master. Because of this (unfortunate) reliance on a master computer, the ring standard includes further protocols for checking that the master computer is still working and for appointing a successor if not. These are implemented using special control messages distinguished by particular values in the frame control byte of the message.

Communication multiplexing

The basic token ring multiplexing scheme is permission-based, and a computer must receive a message representing the token on the ring before being allowed to broadcast a message. To send its first message, the token message is converted into an information-carrying message by changing one bit in the access control byte as it passes through. After sending its message, and receiving it back again when it has passed round the ring, the computer either sends another message or sends a message representing the token to its neighbour in the ring.

As described, this allows fair sharing of the multipeer channel, as long as computers are not allowed to hold the token for an excessive period of time. The default limit on the token holding time is 10 milliseconds. The master computer is responsible for ensuring that the token keeps in circulation. If a computer fails while holding the token, then the token disappears. Given the maximum token holding time, and knowing the number of computers in the ring, the master computer is able to determine a maximum period of time that can pass before it should receive the token for its turn. If this time is exceeded, then it generates a replacement token.

As well as the basic permission-based mechanism, there is also a fairly elaborate reservation-based mechanism to allow computers to have different priorities of access to the ring. The access control byte of a token message contains a three-bit field giving the priority (0 to 7) of the token. When a computer wants to broadcast a priority p message, it must wait for a token with a priority value that is less than or equal to p. Any other tokens must be allowed to pass by, destined for computers with higher priority needs. The access control byte of all messages contains a three-bit reservation field, which allows computers to make claims for future use of the token.

The reservation field is set to zero when a message begins travelling round the ring. If a computer is waiting to transmit a message of priority p, and a message passes through with a reservation field value less than p, then it can set the reservation field to p before passing on the message. The effect is that, when the message arrives back at its original sender, this computer knows what priority is required for the token it reintroduces to the ring. Note that, after a high priority token has been used, this rule may have the effect of reducing the priority again to a lower value.

6.4.3 FDDI and FDDI-II

The Fibre Distributed Data Interface (FDDI) is a message broadcasting system based on the token ring. It makes use of optical fibre channels operating at 100 Mbits per second. Because of the higher data rate, there are several differences from the token ring mechanisms described in the previous section. Also, FDDI allows synchronous communications to take place at regular intervals, which involves a mechanism additional to the token-based mechanism for multiplexing asynchronous communications. This mechanism does not guarantee that synchronous communications will take place at precisely regular times, and a variant known as FDDI-II has been developed to allow this. The description of FDDI below focuses only on the ways in which FDDI and FDDI-II differ from the IEEE 802.5 standard for token rings.

Information

The format of FDDI messages is shown in Figure 6.10. Although the format can be largely understood in terms of bytes, the introduction of **nibbles** is necessary for a full understanding. A nibble is a four-bit unit, that is, half of a byte. The

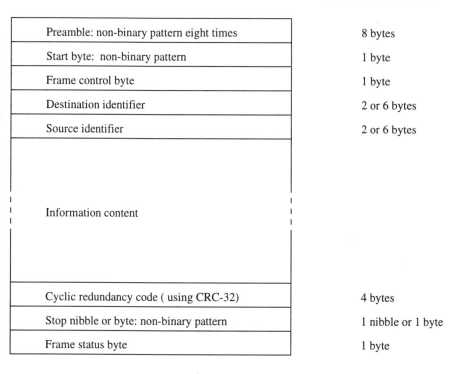

Preamble: non-binary pattern eight times	8 bytes
Start byte: non-binary pattern	1 byte
Frame control byte	1 byte
Destination identifier	2 or 6 bytes
Source identifier	2 or 6 bytes
Information content	
Cyclic redundancy code (using CRC-32)	4 bytes
Stop nibble or byte: non-binary pattern	1 nibble or 1 byte
Frame status byte	1 byte

Figure 6.10 Format of FDDI message

use of nibbles is a significant difference for FDDI: for transmission, each four-bit nibble of a message is encoded as length five on-off pattern of light signals. This is different from Manchester encoding, but the 'five bit' representation of each nibble ensures that there are always transitions between on and off present, to assist synchronization.

The FDDI message has an eight-byte preamble to allow synchronization between transmitter and receiver at the higher transmission speed. There is no need for a token ring-style access control byte, since FDDI does not use priority and reservation fields, as will be seen below. The end delimiter is a stop nibble on normal messages, and is a stop byte on token messages. The maximum allowed message size is 4500 bytes.

Time

The latency of an FDDI network depends on the same type of issues as the latency of a token ring network. A significant factor is that the distances allowed between computers in the ring are significantly larger, ten- or twenty-fold. Thus, the latency is higher. For example, in a maximum size 100 km ring with 500 computers attached, the latency would be around 1 millisecond. The greatly improved time factor is that the data rate is 100 Mbits per second, which is 25 times higher than the basic token ring rate. The basic service is an acknowledged connectionless service, like that of a token ring.

Space

FDDI is intended for use in Metropolitan Area Networks, since the fibre optic technology allows transmission over longer distances. Unlike IEEE 802 LANs, where either 2-byte or 6-byte computer identifiers may be used, but not both at once, FDDI allows a mixture. The frame control byte of each message contains a bit that indicates which identifier scheme is used in that message. This can be viewed as being a simple two-level hierarchical identifier scheme.

Multipeer channel implementation

FDDI uses a ring formed by simplex channels between computers. The standard provides for dual rings, travelling in opposite directions. In normal usage, only one ring — the primary ring — is used, and in the same way as for the token ring. The other ring — the secondary ring — is not used. However, if one of the channels forming the primary ring fails, or one of the computers fails, then the two computers on either side of the failure point can loop back their incoming primary ring channels to their outgoing secondary ring channels, thereby forming an emergency ring with a sequence of secondary ring channels replacing the missing one or two primary ring channels. This is illustrated in Figure 6.11. It is allowable to have computers that have only single ring attachments. However, to ensure reliability, such computers must be connected to a dual ring via hubs that are able to isolate the computers in the event of their failure.

As with the token ring, the way of implementing broadcasts allows an acknowledgement service to be provided easily. Messages carry A, C and E bits, which can be set by computers around the ring. The only difference from the token ring standard is that the E bit is located alongside the A and C bits in the frame status byte at the end of the message.

A significant difference from the token ring standard is that there is no master computer responsible for the token-passing mechanism. This, like the presence of a dual ring in FDDI, gives greater reliability against failures. In general, the monitoring responsibility is distributed among the computers in the ring. This is intimately connected with the way that token-passing works, but a side-effect is that messages which are not removed from the ring by their transmitter can either be discreetly removed by one of the other computers, or are removed through re-initialization of the ring following token loss.

Communication multiplexing

Because the latency of an FDDI ring is significantly longer than that of a token ring, there is a fundamental change to the token-passing mechanism. Rather than re-introducing the token when broadcasting of a message is complete, i.e., the message has travelled round the ring, the transmitting computer sends a token

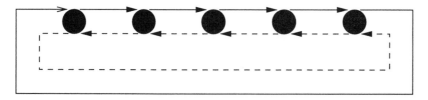

(a) Normal operation

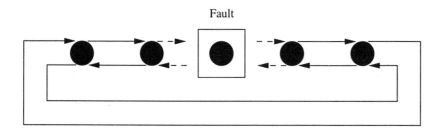

(b) Operation to avoid fault

Figure 6.11 Single FDDI ring formed from dual ring channels

message to its neighbour as soon as it has finished sending its message(s) to its neighbour. Thus, potential transmitters are not delayed by a time corresponding to the latency of the network. In particular, more than one message may be circulating the ring at one time, each message corresponding to a different broadcast message. This relaxation of the multiplexing conditions is possible in chain or ring-style networks because the multipeer channel is based on a collection of independent unicast channels, and so there is scope for multiple simultaneous broadcasting without contention occurring.

The rules that a transmitting computer must follow when holding the token are more complex, since provision is made for synchronous communications to be guaranteed reasonably accurate time periods. Each computer has a time allocation (possibly zero) that it is allowed to use for the transmission of synchronous communication each time it receives the token. Then, the rules for allowing other, asynchronous, communications are set up to ensure that synchronous communications can occur on a regular basis.

An important parameter is the target token rotation time (TTRT), which measures the maximum time that a computer should have to wait between occasions of receiving the token. All of the computers have the same value for the TTRT, which is chosen from the range 4 to 165 milliseconds. The setting of the synchronous time allocations, and the setting of the TTRT value, is carried out by a network management computer. Computers in the network negotiate with it, and are informed by it, using a special protocol.

It can be shown that no computer ever has to wait more than two times the TTRT to receive the token. If this happens, the computer initiates steps to introduce a new token, assuming that the old token has been lost. In fact, it can be shown that the actual time for the token to circulate tends towards TTRT in normal operation. On receiving the token, a computer is allowed to use up to its time allocation for synchronous transmission. Then, if the token arrived before TTRT time units had elapsed since the computer's last use of the token, the computer can use the time difference for asynchronous transmission.

The rules for asynchronous transmission are further elaborated to allow different priorities. This is also based on the use of absolute time measurement, rather than the reservation-based mechanism used on the token ring. There are eight different priorities, each with a threshold value that measures a length of time. When it holds the token, a computer tries to transmit its messages in decreasing order of priority. The computer is only allowed to transmit a message if the remaining time allowed to it for asynchronous transmission is greater than the threshold value for the message's priority level. Thus, a computer may have to pass on the token early, rather than transmit lower priority messages. The effect of this is that higher priority messages can usually be sent in an unconstrained manner, whereas lower priority messages are sent in a fair manner when network capacity is available.

Although the basic mechanism gives a reasonable guarantee for the delay suffered by synchronous traffic (no more than TTRT), this is still not good enough for **isochronous** communications. These are required to convey traffic which is generated at regular time intervals and must be delivered at the same constant rate. The FDDI-II standard supports the above FDDI service as its **basic mode**. However, if all computers attached to the network have been upgraded appropriately, then FDDI-II also supports **hybrid mode**. In this mode, time division multiplexing is used to divide the multipeer channel among a collection of slower channels. Each of these can be used for either isochronous transmission or for normal FDDI-style transmissions.

The time division multiplexing is carried out by a master computer, which repetitively generates bit sequences called **cycles**, one every 125 microseconds. Note that this rate corresponds to the normal sampling frequency for human speech, which is no coincidence, since real-time speech communication is a significant application requiring isochronous communication. Given the 100 Mbits per second transfer rate, this frequency means that each cycle is a string of 12 500 bits. The format of a cycle is shown in Figure 6.12.

Within each cycle, after 29 nibbles of preamble and header field, there are 12 groups that carry information. The first byte in each group is used for a normal FDDI-style channel, and so a total of 12 bytes are carried for this channel in each cycle. The rest of the group is used for 16 different isochronous channels, with

Preamble: non-binary pattern five times	5 nibbles
Cycle header	24 nibbles
Group 0	129 bytes
Group 1	129 bytes
	8 * 129 bytes
Group 10	129 bytes
Group 11	129 bytes

(a) Overall cycle format

Dedicated packet byte	1 byte
Wideband channel 0	8 bytes
Wideband channel 1	8 bytes
	12 * 8 bytes
Wideband channel 14	8 bytes
Wideband channel 15	8 bytes

(b) Group format within cycle

Figure 6.12 Format of FDDI-II cycle

eight bytes carried for each. Thus, a total of 96 bytes is carried for each channel in each cycle. This arrangement allows the FDDI-style channel to carry 768 kbits per second, and each isochronous channel to carry 6144 kbits per second. The capacity of each isochronous channel can be further time division multiplexed to carry many slower channels, for example, one channel can carry 96 ISDN channels running at 64 kbits per second each.

The division between non-isochronous and isochronous channel use is flexible, because the bytes allocated to any isochronous channels that are not required can be reassigned to the FDDI-style channel, thus increasing its capacity in steps of 6144 kbits per second, to a maximum of 99.072 Mbits per second. Part of the header field of each cycle indicates which, if any, such reassignments apply to the cycle.

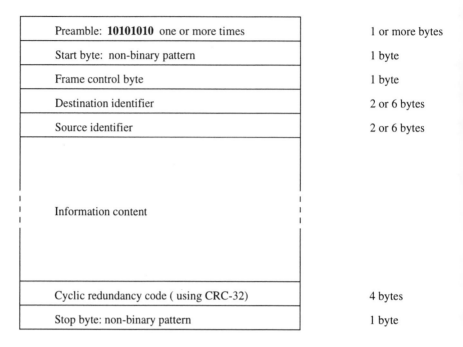

Preamble: **10101010** one or more times	1 or more bytes
Start byte: non-binary pattern	1 byte
Frame control byte	1 byte
Destination identifier	2 or 6 bytes
Source identifier	2 or 6 bytes
Information content	
Cyclic redundancy code (using CRC-32)	4 bytes
Stop byte: non-binary pattern	1 byte

Figure 6.13 Format of IEEE 802.4 token bus message

6.4.4 Token bus

Token bus is a message broadcasting system based on a common guided medium, with a ticket used for permission-based multiplexing. Thus, it merges the channel technology of ethernet with the multiplexing mechanism of token ring. As with the token ring, the word 'token' is used to refer to the ticket. Token bus networks are designed to be useful in situations where bus-style cabling between computers is already available, and guarantees are needed on the maximum delays suffered by any computer wishing to broadcast. This type of network was first developed as part of the Manufacturing Automation Protocol (MAP) project, which is discussed in detail in Chapter 10. The work evolved into the IEEE 802.4 standard for token buses. The following description refers to IEEE 802.4, which is the only significant type of token bus network in existence.

Information

The messages transmitted on an IEEE 802.4 network have the form illustrated in Figure 6.13. This is the same format as used by FDDI, except that there is no frame status byte at the end. Such a byte cannot be used on a token bus because messages are genuinely broadcast to passively listening computers, and there is no scope for computers to adjust frame status bits as messages pass by. The FDDI and IEEE

802.4 message formats can be viewed as the natural evolution and improvement of the token ring format, which was based on earlier pre-existing LAN products.

Time

The basic timing characteristics of a token bus are the same as for ethernet, in that all messages broadcast along a common medium. There are a range of standardized speeds for IEEE 802.4 token buses: 1, 5 or 10 Mbits per second for coaxial cable technology. There are also standards of 5, 10 and 20 Mbits per second for optical fibre technology. The coaxial cable standards specify broadband transmission, using a modulated signal with an appropriate bandwidth for the different speeds. This type of transmission is used so that the same cable can be shared by different transmissions, not necessarily all related to computer communications, at different frequencies. Relatively sophisticated electronics are required to ensure that the signals do not stray outside the proper frequency range. A cheaper alternative is to use carrierband transmission, which uses a modulated signal but does not share the medium with any other transmissions.

The basic service offered on a token bus network is an unreliable connectionless service, as for ethernet, with the transmission of each datagram being continuous over a time period.

Space

Like ethernet, token buses are used as local area networks. However, the fact that broadband signalling is used on coaxial cable media, and over better quality coaxial cable than that used for ethernet, means that longer distances can be spanned, for example, tens of kilometres. Also, it is possible to have more computers connected to a cable segment. The identifier scheme in token bus is the same as for ethernet, as is the way in which unicasts and multicasts are implemented in terms of the provided broadcast service.

Multipeer channel implementation

Broadband transmission is inherently unidirectional, since it is not possible to pass signals of the same frequency in both directions along the same cable. Because of this, a coaxial cable bus for an IEEE 802.4 network can be implemented in two different ways. These are illustrated in Figure 6.14. The first way has a pair of physical buses, one for each direction. All transmissions are made in one direction towards a **headend** at one end of the bus. The headend then repeats the transmissions along the other bus, from which the computers receive the signals. An alternative is to have a single bus, but to use two different signal frequencies. Transmissions are made at one frequency to the headend, and it then repeats the transmissions at a different frequency in the opposite direction, to be received by the computers in the network. This scheme is using frequency division multiplexing to implement the dual bus scheme.

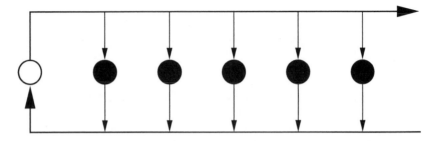

(a) Headend repeater

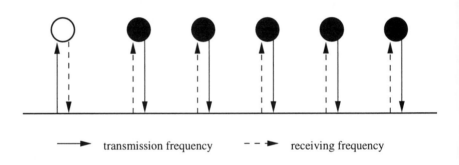

⟶ transmission frequency   - - ▶ receiving frequency

(b) Headend frequency convertor

Figure 6.14 Two implementations of IEEE 802.4 bus

For optical fibre transmission, a single bus is not physically realizable. Instead, a hub configuration is used to form a logical bus. Computers are connected by fibres to the hub, and the only requirement is that the hub behaves like a bus: all computers hear all transmissions. This arrangement is the same as for hubbed ethernets, in particular the 10 BASE F standard. Note that broadband transmission is not used for optical fibre token bus networks.

Communication multiplexing

The token bus multiplexing scheme, as for token ring and FDDI, involves a token-based permission method. The central difference is that there is no physical ring to pass a token around. Instead, a logical ring is formed within the network, using unicast channels implemented on the multipeer channel. This ring need bear no resemblance to the ordering in which computers are physically located on the bus. All that is required is that the token circulates fairly around all of the computers in the network. Whenever the token is passed from one computer to another, all computers receive the transmission. However, just as for normal unicast message transmissions, computers do not accept the token message unless it has their identifier as its destination identifier.

There is no master computer on a token bus network, in order to enhance reliability. Instead, there are complex protocols for managing the logical ring and the token in a distributed manner. This is a major disadvantage compared with ethernet. The essence of these protocols is the use of special types of message by computers to find new successors in the logical ring. When a token holder transmits the token to its current successor in the ring, it then listens for a next transmission. If the token transmission succeeds and the recipient is working normally, then the previous token holder will hear a message transmission by its successor. This will either be an information-carrying message or the token message being passed once again. If nothing is heard, or a damaged message is received, then the original token holder tries to pass the token once more. If this still does not work, the original token holder assumes that its successor has failed, and broadcasts a special message asking its successor's successor to respond, so that the logical ring can be relinked. If no response is received from the successor's successor, a more general procedure is followed to find a new successor.

This procedure is also activated at random intervals by token holders, in order to allow new computers to join the logical ring. It makes use of the fact that computer identifiers can be interpreted as numerical addresses. A special message is broadcast, inviting any computer with an address in the range between the message's source address and destination address to join the ring. Normally, the destination address is the address of the computer's successor in the logical ring. However, after a failure of its successor, and its successor's successor, the token holder puts its own address in the destination field, thus inviting all computers in the network to volunteer to be its successor. If this works, the result is that a two-computer ring is formed, and then it grows by soliciting further successors.

The problem with the invitation process is that more than one computer may respond at the same time. This is dealt with by a little ethernet-style process. If a collision is detected, the competitors wait for a small number of time slots before trying again. This is not a random number, but is determined by the least significant two bits of each computer's address. The first competitor to achieve a collision-free response to the token holder is admitted to the logical ring. A competitive process is also used to decide on the token holder when the network is initialized. In essence, each computer broadcasts for a period of time proportional to the logarithm of its network address. The result of this is that the highest-numbered computer becomes the token holder, and it can then solicit a successor to start forming the logical ring.

Like token ring and FDDI, token bus allows different priorities for message transmissions. The basic scheme used is like that of FDDI. There are four different priority levels, together with target token rotation times for each level. A computer transmits messages in descending order of priority, only allowing each priority

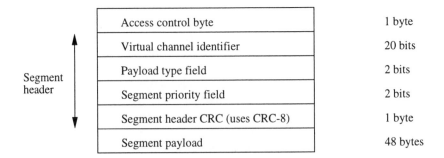

	Access control byte	1 byte
Segment header	Virtual channel identifier	20 bits
	Payload type field	2 bits
	Segment priority field	2 bits
	Segment header CRC (uses CRC-8)	1 byte
	Segment payload	48 bytes

Figure 6.15 Format of DQDB message

level if the time used for transmission so far does not exceed the threshold for that priority level. The overall effect is that absolute guarantees can be made of the delay suffered by top priority messages. Such guarantees cannot be made for ethernet-style bus networks.

6.4.5 DQDB

The Distributed Queue Dual Bus (DQDB) is a message broadcasting system based on a pair of independent unidirectional buses. It is designed to be usable over a variety of types and speeds of physical channel. DQDB qualifies for inclusion in this chapter, because it does not involve any switching by computers. However, strictly speaking, DQDB does not agree with the chapter title, since messages are normally only broadcast in one direction along a bus, not in both directions. However, broadcasts are still possible, just by the sender carrying out two 'half broadcast' communications. DQDB was developed by IEEE as its 802.6 standard for Metropolitan Area Networks, and the typical type of channel would be a high speed offering from a common carrier.

Information

The format of a DQDB message is shown in Figure 6.15. All messages have the same length: 53 bytes. The first byte is used to control the access mechanism for the DQDB message broadcast capability. The remainder of the message is called a **segment**, and has a four-byte header, followed by 48 bytes of information. The choice of the word 'segment' reflects the fact that the information content consists of parts of a larger information unit chopped up to fit into the fixed-sized messages. The DQDB message is very closely related to the ATM cell, which is introduced in Chapter 2: 53 bytes long, with five bytes of control information followed by 48 bytes of real information. The close relationship is no coincidence.

Time

The operation of a DQDB network is controlled by a 125 microsecond clock. Each clock period, a sequence of messages is sent from one end of each bus to the other end. The exact number depends on the bit rate of the physical medium used. For example, over a 34.368 Mbits per second channel, nine messages can be transmitted in each period (including small amounts of extra control information between messages). The clock cycle length is the same as that for FDDI-II, and was similarly chosen to provide support for isochronous services because it is the same frequency as that used for telephone voice services.

The basic service offered by DQDB is an unreliable connectionless service, with the transmission of each fixed-length message being continuous over a time period. There are two classes of service: the pre-arbitrated (PA) service that gives guaranteed access to the multipeer channel for isochronous traffic, and the queued arbitrated (QA) service that gives access to the channel for other types of traffic.

Space

DQDB is primarily intended as a LAN interconnection facility. Thus, a DQDB network and a collection of connected LANs can be used to implement a larger network. The distance covered by a DQDB network may be in excess of 50 kilometres, and so can connect together a collection of LANs across a city. Thus, DQDB networks can be Metropolitan Area Networks (MANs). Because DQDB connects together computers representing LANs, rather than individual end-user computers, the typical number of connected devices is somewhat lower than for the networks already seen.

The mechanism used for identifying the source and destination computers for a communicated message cannot be directly decribed in terms of the basic message format. The actual mechanism varies, depending on the sort of service that is offered on top of the basic connectionless DQDB service for communication of 53-byte messages. There are three types: an isochronous service, a connection-oriented service and a connectionless service (allowing communication of arbitrary length messages).

The four-byte segment header contains a 20-bit **virtual channel identifier** (VCI) field. For the isochronous service and the connection-oriented service, this field identifies a particular connection between two computers, and so computers receiving such a broadcast message can regard this field as being a combined source and destination identifier. For the connectionless service, this field contains an all-ones value, and so does not give any spatial information. To determine the destination identifier of a message, it is necessary to look more deeply inside.

Each unit of data conveyed by the connectionless service may be up to 9188 bytes long, and is transmitted with a header that includes an eight-byte destination identifier and an eight-byte source identifier. Either 16-bit or 48-bit identifiers, as used in other IEEE 802 standards, or also 60-bit identifiers, as used in ISDN, can be carried. The data unit, together with its header (and also a trailer), must be divided up to fit into a sequence of segments. Each segment used carries an identifier value that is unique to the data unit. Thus, computers receiving broadcast messages must look for segments that carry the first part of a data unit, and so contain the header, to check the destination identifier. If the destination identifier is relevant to a computer, then all subsequent segments with the same identifier are also relevant. In fact, the transmission of each data unit is akin to a little connection being established: the first segment gives the computer identifiers involved, then later segments just carry the identifier for the connection.

The exact details of data unit formats are not included here, since they are an issue for the next stage of implementation above the basic DQDB message transfer service. It is unfortunate that the connectionless DQDB provision has to be aware of the next-level implementation detail in order to determine the destinations of its messages.

Channel implementation

A DQDB network has two independent buses, one for transmission in each direction, so there is double the carrying capacity of a normal bus. At the beginning of each bus, there is a computer called the **head**, which is responsible for generating the messages that are sent along the bus. One difference from the other bus-style networks considered earlier is that computers can change messages as they pass by, rather than just passively read them. In this sense, 'dual chain' would convey a more accurate description than 'dual bus'.

Computers transmit information by placing it inside empty messages that pass by. A message is empty when it leaves the head, and until it reaches a computer that fills it. Thereafter, the message is full, and its contents can be read by any computer downstream on the bus. Thus, a half-broadcast effect is achieved. A source computer must ensure that information is transmitted on the appropriate bus(es), in order to reach its intended destination(s).

In an **open bus topology** network, the two heads are at opposite end of the network. In a **looped bus topology** network, the same computer is the head for both buses. Thus, the layout of the network is similar to the dual ring arrangement used for FDDI. The difference is that the head computer does not relay messages from one side to the other, and so there is no ring effect. The two topologies are shown in Figure 6.16. The advantage of a looped bus topology is that, should a failure occur in the network, the computers on either side of the fault can be configured as heads, with the normal head relinquishing head duties, and so an open bus topology network can be formed. This is also shown in Figure 6.16.

Multiplexing

To support isochronous communications using the pre-arbitrated (PA) service, some of the messages transmitted by the head are pre-reserved for use by computers with PA connections established. Because messages are transmitted at a regular frequency, this gives the effect of time-division multiplexing. For other types

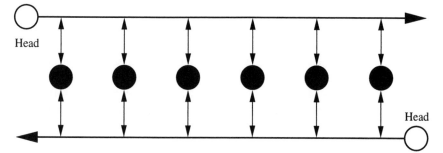

(a) Open bus topology

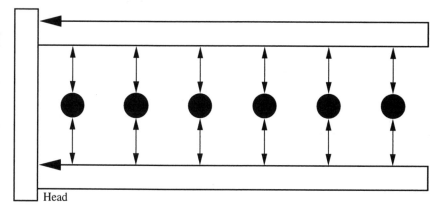

(b) Looped bus topology

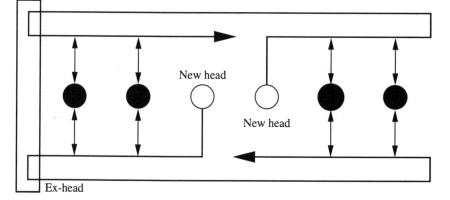

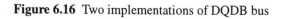

(c) Looped bus topology reconfigured around failure point

Figure 6.16 Two implementations of DQDB bus

of communication, using the QA service, a reservation-based scheme is used to multiplex each bus. As the name Distributed Queue Dual Bus may suggest, this involves a distributed queue scheme, based on the ideas described earlier in the general discussion of reservation-based methods. In essence, each message transmitted by the head that is not pre-reserved can be regarded as having a ticket piggy-backed on it. When a message is used by a computer to transmit information, the ticket is removed before the message is passed on down the bus. This appears to be a scheme that is extremely biased towards computers that are near the head. However, the distributed queue reservation scheme stops this happening.

It is simplest to describe the reservation scheme for one bus only; a symmetrical and independent scheme applies to the other bus. Four bits in the access control byte at the beginning of each message are relevant. The 'busy' bit indicates whether or not the message is currently carrying information. When a message passes a computer with the busy bit zero, it is interpreted as being a ticket permitting transmission of a segment. Transmission is achieved by inserting the segment into the message and setting the busy bit. However, to achieve the distributed queuing effect, computers only make use of a ticket when it is their turn, as described earlier.

Three request bits are used to make reservations for the bus in the other direction. There are three bits because three levels of priority can be used, with a distributed queue for each priority. All queued reservations for a higher priority get precedence over queued reservations for a lower priority. Each computer can only make one request at any priority level at a time. To make a reservation for the bus at a particular priority level, a computer waits until a message passes on the other bus with the appropriate request bit zero, and then sets this request bit. After this, it must then allow enough empty messages to pass through, to satisfy all pending higher priority requests and any same-priority requests made earlier than its own.

In fact, the rule is a little more complicated than this. With the basic distributed queue scheme, it can be shown that, on highly loaded lengthy networks, computers nearer the head end of a bus get preferential treatment. To deal with this, a technique called **bandwidth balancing** is used. Instead of always exerting its right to use a passing empty message, a computer lets a certain proportion pass by. A parameter β is fixed for each bus. After using β messages, a computer is obliged to let the next empty message pass. The value of β must be in the range 1 to 64, with the default being 8. The IEEE 802.6 standard recommends that bandwidth balancing is enabled on any bus that spans a distance greater than the distance needed to accommodate a 53-byte message, for example, 2 km when a 44.376 Mbits per second transmission rate is used.

6.4.6 100 BASE VG-AnyLAN

In earlier examples, such as ethernet and token ring, one possible physical arrangement for the network is to use a hub, with individual computers connected by paired channels to the hub. This has become a very widespread arrangement, since it allows centralized management and maintenance. The technology is sufficiently reliable that having a single central component is not a major concern — in the event of an occasional failure, one hub box can be speedily substituted for another. Given this, the IEEE 802.12 committee, which developed one of the higher speed follow-ons to the IEEE 802.3 10 BASE T ethernet standard, took the natural step of modifying the logical topology to align with the physical topology, i.e., having the form of a star or tree. This resulted in the **100 BASE VG-AnyLAN** standard. The '100' refers to a 100 Mbits per second transfer rate, and the 'VG-AnyLAN' refers to the use of Voice Grade cable (UTP twisted pair, used in existing 10 BASE T networks) and the ability to inter-work with existing 802.3 ethernet LANs and 802.5 token ring LANs. Thus, the standard offers a high speed upgrade path that can make use of existing LAN cabling, and also inter-work with existing LANs that are left intact. When older style networks, LANs in particular, are incorporated into newer networks, they are usually referred to as **legacy networks**. For brevity, the rest of the description will abbreviate the inelegant name 100 BASE VG-AnyLAN to just 'AnyLAN'. This is a shade more humanized than the alternative numerical name '802.12'.

Information

The messages broadcast in an AnyLAN network may have the format of either IEEE 802.3 ethernet messages or IEEE 802.5 token ring messages. However, one or other format must be agreed upon, and used throughout the network. The choice will be affected by the type of other LANs, if any, that are connected to the network.

Time

The raw data transfer rate of an AnyLAN network is 100 Mbits per second, as its full name indicates. In most existing hub networks, the maximum length of cable between a computer and a hub is 100 metres, so the typical latency between two computers via the hub will be less than 1 microsecond.

The basic service of AnyLAN is an unreliable connectionless service. Any communication time period involves the continuous transmission of a message. The bits of each message are transmitted synchronously, with each group of five bits being transmitted as a group of six bits. Each transmitted sextet has the same number of zeros and ones in order to maintain synchronization.

Space

In overall effect, the space characteristics of AnyLAN are similar to those of ethernet and token ring. A flat identifier scheme is used, and unicasts or multicasts are delivered to any computers that match the broadcast message destination identifier with their own identifier. Genuine broadcast messages have an identifier that is matched by all computers in the network.

The important difference is that genuine broadcasting of all messages does not happen. Although all messages are available to all computers in principle, the network prevents the delivery of irrelevant unicasts or multicasts to computers, rather than the computers themselves filtering out unwanted messages. This gives an added degree of security, since the computers are no longer required to be trustworthy. Because of this behaviour, there is not a true multipeer channel involved in implementing the network service. As for DQDB, redundant broadcasting is suppressed.

Channel implementation

In its simplest form, an AnyLAN network is connected as a star, with a switching device called a **repeater** in the centre. Note that this use of the word 'repeater' is different from its use in an ethernet context. Individual computers are connected to the repeater using voice grade cable. This cable in fact contains four UTP twisted pairs, and is of the same type described earlier in the brief summary of the 100 BASE 4T ethernet standard. An overall 100 Mbits per second transfer rate is achieved using splitting over the four pairs, each one having a 25 Mbits per second transfer rate.

In a more complex form, an AnyLAN network can be organized as a tree, with repeaters as nodes and computers as leaves. The connections between repeaters may use a variety of cabling technology, since these are not constrained by the need for backward compatability with earlier hub-style LAN cabling of computers. This can include either higher quality twisted pair or optical fibre. These types of cable allow distances of greater than 100 metres to be spanned.

When a computer wishes to broadcast a message, it sends it to its parent repeater. Then, the exact operation of the repeater depends on whether or not it is at the root of the tree (always the case when the network is a simple star). A root repeater selectively copies input messages received from one child to all of its children. All messages are copied to children that are repeaters. Only relevant messages, i.e., messages with appropriate destination identifiers are copied to children that are computers.

A non-root repeater copies input messages received from its children to its parent, and copies input messages received from its parent to all repeater children and to relevant computer children. Thus, although it may be physically organized as a tree, the network always functions logically as a star, with the root repeater at the centre. All messages are received by all repeaters, with selectivity of reception only occurring for individual computers.

Multiplexing

It is possible that more than one computer in an AnyLAN network might want to transmit a message at the same time. Permission-based multiplexing is used to ensure that only one computer transmits at a time. This does not need any distributed scheme for issuing tickets: repeaters use round-robin polling to arbitrate between their children. As with the actual transfer of messages, polling is carried out on a logical star basis, centred on the root repeater. The root repeater polls each child in turn, to see if it needs to transmit. If the child being polled is a repeater itself, then it polls each of its children in turn. Thus, the overall effect is of the root repeater polling the leaves of the tree in turn.

In fact, a repeater does not explicitly poll each child in turn by asking whether the child wants to transmit and then awaiting a response. During, or at the end of, each message transmission, each child that wishes to transmit turns on a control signal to the repeater, indicating a request. Then, polling just consists of looking for the next request signal in the round-robin order. In a sense, each child is making a reservation; however, each time round, only one child has its reservation honoured, and the others must try again later.

AnyLAN allows two priorities of message transmission — normal priority for most messages and high priority for delay-sensitive messages. Therefore, there are actually two round robins. On each poll, a repeater checks the high priority round robin for requests and, if there are any, it grants the next request in order. If there are no high priority requests, the repeater checks the normal priority round robin for requests. Clearly, if there are many high priority messages, it is possible that normal priority messages might be severely delayed. To prevent this, there is a timeout of between 200 and 300 milliseconds set on the delay suffered by a normal priority message. If this expires, the message is treated as a high priority message by the repeater.

6.4.7 Discussion of examples

The examples of guided technology message broadcast networks illustrate a range of different multipeer channel implementations and communication multiplexing methods. Overall, the evolution of local area networks shows some evidence of a wheel turning full circle. Before the advent of LANs in which all computers are peers, such as ethernet and token ring, a 'local network' typically consisted of one powerful master computer managing a collection of smaller computers or terminals, in a logical star arrangement. With the introduction of hub-style cabling for LANs, the physical connections moved back towards the star style. Now, in examples like the 100 BASE VG-AnyLAN network, and switched ethernet, described in Section 7.4.1, the logical star arrangement has reappeared. However, the essential difference is that the centre of the star is no longer a master computer that is communicating with slaves. It is now just a dedicated switch with the sole purpose of facilitating communication. This means that the LAN still consists of a community of peer computers, and the dedicated centre can be made far more reliable than a general-purpose computer with extra networking roles could ever be.

Meanwhile, the traditional LAN technologies find applications in Metropolitan Area Networks, where the distances involved make distributed control mechanisms still desirable. Thus, FDDI is a natural extension of the token ring mechanism. Ethernet is not particularly suited to extension over long distances, because its mechanism hinges on global sensing of the network state; thus, delays in acquiring sole use of the medium increase with distance. This is why DQDB has emerged as an alternative mechanism for bus-style MANs.

The other major trend in the development of message broadcasting networks has been the need to cater for traffic that is not asynchronous computer communication. Token ring and token bus allow some guarantees to be made on communication delays. However, these are not good enough for isochronous traffic, such as real-time speech and video. Therefore, a very traditional multiplexing scheme — time division multiplexing — has become popular again. As the examples of FDDI-II and DQDB show, modern time division multiplexing is blended with other multiplexing schemes, so that a mix of isochronous and asynchronous traffic can be supported efficiently.

6.5 UNGUIDED TECHNOLOGY NETWORK IMPLEMENTATIONS

The examples of message broadcasting networks given above all involved guided media — electrical cable or fibre optic cable. Wireless networks make use of unguided media, and have the dual advantages of avoiding cabling costs and of allowing the easy interconnection of portable computers. The media used are either radio waves or infra-red signals. These are inherently broadcast media, and so are directly suited to message broadcasting.

A variety of manufacturers have developed their own wireless LANs, each one rather different from the others. However, international standards began to emerge in the mid 1990s. The IEEE 802.11 standard — an addition to the well-established set of IEEE 802 LAN standards — is concerned with wireless LANs. In addition, and in consultation with IEEE, the European Telecommunications Standards Institute (ETSI) formulated a standard called HiperLAN (High Performance Radio LAN) for wireless LANs operating at around 10 times the rate of 802.11 standard LANs. In due course, as with wired LANs, these standards will prevail over proprietary solutions, allowing inter-operability of different manufacturers' equipment.

Broadcasting messages using radio is not a new phenomenon. Indeed, the multiplexing scheme used by ethernet had its root in the scheme used in a pioneering radio-based network of the University of Hawaii. There is also a long-standing standard protocol (AX.25) used by amateur radio enthusiasts for communication between computers. The difference with modern wireless LANs, compared with earlier radio networks, is that they provide much higher speeds which are comparable with guided media LANs, and also have much higher reliability. The higher speeds are facilitated by the use of higher frequency radio or by the use of infra-red. The higher reliability is achieved by the use of more subtle signalling techniques to lessen radio interference and by the use of error detection and correction techniques.

Information

The messages used for wireless broadcasting do not need any particular additional control fields. Starting and ending delimiters, together with source and destination identifiers, are sufficient along with the information itself and an error detection code. If the bit error rate suffered by messages is relatively high — for example, rates of up to 1 in 1000 bits being corrupted are not unknown where precautions are not taken to guard against radio interference — then it is necessary to keep messages short. Otherwise, it is probable that a high percentage of messages will suffer damaged bits. If such a consideration does not apply, then there may be an incentive to make messages as long as possible. For example, if CSMA/CD is used to multiplex the medium, then larger messages are better to minimize contention periods.

Time

The IEEE 802.11 standard offers raw data rates of 1 or 2 Mbits per second using radio or baseband infra-red, and also rates of 4 or 10 Mbits per second using broadband infra-red. The HiperLAN standard offers up to 20 Mbits per second using radio, with a genuinely high-speed rate of 155 Mbits per second planned eventually. The basic service of wireless LANs is an unreliable connectionless service. A communication time period involves the continuous transmission of the bits of a message. The timing of the individual bits depends on the transmission mechanism used.

For radio, **direct sequence spread spectrum** transmission involves the data bit stream being exclusive OR-ed with a higher rate pseudo-random bit stream, so that the resulting transmitted bit stream has the higher rate. This is illustrated in Figure 6.17. An alternative is to use **fast frequency hopping** transmission. This involves changing the signal frequency pseudo-randomly several times during each bit time. Both of these types of transmission mode are allowed by the 802.11 standard. They do not feature as part of the HiperLAN standard. Both types involve a decomposition of each bit time period. The reason for doing this is to lessen interference from other radio sources. Roughly speaking, the idea is to widen the frequency band used for transmission, but the technical details are not of mainstream importance here.

Space

The range of a 20 Mbits per second radio LAN is only 50 metres, which is distinctly low, compared with guided media LANs. However, a range of 800 metres is possible for a 1 Mbit per second radio LAN. Infra-red LANs are restricted to a single room, for the simple reason that infra-red does not pass through walls. The shorter distances possible with high-speed radio LANs and infra-red

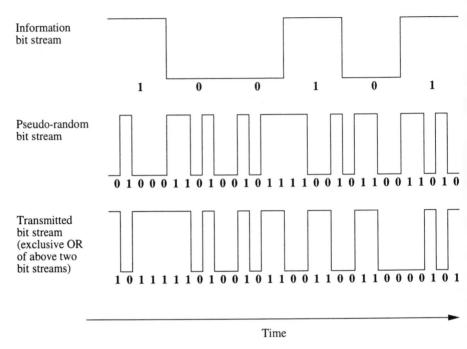

Figure 6.17 Example of direct sequence spread spectrum transmission

LANs are adequate to allow mobile communications. A portable computer can communicate with other computers in the same room or corridor that are connected to guided media LANs. When a more extensive radio-based network is required, then intermediate relay stations must be used to pass on radio transmissions from one computer to another. However, this is not really message broadcast networking any more (although, of course, messages are being physically broadcast); rather, it is message switching, which is the subject of Chapter 7.

Mobility is an issue that relates space and time, since computers may actually be in motion when communicating. The IEEE 802.11 and HiperLAN standards are both intended for slowly moving computers only, for example, at speeds no more than 10 metres per second. This is acceptable for LANs, where the assumption is that the computers are in a small area, probably indoors, and so the scope for movement is limited. Other arrangements are needed for WANs, where computers might be in fast moving vehicles.

No special identifier schemes are necessary for wireless LANs. A flat identifier scheme, as used in wired LANs, is perfectly adequate. In amateur message radio networks, one particular identifier scheme is used: computers are identified by the radio call signs of the radio hams who are carrying out the communication for the computers.

Channel implementation

Radio provides a natural broadcast technology. Infra-red is more natural as a point-to-point medium. However, the output of an infra-red source can be optically diffused, so that it is spread over a wide angle. By this diffusion, and then reflection from walls of a room, it is possible to achieve broadcasting. An alternative, for both radio and infra-red, is to use directional transmission in connection with a fixed roof-mounted reflector. A passive reflector just reflects the waves so that they reach the receivers of all computers. An active reflector repeats signals received by transmitting them back to the receivers. This means that the transmitters can use lower power signals than are necessary with passive reflectors.

Multiplexing

Wireless LANs suffer from the same multiplexing problem as wired LANs, and the main solutions either involve some form of contention avoidance or detection, or physical division. Permission-based or reservation-based schemes are less common, but an optional feature of the IEEE 802.11 standard involves a permission-based system operated by a master computer.

CSMA/CD schemes for wireless LANs are similar to the scheme used for ethernet. The only complication is that it is not possible to transmit and receive at the same time, and so a transmitter cannot hear collisions. To deal with this, at the beginning of message transmission, computers rapidly switch between transmitting and receiving, in a pseudo-random manner, for the first few bit time periods. If another transmission is heard during one of the receiving periods, a collision has been detected.

The multiplexing scheme used in HiperLAN is called **EY-NPMA** (Elimination Yield — Non-pre-emptive Priority Multiple Access), which is essentially a variant of CSMA/CD. The elaboration comes in the collision detection method. Rather than any computer immediately starting to transmit its information when the channels becomes free, computers transmit control messages of differing lengths. The general idea is that the computer that has chosen the longest length wins, since other competitors back off. The lengths are chosen pseudo-randomly, so the mechanism is fair. In fact, matters are a little more complex than this, since messages can have five different priority levels.

An elaboration of CSMA/CD is **CSMA/CA**: CSMA with Collision Avoidance. This involves a little extra politeness before transmitting. After sensing that the medium is free, computers wait for a short random period of time before transmitting. This is designed to lessen the problem of clashes when several computers are waiting for the medium to become free before transmitting. CSMA/CA is used in the IEEE 802.11 standard, in which it forms part of an overall multiplexing protocol: **Distributed Foundation Wireless Medium Access Control** (DFWMAC), which also deals with other problems suffered by wireless communication.

A main problem solved by DFWMAC is the fact that wireless communication might not involve a genuine multipeer channel. There may be computers that are not able to hear 'broadcasts' made by other computers that are too distant. Thus, strictly speaking, the wireless network implements a set of 'multi-multicast' channels. This leads to the **hidden station problem**, which is illustrated in Figure 6.18. The problem occurs when one computer cannot hear that another computer is already broadcasting to part of the network, and so it begins to transmit. This transmission will collide with the one already in progress, in those parts of the network that can hear both transmissions.

DFWMAC involves a four-stage handshake for message transmission, the first two stages of which deal with the hidden and exposed station problems. The four stages are:

- message sender sends a Ready To Send (RTS) control message, containing the length of the actual information message, to that message's recipient(s);
- message recipient(s) send back a Clear To Send (CTS) control message, also containing the information message length;
- message sender sends the information message; and
- message recipient(s) send back an acknowledgement control message.

An alternative at the second stage is to send back a control message indicating the intended recipient is not able to receive; obviously, if it is completely switched off, no message will be returned. The protocol deals with the hidden station problem because, although a hidden computer will not hear the RTS message, it will hear the CTS message. The fact that the RTS and CTS messages contain the length of the actual message to be transmitted means that all listening computers are given warning of the length of the transmission to follow.

Further, but optional, complication is introduced in DFCMAC, in order to deal with communications that require guarantees on their starting time. A master computer can be present, and it is given priority access in the CSMA/CA scheme, that is, it can acquire channel use rights before any other computers make bids. Having acquired the channel, the master computer then issues polls to computers that have been put on its list of those requiring regular time periods for communications. When polled, each such computer can transmit without encountering contention. After polling is finished, the channel is made available for normal communications.

As alternatives to isolated multiplexing methods, time division multiplexing and frequency division multiplexing are two possible physical division schemes which can be used when all transmissions occur via a central base station. That is, computers transmit messages to the base station, which then broadcasts them to all of the network. Assignment of time slots or frequency bands may be done statically by the base station, or dynamically in response to requests.

A further possibility is **code division multiplexing**, for example, where frequency hopping is used. A pseudo-random sequence is used to control the hopping and, if a different sequence is assigned to each computer, and all the assignments are made known to all of the computers, then multiplexing of point-to-point channels is possible. A transmitter selects the pseudo-random sequence of its receiver, and uses it for the transmission. With care in the choice of pseudo-random se-

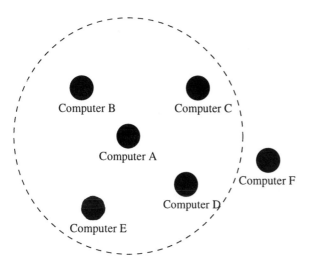

(a) Computer A is heard by Computers B, C, D and E, but not F

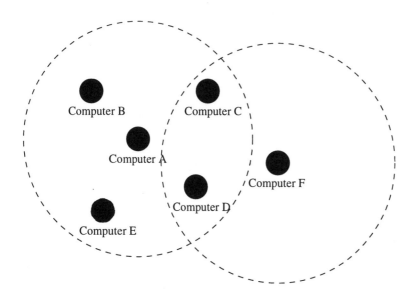

(b) Computer F collides with A at Computers C and D

Figure 6.18 Hidden station problem in wireless network

quences, it is unlikely that competing transmissions will use the same frequency at the same time. The disadvantage is the need for all computers to have complete knowledge of the others' sequences. Also, of course, this technique does not work for broadcasting to multiple receivers.

6.6 CHAPTER SUMMARY

Networks in which computers perform no switching, or only non-selective switching, generally offer a connectionless message broadcasting service. In practice, such networks occur as Local Area Networks or Metropolitan Area Networks. These networks are formed by directly using physical channels between computers. The physical media are normally guided at present, but use of unguided media is increasing to give wireless communication. Communication rates are in the megabits per second range: 10 or under for earlier networks, 100 or over for more modern networks.

One main implementation problem is constructing a multipeer channel using the underlying channels of the network. This might involve using an existing common medium directly, or constructing regular topologies from existing unicast channels. Topologies include chains, rings, stars and trees. The other main implementation problem is multiplexing broadcast channels onto the multipeer channel. This might involve a standard flow control technique, such as use of permissions or reservations. Alternatively, computers may act independently and aim for contention avoidance and/or detection. Finally, physical division multiplexing of time or frequencies may be used.

Various standard message broadcasting networks have been developed, to take account of differing channel availabilities, and different qualities of service required. The standardization efforts of IEEE have been a significant influence on what is used in practice. Examples of different networks types are: ethernet; token ring and its successor, FDDI; token bus; DQDB; 100 BASE VG-AnyLAN; and 802.11 wireless. There are many other variants, both developed experimentally and used in practice.

6.7 EXERCISES AND FURTHER READING

6.1 Explain how a room full of air can act as a message broadcasting network for a group of people within the room.

6.2 Discover whether you have access to any (a) LANs, and (b) MANs, and, for each, what computers are connected together.

6.3 In what ways might you find a Home Area Network useful?

6.4 What common features does the outline message format given on page 169 have with a letter sent through the postal system?

6.5 Suppose a message is transmitted over a phsical channel. Give an intutitive argument why the minimum length of the communication time period is equal to the bandwidth*delay product of the channel plus the message size in bits, divided by the channel rate.

6.6 If you are familiar with the internal architecture of computers, discuss the differences and similarities between an internal computer bus and a bus used as a common medium for a message broadcasting network.

6.7 Give examples of how bus, chain, ring, star and tree arrangements might underpin human communication systems.

6.8 Hub arrangements for common medium or chain/ring networks are popular with maintenance engineers and network managers. Why do you think this is so?

6.9 Compare the contention avoidance/detection strategy for isolated multiplexing with the ways in which people within a group would try to avoid speaking at the same time.

6.10 Find out about the pioneering University of Hawaii radio network that led to the ALOHA method for isolated multiplexing.

6.11 Suppose that three computers are in contention for transmission rights, and the randomized exponential backoff algorithm is being used to underpin isolated multiplexing. Either by evaluating probabilities, or by computer simulation, investigate how many time slots are likely to pass before all three transmissions are successfully carried out.

6.12 Give practical examples of network use patterns for which an isolated multiplexing approach is likely to be (a) effective, and (b) ineffective.

6.13 If you have read the book *Lord of the Flies* by William Golding, describe the permission-based mechanism used for multiplexing communications among the shipwrecked boys.

6.14 Repeat Exercise 6.12, but for a permission-based multiplexing approach.

6.15 A shop can serve only one customer at a time, and has an area for customers waiting to be served. If customers can stand/sit where they like in this area, but the aim is that customers are served in order of arrival, describe a reservation-based approach to multiplex the communications between customers and the server.

6.16 The term 'channel hopper' is sometimes used to describe a television viewer who continually switches from one channel to another. Explain why this process is more akin to time division multiplexing than to frequency division multiplexing.

6.17 Find out the main differences between the original Ethernet design and the final IEEE 802.3 ethernet standard.

6.18 Work out the minimum length that a communication time period can have on a 10 BASE 2 ethernet.

6.19 In an **empty packet ring** (sometimes called a 'slotted ring'), small empty messages circulate, rather than a token circulating. When wishing to transmit, computers wait for an empty message, then fill it. Discuss the relative merits of the token ring mechanism and an empty packet ring mechanism.

6.20 Find out how a **register insertion ring** operates, and explain how it is contention-free, but does not involve either a permission-based or reservation-based approach.

6.21 Write down a list of differences between the 802.5 token ring mechanisms and the FDDI mechanism, and explain why each difference exists.

6.22 Give informal arguments to show why the typical token circulation time in an FDDI network tend towards the TTRT, and the maximum circulation time is never more than double the TTRT.

6.23 Suggest reasons for the format of an FDDI-II cycle, in particular, for the existence of the 12 groups within each cycle.

6.24 List the properties of a token bus network and, for each property, say whether it is inspired by ethernet or by token ring (or by neither). Explain why this inspiration is likely to have occurred.

6.25 Find out more about how the logical ring is maintained by the IEEE 802.4 protocol.

6.26 Investigate the history of the IEEE 802.6 standardization process that led to the selection of DQDB as the MAN standard.

6.27 Give an intuitive explanation of why bandwidth balancing is needed as an extra mechanism on length DQDB networks.

6.28 The 100 BASE VG-AnyLAN standard is designed to harmonize with existing legacy networks that, in particular, use IEEE 802 identifiers for computers. Discuss the benefits that would arise if a more hierarchical identifier scheme could be used instead.

6.29 Why might the timeout period for AnyLAN to upgrade normal priority messages to high priority have been chosen to be in the range 200–300 millseconds?

6.30 Conduct a survey of the different radio-based LAN technologies that are currently available.

6.31 Discuss the difference between contention avoidance and detection in networks based on guided, and on unguided, technologies.

6.32 The **exposed station problem** occurs when a computer hears a transmission in a wireless network, and needlessly holds off transmitting since its target recipient is outside the range of the transmission in progress. Explain how the four-stage handshake mechanism used in DFWMAC gives scope for avoiding the exposed station problem.

Further reading

The topics covered in this chapter are predominately associated with the topics covered in the LAN and MAN literature. One possible specialist textbook for further reading is *Local and Metropolitan Area Networks* by Stallings (Macmillan 1997), although there are several others. The classic paper on Ethernet is "Ethernet: Distributed Packet Switching for Local Computer Networks" by Metcalf and Boggs, published in Communications of the ACM, **19**, July 1976. For information about a major influence, the University of Hawaii radio network, see "Development of the ALOHANET" by Abramson, published in IEEE Transactions on Information Theory, **IT-31**, March 1985. A full understanding of the issues behind the workings of more recent, and considerably more complex network types, such as FDDI and DQDB, requires a plunge into the communications research literature.

SEVEN

MESSAGE SWITCHING NETWORKS

The main topics in this chapter about message switching networks are:

- information, time and space basics of message switching
- routing of messages between switches
- congestion avoidance and detection
- examples of switching networks: switched ethernet, POTS, X.25, frame relay and ATM cell relay
- public switching network services: PSTN, CSPDN, PSPDN, ISDN and SMDS
- the Internet electronic mail network

7.1 INTRODUCTION

In this chapter, the computers forming a network are able to carry out selective switching of messages. That is, a computer can receive a message from one computer, and then may pass it on to one or more others. The switching is normally sufficiently selective that unicast channels can be implemented between arbitrary pairs of computers in the network. That is, viewing the network as a graph, there is a path between every pair of vertices in the graph. Although multicasting and broadcasting can be implemented, these features do not occur as a natural side-effect of the switching, unlike in Chapter 6, where they are a natural side-effect of the broadcasting.

The term 'message switching networks' is an accurate description of what happens in these networks. Remember, however, that the channels being used to form the networks of this chapter are assumed to have similar information, time and space characteristics. Thus, switching does not need to involve translation work as well. Networks that involve such work are the subject of Chapter 8.

The terms **switching** and **relaying** are both used in connection with message switching networks. Sometimes, these terms are used synonymously. However, where there is a distinction, it is one of quality. Message switching is reliable, in that messages input to the network are switched through appropriate intermediaries and are then output at their intended destination. Message relaying is 'best effort', in that the network does its best to relay messages through intermediaries, but there is a possibility of messages being lost or damaged.

This distinction is rooted in a historical difference between two types of switching: **circuit switching** and **packet switching**. The first, which comes from the traditional telephone system, involves setting up an electrical connection between two end-points via a sequence of switches. This connection acts as a dedicated path that ensures input to the network goes physically to the correct output. The second, which comes from the first computer data networks, does not involve a dedicated physical connection. Instead, packets of data (i.e., messages) are input to the network, routed through the network, and then appear at the correct output.

The essential difference is one of structuring time. Circuit switching allows a continuous communication over time; packet switching allows a communication that is partitioned into asynchronous components (packets) over time. The latter can be arranged to simulate the former by providing a reliable connection-oriented service. However, another option is to supply a less reliable service, which is where message relaying becomes an alternative.

This chapter is concerned with networks that switch messages, that is, the packet switching style rather than the circuit switching style. However, note that true circuit switching is becoming obsolete with the advent of digital telephony. The illusion of a dedicated circuit is still created, but by carrying out sufficiently frequent sampling and packetization of the data carried by the circuit. Thus, circuit switching can be viewed as a service provided by message switching. This inverts the old order, where message switching services were provided using circuit switching, i.e., the telephone system.

The operational principles of message switching networks are easier to describe than those of message broadcasting networks. Essentially, it is only necessary to describe the local switching mechanisms used in order to ensure that messages travel from source to destination(s) by an appropriate route. This avoids the need for more global concerns, such as implementing a multipeer channel and then multiplexing communications onto it. The fact that all message switching networks are basically the same in general principles is often easy to forget, because different families of networks have different jargon words to describe their features, creating an illusion of fundamental differences.

Message switching networks are most common as **Wide Area Networks** (WANs), where computers are located in geographically appropriate locations, and physical channels between particular computers reflect geographical adjacencies. The use of message switching technology for more local areas was superceded by the advent of message broadcasting LANs and MANs. However, this is changing with the advent of very high speed switching elements that make it practicable to employ message switching networks with similar, or higher speeds, than message broadcasting networks.

Message switching can also be used over channels provided by other networks, rather than just by physical channels. The term **tunnel** is sometimes used for such logical channels, suggesting the idea that they are tunnelling under the intervening network. Most of the ideas in this chapter are applicable to networks composed of either type of channel, physical or logical. This is in contrast to the ideas in Chapter 6, which largely apply to physical channels only.

7.1.1 Information basics

The messages transmitted by switching networks over physical channels are traditionally called **packets**. However, the terms **frame** and **cell** are also used in more modern, higher speed, networks, where the mechanisms are rather lighter weight than those used in packet switching networks. In switching networks over higher-level channels, the general term **message** itself is usually used.

In a switching network that offers a connectionless service, the basic components of a message are the same as those for a broadcasting network: source identifier, destination identifier and the information. However, most switching networks offer a connection-oriented service — reflecting a background in circuit switching — and so information-carrying messages have two basic components:

- connection identifier: uniquely identifies the connection that the message is travelling within;
- information: the information being communicated by sending the message.

The **connection identifier** is included so that intermediate computers can tell where the message originated from and where it is destined for. When a connection is established, its identifier must be selected, and then associated with the identifiers of the computers involved in the connection. Thus, a connection-establishing message for a point-to-point connection must have at least three components:

- connection identifier
- source identifier
- destination identifier.

For multicast services, an identifier is associated with a multicast group: the group of computers involved in a multicast connection. Usually, there are mechanisms to allow computers to join or leave the connection during its existence, so the identifier denotes a current group of participants.

The connection identifiers, and the source and destination computer identifiers, are of a fixed size agreed for the network. Connection identifiers are just integer values, for example, 12-bit integers in many public packet switching networks. In these networks, computer identifiers consist of 14 binary-coded decimal digits, which are interpreted in a hierarchical manner according to the ITU-T X.121 standard mentioned in Chapter 5. One advantage of using connection identifiers instead of pairs of source and destination identifiers in messages is that the overhead of control information is reduced: from 112 bits to 12 bits in this example.

As with message broadcasting networks, the information content of a message is just regarded as a sequence of bits. Traditional packet switching networks allow variable lengths for this sequence, but with a fairly low upper bound on length, for example, 128 bytes. More modern message switching networks change this assumption in two ways: either a greatly increased upper bound on length, for example, 9188 bytes, or a reduced and fixed length, for example, 48 bytes. The reason for imposing the various restrictions on message size is to simplify handling of messages by intermediate switching computers.

The standard network message format, which may include other components in addition to the above, has a standard representation that is the subject of an absolute agreement, fixed for all computers participating in the network. Where physical channels are used to implement the network, this representation is in terms of bits. However, the representation of bits on the physical channels used to connect computers in the network may vary from channel to channel.

In summary, from an information point of view, messages in switching networks have similar characteristics to those in broadcasting networks. There is absolute agreement on the type of information communicated: the standard network packet, frame or cell. There is also absolute agreement on how this is represented.

7.1.2 Time basics

Two absolute time measurements are of interest for message switching networks. The first measurement is the rate at which transmissions can take place over the channels within the network. There is no need for every connection to operate at the same speed. For example, more heavily loaded parts of the network may have channels operating at higher rates, or some parts of the network may offer a higher quality of service than others.

The second absolute time measurement is not usually expressible as a constant value. This is the switching time for messages passing through intermediate computers. There is a fixed component of this time: the time to receive a message, decide its fate and then transmit it onwards. However, a variable component is introduced by any other messages that might be present simultaneously. These may slow down the overall operation of the computer, or might compete for the use of outgoing transmission channels. The latency is determined by the sum of delays introduced by the channels, plus the delays introduced by any intermediate computers. Thus, the latency and rate experienced by a message is not nearly as easy to calculate as for the situation in message broadcasting networks.

The duration of a communication time period when a message is sent between two points is determined by the latency and transmission rate. The maximum duration is not easily worked out. However, in real networks, crude constant upper bounds, or upper bounds based on measurement of earlier communications, usually exist. These are needed for things such as timeout mechanisms.

There is a natural way in which to implement each communication over a channel provided by the network: as separate communications, one for each leg of the journey. The characteristic of these communications is that the information is the same, the spaces are different, and the time periods are sequential for the legs of the journey. In a **store-and-forward** message switching model, which is the norm, intermediate computers receive complete messages before forwarding them onwards. With this model, the communication time periods will be sequential and disjoint.

In an alternative **cut-through** message switching model, intermediate computers may begin forwarding messages before they have been fully received, using identifiers carried at the beginning of the messages. With this model, the communication time periods may overlap. One particular implementation of this model is **wormhole routing**, in which message communications are segmented into sub-communications of **flits** — fixed length messages. Flits are communicated using the store-and-forward model, but the overall effect is to implement a cut-through switching style.

As already remarked, in terms of time packages, the service offered by message switching networks is usually a connection-oriented service. Information is presented to the service in message-sized units, and each results in the transmission of a message. However, the transmission of other messages (and possibly the retransmission of the information-carrying messages) will be necessary in order to implement the required connection-oriented protocol. The complexity of the protocol depends on the quality of the channels that are used to form the network.

In message switching WANs, distances may be long and the quality of physical channels may be less good, leading to lower reliability than in LANs or MANs. The situation has improved markedly, with the introduction of fibre optic media and digital transmission techniques, but problems may still be introduced by over-worked switching computers needing to jettison excess messages trying to pass through. When offering a reliable connectionless service, the network itself takes responsibility for shielding the service user from unreliabilities. This may be achieved on an **end-to-end** basis, where individual network channels may be unreliable, but any problems are fixed up by protocols between the sender and recipient computers. Alternatively, this may be achieved on a **point-to-point** basis, where the individual channels are made reliable and the intermediate computers are made reliable.

Alternatively, some network types only offer an unreliable service, in which case loss, damage or duplication of messages might occur. Instances of damage can be turned into losses if the network includes an error-detection code in messages, with received messages just being thrown away if they fail an error-detection test. Duplication is a possibility, either if multiple copies of the same message are sent by different paths through the network or if the same message is retransmitted along the same path through the network.

A further component of a connectionless service may be flow control between the sender and recipient(s) of messages. This feature may also be of benefit to the network since, because the rate at which messages are presented to the network for transmission is controlled, the internal demands on switching computers become more predictable. In some cases, such as cell switching networks, the fact that the flow of information over connections is known is essential to the efficient operation of the network. Within the network, flow control is needed over the channels between computers, unless it is acceptable for unreliability to occur on individual legs of a message's journey.

In summary, from a time point of view, there are no inherent absolute bounds on the duration of communication time periods, since these vary with the communication space and the structure and operation of the network. However, the sub-communications for each leg of a message's journey have more predictable absolute bounds on their time periods if they occur over physical channels. These sub-communications are not decomposed over time, and the bits of a message are communicated synchronously. Time-related error correction or flow control may be provided by the network in order to deliver a higher quality service. Time-related flow control may also be used for mutiplexing communications within the network. This is discussed in detail in Section 7.3.

7.1.3 Space basics

The message switching style of network is very scalable, and there is no inherent need to impose absolute physical constraints on the area occupied. In fact, this might range from one desk to the entire world. Constraints are only needed if the latency of the network must be kept under some limit, or if long-distance channels are unable to deliver the performance needed for high transmission rates. As already mentioned, message switching networks have traditionally been used as Wide Area Networks (WANs).

The identifer scheme for switching networks is usually hierarchical. One reason is that, given such networks are often WANs, it is useful to have a scheme that allows local management of the allocation of identifiers. Another reason is that the structure of identifiers is often useful to switching computers in deciding how messages should be routed through the network. As a simple example, suppose that the first part of an identifier indicates the country in which the computer is located (as is the case in X.121 identifiers). Then, a switch might use the same route for all messages destined to a particular country, rather than attempting to be any more discriminating.

The network map known by users will be a partial one. In general, a computer will not be aware of all the other computers present in the network, particularly in a WAN. If a computer requires information about computers in the network, then it must be configured absolutely, or acquired from incoming messages from other computers, or acquired by making enquiries of other known computers.

Overall, a switching network offers at least the ability to use $n(n-1)$ different unicast channels, where n is the number of computers in the network. Each of these channels involves a different pair of computers. The information flow is from one computer to the other. This service could be used as a basis for a multicasting service. However, some switching networks offer multicasting and/or broadcasting as a service, which means that the internal operation of the network needs to be extended appropriately.

Since message switching networks usually are larger, and more widely distributed, than message broadcasting networks, there are likely to be many simultaneous communications being presented to the network for service. A key difference from most message broadcasting networks is that message switching networks are capable of implementing more than one communication simultaneously. In principle, each channel within the switching network can be active at the same time, each involved in implementing a different communication presented to the network.

The implementation of broadcasting networks was presented as having two components: first implementing a single multipeer channel, and then implementing a mechanism for sharing it between a collection of independent broadcast communications. This reflected the global cooperation found in such networks. The implementation of switching networks can be viewed as being almost the opposite way round. First, it involves implementing a mechanism for sharing each existing unicast channel between independent sub-communications. Then, it involves implementing a collection of independent unicast (and possibly multicast and broadcast) communications directly by defining switching computer behaviour. This reflects the more localized operation of such networks.

However, this is not the full story. Multiplexing of sub-communications over channels is not necessarily completely localized. Sub-communications over one channel are not completely independent of sub-communications over another channel. Thus, multiplexing can involve a variable amount of cooperation within the network, in order to improve network performance as a whole.

At one extreme, the multiplexing of sub-communications could be completely controlled by the multiplexing of communications onto the network service. For this, a model similar to that in the previous chapter would be appropriate: the network implements a unicast-style multipeer channel, which is then multiplexed. However, this does not correspond well with practical reality, because there is not global control of the independent communications being implemented by the network.

Nevertheless, because multiplexing issues are intimately connected with how communications are implemented as sequential sub-communications, channel implementation is described here first before multiplexing.

7.2 CHANNEL IMPLEMENTATION

The ability to carry out arbitrary unicast communications (and possibly multicasts and broadcasts too) depends on making appropriate use of the actual unicast channels that exist between particular pairs of computers. In most message switching networks, the graph representing the computers and the existing channels does not have a regular structure, unlike the chains, rings, stars and trees seen in the previous chapter. This is because edges exist for geographical or historical reasons, rather than topological reasons. To function as a message switching network, the graph must have a path between every pair of distinct vertices. That is, there is a sequence of one or more edges that connect each pair of vertices via intermediate vertices if necessary. If this is not the case, then the graph is not fully connected — in other words, the 'network' in fact consists of two or more disjoint sub-networks.

Conceptually, there are two ways in which a network can be viewed, as illustrated in Figure 7.1. One way, consistent with everything said so far, is to regard it as a collection of computers wishing to communicate, with some channels available between some computers. An alternative is to view the network as consisting of specialized switching computers, with each 'real' communicating computer being connected to its nearest switching computer. This is a black box view of the network, and is the model used for many public message switching networks. In essence, this black box has the same role as a shared medium in message broadcasting networks. Some authors use the term **subnet** to refer to the black box, and 'network' to refer to the black box and its attached computers. Another, less technical, term used for the black box is **cloud**, to suggest that it is something opaque to outside viewers.

A black box model is convenient for use here, since the channel implementations concern what happens inside the black box, and the description can focus on the requirements of switching computers. There is no need to complicate matters by worrying that the same computers might also be users of the network service as well (although this might indeed be the case, but the two activities can be separated within the computer). For brevity, a switching computer will be referred to just as a **switch** in the following description.

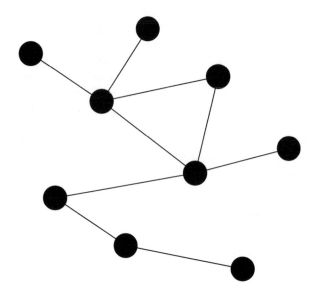

(a) Graph view of network with some computers doing switching

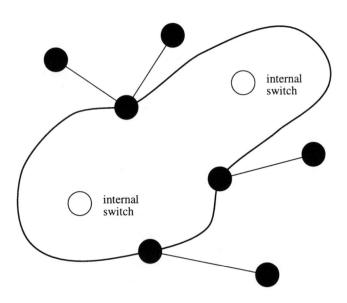

(b) Black box view of network shown in (a)

Figure 7.1 Two ways of viewing a network

The essential feature of a switching network is the behaviour of a switch when it receives an incoming message. First, it must ascertain where the message is heading to. Then, it must assist the message on its way. The first of these operations is relatively straightforward. If messages carry explicit destination identifiers, it is trivial. If messages carry connection identifiers, the switch must look up a table. For a particular connection established between two computers, a different identifier is usually used on each hop of a message's journey. The identifiers are chosen when the connection is first set up, in such a way that each switch can unambiguously look up the details of the connection being used by an incoming message using its connection identifier. When a switch forwards a message, it changes its connection identifier to the appropriate value for the next hop of the journey.

Given knowledge of a message's destination, there are two possibilities for a switch. The message has arrived at its destination if its recipient is a computer attached to the switch. In this case, the switch just forwards the message to its rightful owner. Otherwise, the message needs to travel further. In this case, the switch must decide on which one or more of its available outgoing channels the message should be transmitted.

More than one channel might be used when implementing a multicast or broadcast communication, to promulgate separate copies of the message to different parts of the network. Multiple channels might also be used for unicasts, if the switch believes it useful for two copies of a message to race each other through different network paths. A further possibility is the use of splitting, with the communication of a message being implemented by several sub-communications of message components using separate channels. Therefore, deciding on a choice of outgoing channel(s) represents a major technical problem for implementing efficient switching networks.

A fundamental property of switch behaviour is to ensure that messages progress towards their destination, preferably in all cases, but certainly in most cases. This encompasses not only assisting the correct delivery of messages, but also the efficient delivery of messages. Efficiency applies to both messages and the network itself. Messages should ideally be delivered with as low a latency and at as high a rate as the network user requires. However, the resources of the network should be used as efficiently as possible. There is likely to be a trade-off between message and network efficiency. Further, there is a trade-off between these factors and the efficiency of switches themselves. Switches are simplest, and so smallest and fastest, when their functions are simple. That is, simple methods are used to choose outgoing channels for messages. Unfortunately, simplifying switch functions restricts the scope for subtle approaches to increasing message and network efficiency.

The biggest simplification of switches can come when **source routing** is used for messages. With this approach, the network is most definitely not presented as a black box to communicating computers. In fact, a computer that sends a message to some destination must attach a list of the switches which the message is to pass through. On receiving a message, a switch just obtains the identifier of the next switch to be used from the beginning of the list, and then removes this first item before forwarding the message along the output channel leading to the next switch.

When used in practice, source routing can be very effective, because sending computers can choose a route that is best for their needs. However, the cost of this is that, prior to sending any messages, computers must probe the black box to discover different routes that can be compared. This can generate substantial extra communication traffic. If a network has a fairly static structure and loading, then the cost of discovering good routes can be amortized over many communicated messages.

Without source routing, switches need to make their own decisions. Assuming that messages carry connection identifiers, the basic operation of a switch is to look up that identifier in a table (usually, identifiers are sufficiently short that they can be used to index directly an array of connection information). The table entry contains two things:

- the number of the output channel to be used for the message; and
- the connection identifier to be used on the next leg of its journey.

This entry is set up when the connection is established, and normally does not change thereafter. That is, all messages belonging to a connection follow the same route through the network. When this is the case, the channel used by the connection is often referred to as a **virtual circuit**, reflecting the fact that it has similar properties to a channel through a circuit switched network. On the establishment of a connection (or for each message if messages carry destination network identifiers), the 'best' output channel for forwarding on has to be identified. There are a variety of ways in which this can be done, and five main methods are outlined in the following sections. Each has advantages and disadvantages when compared with the others.

Non-adaptive isolated routing

This involves switches having fixed tables that give, for each destination in the network, the output channel that should be used. The advantage of this method is its simplicity. The major disadvantages are that the tables may be very large, and that the method is non-adaptive: that is, routes do not change in response to changes in the network structure or loading.

The first problem is not unique to this routing method, but applies to all table-driven schemes. One solution to the problem is to exploit hierarchical identifier schemes, and use the same channel for all destination computers within the same part of the hierarchy. A partial solution to the second problem is to store more than one channel number for each destination, and then make a random choice between the possibilities, or cycle through the possible choices, when determining the channel actually to be used. An extreme form of this method is for a switch just to choose randomly one of its outgoing channels, dispensing with the guiding tables altogether. However, in general, this is not likely to ensure that messages progress towards their destination.

Broadcast isolated routing

An alternative to making a choice of outgoing channels is to use all of them. That is, send a copy of each message on all of the outgoing channels, except the one back to the computer the message arrived from. This is a form of broadcasting, and is called **flooding**. It solves the various problems associated with the previous method, since no tables are involved and, given a changing network structure and loading, it will find the most efficient path for each message.

Of course, the major flaw in this is that numerous other paths will be found too, and the network will become congested by the multiple copies of each message. To avoid infinite numbers of copies, it is necessary to prevent computers from forwarding extra copies of messages that arrive after a first copy has been distributed. Another way of stemming the tide is to limit the number of hops that a message can make, on the basis that another copy should have reached the destination in a shorter number of hops. A further way is to be selective about which outgoing channels are used, for example, to use only channels that are heading in roughly the right direction. This sort of selective flooding is a way of implementing multicasts and broadcasts.

Adaptive isolated routing

Adaptive isolated methods attempt to find paths for messages that take network structure and loading into account, but without introducing the speculative copying found in flooding type methods. As the name 'isolated' suggests, the switch must base any decisions only on its observations of messages received and transmitted in recent history, rather than explicitly consulting any other switches (or computers, in general) in the network.

One possible observation from incoming messages is to note the source computer identifier and the arrival channel. A plausible guess is that it is good to send messages destined for that computer back in the same direction. This, at least, gives one plausible path, although it may not be the best. One possible observation from outgoing messages is to note the channels on which flow control has been allowing messages to be passed on most quickly. This may suggest paths that will cause messages to be delivered more quickly than others.

There is no guarantee that adaptive isolated routing will lead to good results, since the information used as a guide is so localized, and does not look beyond the horizon of a single switch to see the overall condition of the network. A further problem is that even the limited information that is gleaned from local observation may become out of date. Therefore, switches must arrange expiry times for their stored routing information.

Adaptive centrally-controlled routing

Isolated adaptive routing schemes rely only on local knowledge at switches. However, what look like good decisions at one point in the network may not prove to be good decisions, either over the whole path followed by a message or for efficient operation of the network generally. An opposite alternative, which introduces something of the global state sensing found in most message broadcasting networks, is to have a single computer that specializes in finding good paths through the network. Acting on information supplied periodically by the different switches in the network, this computer can compute good choices for each switch based on a global view. As explained on page 157 in Chapter 5, there are standard algorithms that, given a description of the vertices and edges of a graph, can find the best path between any given pair of vertices. In the context of a switching network, 'best' might mean shortest distance, fastest or cheapest.

Ideally, a switch would ask the central computer for advice for each message that has to be routed. However, this is not practicable due to the extra communication and delays introduced by a dialogue between switch and central computer. Therefore, instead, the central computer periodically sends back details of best routes to the switches, which can then store the information in tables until the next update. The major problem with this method is its reliance on a single computer — clearly, if this computer fails, the switches will be relying on increasingly out of date routing information. As a precaution, switches must have expiry times for routing information, just as in the adaptive isolated approach. A further problem with centrally-controlled routing is that the central computer becomes a hot spot of network activity, as switches send in updated local information and then receive back global information in return.

Adaptive distributed routing

Most practical adaptive routing schemes involve a compromise between the isolated and the centralized schemes. There are two notable classes of compromise scheme — one that is at the isolated end of the scale, but which introduces limited communication with other switches, and the other that is at the centralized end of the scale, but has each switch acting like a little centralized computer. Both classes perform well in most practical circumstances.

The first class is called **distance vector routing**, and switches exchange their entire routing tables with their immediate neighbours in the network. On receipt of such information from a neighbour, a switch checks to see whether it can find any better routes than already known by sending messages via that neighbour. The benefit is to restrict communication about choices of path to local parts of the network. Specifically, each switch can exchange information only with the immediate neighbours to which it is connected by channels. The overall effect of this is that information about choosing paths gradually propagates through the network. Each switch gets direct information about its neighbours' view, which incorporates indirect information about its neighbours' neighbours' view, which incorporates indirect indirect information etc. etc. This is less accurate than centralized route computation, and can lead to inconsistent views across the network, but avoids the central potential point of failure and communication bottlenecks.

The second class is called **link state routing**, and switches broadcast to all other switches information about the cost of their channels to immediate neighbours. Flooding can be used to carry out this broadcasting. Thus, each switch sends a small quantity of information to all other switches, in comparison with distance vector routing, where each switch sends a large quantity of information to only neighbouring switches. With link state routing, each switch can perform its own shortest path calculations, in the same way as a centralized routing computer would, in order to generate a local routing table. The disadvantage is the amount of extra traffic imposed on the network, and the computational resources required to carry out the routing table calculations.

A more specific type of adaptive distributed routing has arisen to deal with the advent of mobile computing. The problem is that computers may no longer be semi-permanently located at the same point in the network. In principle, if an adaptive routing method gives good routing information for individual computers, then it can cope with mobility. However, in practice, the routing information typically refers to groups of computers that have co-located identifiers in a hierarchical identifier scheme — this dramatically reduces the size of routing tables. The expectation is that computers with co-located identifiers are also co-located in the network. If a computer is mobile, it is desirable for it to always have the same identifier, otherwise all communication partners of the computer will have to learn its new identifier each time it moves. This is in direct conflict with the exploitation of a hierarchical identifier scheme in order to reduce routing table size.

A solution is for a mobile computer to have a 'home' switch, which is its closest switch in terms of the routing information for its position in the identifer hierarchy. This means that messages bound for the mobile computer are routed to the home switch. However, depending on its location, the computer also has a 'foreign' switch, which is its current closest switch. There is a specific transfer of routing information from the foreign switch to the home switch. This tells the home switch where messages for the mobile computer should be forwarded to. This mechanism is needed, otherwise communication with the mobile computer would not be possible at all. A further elaboration, to improve efficiency, is for the foreign switch to supply direct routing information to switches at which messages for the mobile computer originate.

7.3 COMMUNICATION MULTIPLEXING

The issues of making efficient use of a network and taking account of network loading featured in the discussion of Section 7.2. These arise because switching networks allow many different communications to take place simultaneously. The communications arise not only from different users of the network, but also from the control communications needed for the exchange of routing information and for network management functions in general. The result of this, at a network channel level, is that sub-communications of these communications must be multiplexed. There is a problem if two (or more) sub-communications need to use the same channel at the same time.

The essential problem in implementing the network is to decide what a switch should do when it has two or more messages that need to be transmitted on the same output channel. A luxurious option is to multiply the capacity of each channel so that it is able to carry more than one message simultaneously, and the contention problem does not arise. This is a rather expensive solution if contention does not arise very often, and it could only be considered for a few specific channels that had persistently high loadings. The normal solution is for a message to be temporarily queued in the switch if its outgoing channel is already in use, or indeed if flow control for the channel prohibits the immediate transmission of the message. Given this, a further desirable property for routing schemes is to ensure that any such queues are always short or, better still, non-existent.

The usual type of within-switch queuing is **output queuing**, where a switch has a queue for each output channel. An alternative is **input queuing**, where a switch has a queue for each input channel, with messages having to queue up for entry to the switch if their output channel is currently busy or unavailable. Output queuing makes more sense from an intuitive point of view, since the queue is directly associated with the resource that is required. Input queuing may penalize other messages caught behind a message unable to proceed, since their own output channels may be available. For this reason, input queuing methods are usually impure, in the sense that such messages are allowed to overtake the messages that are blocking them. Input queuing methods are mainly of interest because they allow finer control of the causes of queues forming, as opposed to dealing with the effects.

If the overall loading of a network is within the physical capacity of the network, it is desirable for each switch and physical channel to be evenly loaded. However, when non-trivial queues form at any switches, then the network is experiencing **congestion**. The result of this is that messages are delayed, due to the time spent in queues. Once congestion occurs at one switch, its effect is likely to spread backwards to switches that are sending messages to that switch, and eventually in turn to the computers that are generating the messages. Thus, control of congestion requires distributed action across the network and its client computers. It is not just a multiplexing issue localized to individual channels.

Essentially, to prevent congestion occurring, the net inflow of messages to each switch must match the net outflow, and so some sort of flow control is needed. This may take place entirely within the network, i.e., between switches, or it may affect the interaction between the network and transmitting computers as in broadcasting network contention control, or it may be achieved as a side-effect of flow control between transmitting and receiving computers that are making use of the network. These three main classes of congestion control techniques are outlined below, all involving concepts seen earlier.

7.3.1 Isolated

An aim of any routing scheme for a switching network is to avoid congestion as much as possible. However, it is possible to make congestion avoidance a more dominating factor in the choice of routes than is typically the case in practice. One example is the use of adaptive isolated routing schemes that attempt to reduce the threat of congestion at switches, in the hope that the results will be for the good of the network as a whole. In a **hot potato** scheme, switches select the output channel with the shortest queue for each message, in order to get rid of the message as quickly as possible. Of course, while lessening the likelihood of congestion, messages may be sent in directions that are far from optimal. Other schemes, such as making a controlled random choice of output channel have similar benefits and drawbacks.

No scheme that is based purely internally in the network can prevent congestion occurring completely, if users can impose arbitrary traffic flows on the network. Thus, detection of congestion, and attempting to deal with it, is needed also. It is easy for an individual switch to detect congestion within itself — its queues have become excessively long. One simple option for dealing with this local congestion is just to throw away any excess messages. This may be acceptable if the network is offering an unreliable service, and the proportion of messages lost in this way stays within reasonable bounds. However, there may be unfortunate side-effects. For example, where a reliable service is being built upon an unreliable network service, messages may be soon retransmitted, and again home in on the hapless switch.

As a more general solution, a switch must share information about its congested state with other relevant switches and, perhaps, with message-generating computers too. Clearly, this sharing must avoid further increasing congestion in the network. If appropriate, it might be performed as part of the regular exchange of routing information. A congested switch will transmit bad news about its routes to other switches, which will respond by choosing routes for messages that avoid the switch, if alternative routes are available. Alternatively, sending explicit congestion warning messages may be necessary to achieve a more immediate effect, either routing around the problem or at least reducing the flow of messages into the congested spot.

Ultimately, if the network is overloaded, congestion problems cannot be solved just by switches attempting work-arounds. Instead, transmitting computers must either slow down or stop completely. This involves imposing temporary extra flow control between the computers and the network. One mechanism is the use of **choke messages**. On receiving a message for forwarding, a congested switch sends a choke message back to the message's original sending computer, asking it to slow down or stop transmitting until further notice. This relies on the goodwill of the sending computer in taking voluntary action to protect the network. It also means that there will be a delay before any change in the inflow occurs, since further messages will already be in transit towards the congested switch before the choke message reaches its destination. An elaboration of this scheme involves all of the switches on the choke message's route back taking note of the choke message. This results in more immediate relief for the congestion point.

7.3.2 Permission-based

With permission-based schemes, congestion is prevented by not allowing arbitrary traffic flows within the network. Between switches, point-to-point flow control can be used over the network channels to control the rate at which new messages arrive at each switch. In effect, this means that switches are performing input queuing, with the queues being distributed backwards to their neighbour switches. This may cause unnecessary delays for messages that could pass speedily through the switch that is imposing the flow control.

The use of choke messages, described above, is an example of flow control between a network switch and a transmitting computer, for use in emergencies. As a more permanent permission-based scheme, there can be flow control between a transmitting computer and its entry switch into the network. This will control the rate at which new messages can be introduced into the network. In order for this to be fully effective, the entry switch must have up to date knowledge of how congested the switches along the path to be followed by the message are.

Simpler alternative solutions impose flow control to protect the network as a whole. One possibility is to allow each user computer to have a fixed number of messages in transit at any one time. This is somewhat crude, and will lead to trouble if all computers decide to use their allowance at the same time. A more subtle possibility is to bound the total number of messages allowed in the network at one time. Such a scheme can be implemented by using a global pool of tickets, with spare tickets that appear at exit switches promulgating in a fair fashion to entry switches with needy clients. *In extremis,* this scheme can be combined fruitfully with a network service offering flow control between the communicating computers. The network then allows its own needs to influence the management of this end to end flow control.

7.3.3 Reservation-based

Reservation-based schemes are focused on networks offering connection-oriented services. At the time connections are established, network resources are reserved to ensure that messages sent within the connection do not lead to congestion at switches *en route.* Introducing extra communication mechanisms to reserve resources for a single message sent by a connectionless service does not really confer any advantage, compared with just trying to send the message itself. As already seen, the route followed by all messages sent within a connection is usually the same, and the switches involved maintain tables showing each currently active connection passing through.

The reservation aspect of connection establishment is that all switches involved only accept the connection if they are able to reserve an appropriate potion of their outgoing channel capacity. That is, given the rate at which messages will flow over the connection, the increased load on each switch will not cause it to become congested. If one path through the network is ineligible because of possible congestion concerns, then alternative paths may also be investigated, rather than just refusing the connection straight away.

One type of reservation is for the computers establishing the connection to specify the maximum number of messages that can be within the network at one time. Then, each switch might accept the connection only if it had enough spare capacity to queue efficiently this number of extra messages simultaneously. Alternatively, under the assumption of a regular flow of messages, a switch might accept the connection if it is able to deal with the maximum number of messages divided by the total number of switches involved in the connection's path through the network.

A more precise type of reservation is to agree the maximum rate at which traffic can be presented to the network, rather than just giving a bound on the amount of traffic present at any one time. Switches then accept the connection only if they have the spare capacity to handle the additional rate of message flow. Such schemes have to be managed with care. Transmitting computers must ensure that messages are presented to the network at no greater than the agreed rate. Switches must try to avoid a too conservative approach to accepting reservations. For example, reserving adequate resource that allows all active connections to operate at their maximum rates may be wasteful if it is unlikely that all will be highly active simultaneously. Congestion will be avoided, but the capacity of the network will be under-used.

7.4 EXAMPLES OF PHYSICAL SWITCHING NETWORK IMPLEMENTATIONS

In this section, the main examples of different styles of message switching network based on physical channels are covered. The description of each includes the level of service provided by each type, in terms of information, time and space properties, as well as the channel implementation methods used and the ways in which communications are multiplexed on network channels. Following this, Section 7.5 outlines some examples of public networking services that are founded on these message switching networks. Finally, Section 7.6 contains an example of a higher-level message switching implementation. There are further examples of switching network implementations in the Chapter 8, but there the switches have extra functions as well as just assisting in routing.

7.4.1 Switched ethernet

Ethernet message broadcasting networks are described in Section 6.4.1. Originally, such networks involved a physical bus connecting together the computers. However, this arrangement evolved to become a hub containing a short physical bus, with each computer physically connected to the hub. **Switched ethernet** is a further development of this idea, in which the hub contains an active switch rather than just a passive bus. The switch connects a number of interface cards using a high-speed backplane; each of these cards has some of the network's computers attached to it.

The physical arrangement is a two-level tree. Computers attached to the same interface cards can communicate locally, without the switch being involved. This sort of local communication may involve the standard ethernet multiplexing mechanism, or the interface card may prevent contention itself. For non-local communication, the switch enables the routing of messages from one interface card to another over the backplane. The backplane runs at a fast enough speed that there is sufficient capacity to deal with multiple simultaneous non-local communications. A further facility that can be provided by the switch is **PACE** (Priority Access Control Enabled), in which the switch prioritizes messages. This adds a priority mechanism, something that is not possible with standard ethernet.

Switched ethernet is an example of the simplest possible kind of message switching network: there is a single switch involved and, moreover, it can cope with all of its input channels being active simultaneously. In most respects, this technology is more akin to a piece of computer hardware rather than to a computer communications system.

7.4.2 Telephone system circuit switching

Circuit switching was discussed at the beginning of this chapter. It involves the direct physical implementation of a point-to-point communication. This is intended to allow the continuous transmission of speech from one telephone to another. The capability can also be used for one computer to talk to another. Information bits are transmitted at a fixed rate from the sending computer to the receiving computer along the direct connection between the computers.

Each switch in the network is capable of routing signals from its input circuit directly to its output circuit. When a connection is established, switches along its path are positioned appropriately so that a physical circuit is formed from sender to receiver via intermediate switches. There is no scope for congestion to occur, since every input channel to a switch is directly connected to its personal output channel. A connection is not accepted initially if no dedicated path can be found for it. Thus, circuit switching is the ultimate form of reservation-based congestion elimination. Of course, it is very wasteful if the computers do not actually require the ability to communicate continuously during the lifetime of their connection.

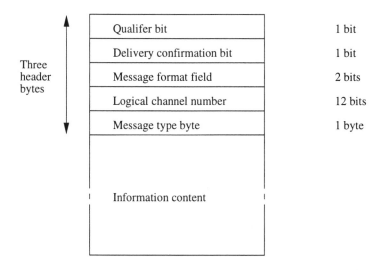

Qualifer bit	1 bit
Delivery confirmation bit	1 bit
Message format field	2 bits
Logical channel number	12 bits
Message type byte	1 byte
Information content	

Three header bytes

Figure 7.2 Format of X.25 message

7.4.3 X.25 packet switching

The most prevalent type of packet switching networks is based on the ITU-T X.25 protocols. These protocols are very much founded on the black box view of a network, and describe the interaction between a computer and a switch on the perimeter of the network. The computer and the switch are known as the Data Terminal Equipment (DTE) and the Data Circuit-terminating Equipment (DCE) respectively, in ITU-T parlance. The internal workings of the network are not prescribed by the X.25 standard. However, it is normal for a variant of the X.25 protocol to be used between internal switches also. The overall effect of the packet switching network is to offer what appears to be a direct circuit between two computers. However, this is not continuously active as with circuit switching, but only comes to life when a packet is transmitted or received. Thus, it is asynchronous in behaviour.

Information

The standard X.25 message is shown in Figure 7.2. It has a three-byte header, followed by information bytes. The header of all messages contains a 12-bit connection identifier, called the **logical channel number**. This allows the recipient of the message to deduce the source identifier and destination identifier of the message. The third byte of the header indicates the type of the message. As well as information-carrying messages, there are control messages, for establishing, resetting and closing connections, and also for flow control of information-carrying messages. Control messages for establishing connections carry the explicit source and destination identifiers of the two communicating computers. Information-carrying messages carry sequence numbers to support a sliding window flow control mechanism; this includes piggybacking of acknowledgements. The length of the information content is often limited, for example, to only 128 bytes. This restriction, together with the very short message header, is to keep down the storage requirement for messages that have to be queued in network switches.

Time

X.25 network rates are often relatively slow, with 64 kbits per second maximum rates within a connection being typical for public networks. However, much higher maximum rates are possible in some private networks, for example, 10 Mbits per second. However, regardless of the actual limit, these are maximum rates, and the normal expectation or requirement is that messages will not be continuously transmitted at this rate. The latency depends on the distances involved, but might be non-trivial, particularly if a satellite channel is involved at some stage of the path through the network.

The standard service offered by an X.25 packet switching network is a connection-oriented service. Over time, this involves a handshake for connection establishment, then the flow-controlled exchange of information messages, and finally a handshake for connection closure. The X.25 packet protocol makes no provision for error correction or sequence correction, since the network is required to provide a reliable path for messages. This is achieved by building reliable point-to-point connections over physical point-to-point channels, using the LAPB protocol described in Chapter 4. Thus, X.25 messages are encapsulated inside the HDLC-style messages used by LAPB. Strictly speaking, LAPB need only be used on the channel between a DTE and a DCE, but it is also used within X.25 networks.

X.25 also offers an acknowledged connectionless service. In essence, this is just the initial connection establishment handshake of the connection-oriented service on its own, and does not involve the introduction of any different features.

Space

There is no fundamental limitation on the area covered by an X.25 network. However, in practice, national boundaries govern the maximum extent, since legislation usually prevents single networks spanning more than one country. In wide area networks, the computers, switches and physical channels are located with geographical significance to suit human convenience. These days, it is unusual for X.25 networks to be used over areas smaller than those serviced by WANs. Fourteen-nibble (seven-byte) identifiers are used in X.25. These might be used as a flat identifier scheme in a single organization but, in public networks, the hierarchical X.121 identifier scheme is used.

Channel implementation

As already stated, X.25 does not prescribe the internal organization of the network. Bidirectional channels are established through switches in the network. However, only the interface between the source and destination computers and the network is specified. A sub-connection is established between each computer and its network switch. In practice, the same interface is used within most X.25 networks. When a connection is established, a series of sub-connections is established along the path to be used by the connection, one sub-connection communication between each pair of switches involved. Messages are sent within the series of sub-connections, with the connection identifier being changed at each stage of the route, so that it is recognizable by the next switch to receive the message. Ultimately, the message is delivered to the destination computer, containing a connection identifier that it recognizes.

The way in which paths are chosen for connections is an internal matter for the network. Historically, in public X.25 networks, this has been done using fixed routing tables, updated manually. This reflects the relatively simple and fixed structure of such networks, together with a fairly predictable loading. Hiding the internal details is consistent with the black box view of a network, an idea very familiar from the telephone system, which is also standardized by ITU-T.

However, where there is a more computer-centred emphasis on standards, matters are somewhat different. Explicit routing methods are specified for use in various proprietary packet switching networks that are alternatives to X.25 networks. For example, in IBM's System Network Architecture (SNA), fixed routing tables are used. These can give alternate routes where possible, but these alternatives are only used when a preferred route fails completely, rather than in response to network loading.

In Digital's DECnet Phase V architecture, an adaptive link-state routing protocol is used. This involves a two-level hierarchy for the network, so flooding of route information is limited, at one level, to local parts of the network and, at the other level, to switches linking local parts of the network. This protocol has formed the basis for an ISO standard routing protocol, which is briefly discussed in Section 8.3.3.

Communication multiplexing

On the interface between a DTE and a DCE, the presence of the connection identification field in messages allows multiplexing of many communications over the same physical channel. Since this field has 12 bits, 4096 different communications might be multiplexed at one time. In practice, some of the allowable values are reserved for special purposes, but there is still scope for thousands of simultaneously existing connections, as long as the sending/receiving computer is able to cope with them all.

Within the network, a collection of communications between various different endpoints are multiplexed onto each physical channel between switches. The normal criterion for defining a switch to be congested is that it has run out of space to queue messages that are waiting for transmission on output channels. Since the X.25 service has to be reliable, just dropping surplus messages is not an option. When connections are established, a limit on the flow control sliding window for each direction of transmission is agreed — this window can be of size between one and seven messages. To be entirely safe from overflow congestion, each switch need only allocate enough buffering space to contain a number of messages equal to the sum of the window sizes for the two directions. When a switch is already fairly highly loaded, a further possibility is for the switch to reduce the window size allowed for the connection, rather than refusing it outright.

This general approach is very conservative, since it assumes that all messages associated with a connection might be present simultaneously at one point of the network. Of course, as with routing policy, X.25 does not prescribe how congestion should be avoided in the network, so this is just one possible option that is made easy by the way the X.25 packet protocol works.

7.4.4 Frame relay

Frame relay is a younger sibling of X.25 packet switching, with a lighter weight approach more suited to the high speed networks of today than the slow WANs of yesterday. The basic service is also connection-oriented, but dispenses with flow control and the requirement for total reliability. Error detection is performed as messages are relayed through the network, but any messages found to be in error are just discarded, and no recovery action is taken. It is assumed that the user of the frame relay service will implement error correction if it is required. Thus, just as packet switching formed a natural evolution from circuit switching, so frame relay forms a further natural evolution, away from a telephone system view and towards a computer system view. That is, instead of complex protocols being used within the network, responsibility for such matters is offloaded to the computers using the service.

Information

The general format of a frame relay message is shown in Figure 7.3. It is easily recognizable as being derived from the classic HDLC frame format. The difference is in how the two-byte header is used. In the header, ten bits are available for a connection identifier called the **logical channel identifier** to identify the connection to which the frame belongs. However, this can either be shortened to six bits, or extended in multiples of seven bits by adding extra header bytes. This is accomplished using the extended address bit in each header byte. It is set to zero when there is a further header byte containing part of the connection identifier, and set to one in the final such header byte. Figure 7.4 shows the layout of a basic two-byte header, and a header that has been extended to four bytes, to carry a 24-bit connection identifier.

Header byte 1	1 byte
Header byte 2	1 byte
Information content	
Cyclic redundancy code (using CRC-CCITT)	2 bytes

Figure 7.3 Format of frame relay message

Unlike X.25, there is no type field in the standard message: essentially, all messages are used for carrying information. There is a bit available for distinguishing commands from responses, but this plays no role in the frame relay service itself. To effect connection management when necessary, the zero connection identifier is used to denote messages that are for control purposes rather than for information carrying. In the original frame relay service offerings from suppliers, all connections were permanent, and configured statically by the network operator. Availability of services with dynamic connections is a more recent phenomenon. Thus, there was no immediate need for a message type mechanism of the sort found in X.25.

The maximum size of message allowed is 8192 bytes, although many vendors only support a maximum of 2048 bytes, and maximums of 256 bytes are not unknown. The fact that variable-length messages are allowed means that network response times are unpredictable. Each message carries a cyclic redundancy code, to allow the detection of errors.

Time

Some frame relay services offer speeds akin to those of X.25, for example, 64 kbits per second. However, much higher speeds are possible, such as 1.5 Mbits per second, which is considerably higher than for X.25. Since protocols are simpler, and so switches do not have to do so much work, the latency experienced by frame relay messages should be somewhat lower than X.25 messages.

The standard service is a permanent connection-oriented service — such a service is also offered by X.25, in addition to its dynamic connection-oriented service. Thus, frame relay connections conceptually are communications with unbounded time periods. There are some bounds of course, but these are determined by human action to enable and disable the connections, rather than any observable feature of the connection itself. Frame relay services allowing dynamic connec-

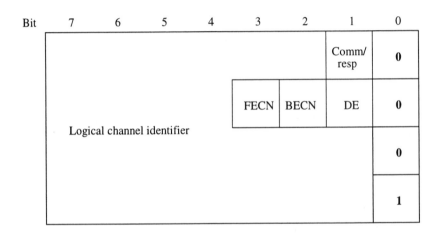

(a) Two-byte header   -   Bit 0 is the extended address bit

(b) Four-byte header   -   Bit 0 is the extended address bit

Figure 7.4 Examples of frame relay header format

tions are also now available, so the overall provision has the same two types of connection as X.25. There is no active flow control procedure, since the communicating computers are assumed to be able to cope with the maximum rate of the connection.

Space

The spatial characteristics of frame relay are similar to those of X.25 packet switching. The differences are in the information and time characteristics. With permanent connections only, there is no need for a network identifier scheme. However, with the advent of dynamic connections, identifiers are needed to specify the endpoints of connections. Either X.121 packet switching identifiers or E.164 ISDN system identifiers can be used. These are both hierarchical schemes.

Channel implementation

As with X.25 packet switching networks, frame relay has a black box view of the network, with no prescribed techniques for how relaying is carried out. However, the basic action of a switch is the same. The connection identifier carried by each message is used to determine the output channel for the next hop through the network. The choice of routes for frame relay connections is a far less dynamic affair than for packet switching connections, when connections are permanent and configured by the network operator. In this case, it is possible to have an accurate central view of the network structure, the pattern of connections and the loading placed by connections on the network. From this view, switches can be statically reconfigured periodically to reflect changes. When dynamic connections are possible, then there is a similar need for routing table updates as in an X.25 packet switching network.

Communication multiplexing

The connection identifier carried by each message allows many connections to be multiplexed on to the same physical channel. Since the standard identifer is 10 bits long, in principle up to 1024 permanent communications could be multiplexed on one physical channel. It is possible for congestion to occur in a frame relay switch, either because too many clients are simultaneously exercising their right to use the maximum transmission rate of their connections, or because several clients are simultaneously sending maximum size messages along their connections.

The use of high speed physical channels is one obvious way of lessening such threats. If congestion does occur, it can at least be cleared rapidly. In extreme cases, switches are allowed to drop messsages, since there is no guarantee of a totally reliable service. Each message carries a Discard Eligibility (DE) bit, which is set if the frame is seen as a better candidate for discarding than others that do not have the bit set. This bit can either be set by the message sender, or by the network if it has detected that the sender is imposing an excessive load.

Although there is no flow control mechanism associated with messages, there is a facility for notification of congestion problems. This uses the Forward Explicit Congestion Notification (FECN) and Backward Explicit Congestion Notification (BECN) bits in the header of each message. When a queue in a switch exceeds a certain length threshold, the switch warns the sending and receiving computers of the congestion. In messages leaving via the congested channel, the FECN bit is set and, in messages being forwarded after arriving via the congested channel, the BECN bit is set. When an end computer receives a message with a congestion notification bit set, it reduces its rate of message generation until the notifications cease.

7.4.5 ATM cell relay

Just as frame relay represents an evolution from packet switching, so Asynchronous Transfer Mode (ATM) cell relay represents a further evolution. However, the motivation behind it is considerably more significant and wide-ranging. This is the true blending together of support for computer communications with support for the telephone system, video transmission, and other telecommunications activities. Ultimately, the plan is that ATM cell relay networking will replace the worldwide telephone system, as the underpinning for Broadband ISDN (B-ISDN) services. However, this will take a long time to come to fruition and, meanwhile, frame relay represents an attractive staging post.

Although ATM appears set to become the dominant networking technology of the future, understanding its operation does not require any major conceptual leaps. The major change is that fixed-sized cells are transmitted, rather than variable-sized frames or packets. This restriction assists in the aim of very high speed transmission and switching. The basic service is connection-oriented but, as for frame relay, with no error-correction facility built in.

The major features of ATM cell relay are described here. A fuller description of some aspects is contained in the case study of Chapter 11, where ATM is used as the basis for the required application: video telephony.

Information

The standard ATM cell was introduced as one of the classic message types in Chapter 2. Figure 7.5 shows the format of the message in more detail. There are two variants: one used between switches and the other used between computers and switches. The fact that there are two variants is rather counter to the idea of having standardized messages to simplify processing by switches. Much of the header is taken up by a connection identifier for the message. This occupies 28 bits for messages sent between switches and 24 bits for messages sent between a computer and a switch. The identifier plays the same role as in the other types of connection-oriented networks, and its use will be discussed fully below. The eight-bit CRC at the end of the header is used for error detection for the header only, not the data content of the message.

After the five-byte header, there are 48 bytes of information. This choice for ATM was a compromise between the telephonic community, which favoured 16 bytes as being most suitable for voice samples, and the computer community, which favoured 128 bytes as being most suitable for computer data. After preliminary compromises from each side to 32 bytes and 64 bytes respectively, 48 bytes emerged as the final compromise, which represents a poor choice for both sides.

Time

The rates for ATM cell relay to underpin B-ISDN are 155 Mbits per second and 622 Mbits per second, with gigabit speeds being supported in the future. Slower

Virtual path identifier (VPI)	12 bits
Virtual channel identifier (VCI)	16 bits
Payload type	3 bits
Cell loss priority	1 bit
Cyclic redundancy code (using CRC-8 generator)	1 byte
Information content	48 bytes

(a) Format of ATM message used between switches

Generic flow control	4 bits
Virtual path identifier (VPI)	8 bits
Virtual channel identifier (VCI)	16 bits
Payload type	3 bits
Cell loss priority	1 bit
Cyclic redundancy code (using CRC-8 generator)	1 byte
Information content	48 bytes

(b) Format of ATM message used between computers and switches

Figure 7.5 Formats of ATM messages

speeds are also possible, of course. The speed categories are familiar from the Synchronous Transfer Mode (STM) standards used in the telecommunications community, and described on page 25 in Chapter 1. The crucial distinction is that the STM service delivers a *continuous* transmission capability at the stated transmission rate. ATM, on the other hand, requires this rate only when there are messages to be sent.

The basic ATM service is connection-oriented, with messages being transmitted within the connection. The standard message corresponds to a standard time period that acts as the unit when segmenting communications. Both permanent connections, as supplied by frame relay, and dynamic connections can be used. The handshakes required for establishing and closing connections take place on a separate connection used for control purposes. This dynamic connection, in turn, is established by making a request on a permanent connection. Thus, for each ATM information connection there is a corresponding ATM control connection. Section 11.3 contains details of how these connections are used and managed.

Each connection has the guaranteed property that messages are not reordered while in transit and so, apart from when messages are lost completely, messages are received in the same order that they are transmitted. When a connection is established, a **flow specification** is given. This specifies the quality of service that is required within the connection, and it is negotiated between the connection user and the network as part of the connection establishment procedure. The parameters can include the cell transfer rate, the cell latency and the cell error rate. As well as ensuring that the connection has the quality required, agreement on the flow specification also enables the network to reserve appropriate resources for the duration of the connection.

Space

ATM networks are intended to be usable at all scales, from wide area networks down to local area networks, and beyond to home or desk area networks. Until the advent of fully ATM networks, an intermediate role for ATM networks is to inter-connecting existing legacy networks of earlier types. Cell switching may feature at all of these scales, although broadcasting is an extra possibility for the more local areas. DQDB is an example of the broadcasting of ATM-style cells.

As well as a unicast service, ATM networks can also offer a multicast service. A multicast has one sending computer and a collection of receiving computers. Connection establishment for a multicast communication involves first the establishment of a connection between the sending computer and one receiving computer. Thereafter, the sending computer can use the associated control connection to add further receiving computers to the connection.

The identifier scheme used for public ATM networks is either 20-byte OSI identifiers allocated by countries (ISO 3166) or by international organizations (ISO 6523), or ITU-T E.164 15-digit decimal ISDN numbers for backwards compatibility. All of these identifier schemes are hierarchical. The identifiers are carried by the messages sent to establish connections.

Channel implementation

The connection identifiers in ATM have two parts, reflecting a two-level hierarchy of identification for connections. The top level is called the Virtual Path Identifier (VPI). It is 12 bits long in messages sent between switches, and 8 bits long in messages passing between computers and switches. This reflects the fact that there are likely to be more paths between switches within the network than there are between one computer and the network. The lower level is called the Virtual Channel Identifier (VCI), and it is 16 bits long.

The idea is that the network supplies **virtual paths** between computers, and then a collection of **virtual channels** shares each virtual path. The advantage of this is that messages travelling on any virtual channel of a particular virtual path all follow the same route. Thus, intermediate switches need only inspect the VPI field to decide the next hop for a message. Given that this field is somewhat shorter than the combined connection identifier, direct indexing into lookup tables is practicable.

As with X.25 and frame relay, an ATM cell relay network is a black box, and protocol standards are just concerned with the interface to the network. However, the basic actions necessary at a switch are the same as in any connection-oriented message switching network. The significant difference for ATM is the increased scale: very high speeds and large numbers of input and output channels. For example, messages might arrive on each input channel at a rate of 155 or 622 Mbits per second and there might be up to 1024 input channels. Thus, there could be around 1000 new messages input every microsecond. A conventional message switch has some type of processor inside, which deals with message arrivals sequentially, forwarding each message to the appropriate output channel. However, there are no processors fast enough to attempt this at ATM rates. Therefore, a parallel approach has to be used: all of the input channels are scanned simultaneously, and messages arriving are switched in parallel to the appropriate outputs.

The fact that messages have a fixed size makes it feasible to design fast hardware to carry out the switching. The necessary technology was largely well-understood, since the problem of connecting together processors and memories inside parallel computers has been much studied since the 1980s and, before that, similar issues arose in circuit switching telephone networks. The simplest solution is the **crossbar switch**. An example of a four-input four-output crossbar switch is illustrated in Figure 7.6. By appropriately setting internal switching elements, parallel paths can be created between each input channel and each output channel. For multicasts, the same input can be sent to several outputs by enabling all of the appropriate switching elements.

The basic crossbar is only adequate if each input channel is being routed to a different output channel, which is unlikely to be the case in an ATM cell switch. Thus, the crossbar has to be embellished with some arbitration facility. Normally, this involves introducing an output queuing mechanism, so that multiple inputs can be switched to the same output. There is no need to cater for the unlikely event that *all* inputs need to be routed to the same output, but the queuing mechanism needs to have enough capacity to handle the likely peaks that will occur under normal loading. In extreme circumstances, excess messages can be discarded, but this has to be fairly rare to preserve the normal limit on message loss of at most 1 in 10^{12} cells lost.

There are alternatives to a crossbar switch with output queuing. A crossbar is wasteful because, for n inputs, there are n^2 switching elements, with only a small fraction playing an active role at any one time. There are alternative architectures, also familiar in the world of parallel computing. One example is a Batcher-Banyan scheme, which requires only $n(\lg n)^2$ switching elements. However, this involves more complex hardware, and makes it harder to deal with output clashes and with

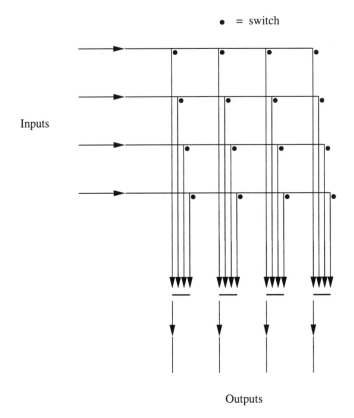

Figure 7.6 Example of crossbar switch

multicasts. In general, input queuing is an alternative to output queuing. However, as remarked earlier, this can unnecessarily delay messages trapped in the queue. Any attempt to allow such messages premature release has to be done carefully, in order to ensure that messages do not become reordered in transit.

Communication multiplexing

As already seen, there are two levels of multiplexing possible for the physical channel between a computer and an ATM network switch. First, communications over several virtual paths can be multiplexed on to the physical channel. Second, communications over several virtual channels can be multiplexed on to each virtual path.

Overall, a set of communications over a collection of virtual paths, and their associated virtual channels, is implemented simultaneously by a cell relay network. Each of these communications involves sequential sub-communications between switches. These sub-communications are multiplexed on to the physical channels between switches.

Given the rate at which messages arrive at switches, and the number of separate input channels that a switch might have, there is significant scope for congestion. This might be of a short-term nature, when chance coincidences lead to a particular output queue being temporarily long. More seriously, congestion might be long term, if the usage pattern is imposing an unwelcome loading on a switch, or switches, in the network. The latter problem is addressed by the use of flow specifications when connections are established, as mentioned earlier. Given that this specification includes a peak required rate for the connection, a reservation-based system can be used, with intermediate switches ensuring that they have sufficient spare capacity to add the connection. As with similar reservation schemes, it may be wasteful to reserve the maximum requirement for all connections, since they may not require their peak rate simultaneously. An alternative is to base the reservation on the average required rate, which is also included in the flow specification. However, this leaves open the possibility of congestion in the worst case.

If congestion does occur, there are dynamic mechanisms to guide a sending computer on reducing its transmission rate, if it is able to. First, each message carries a header bit in the payload type field that can be set by a switch to indicate that it is congested. This is similar to the use of the FECN bit in frame relay, and gives a warning to the cell recipient. Second, a sending computer transmits special control messages along the connection periodically. These contain, as information, the rate at which the sender wishes to transmit currently. The message is sent along the connection, then back again, with each switch allowed to reduce the rate if it is too high. The result is that the sending computer is advised of the rate that is acceptable to avoid network congestion.

To guide switches, should it be necessary to discard messages, each message carries a cell loss priority bit in its header that is set to indicate that it is a better candidate for discarding than any messages that do not have the bit set. This is similar to the provision in frame relay.

7.5 EXAMPLES OF PUBLIC SWITCHING NETWORK SERVICES

The previous section gave examples to illustrate the main types of message switched networks over physical channels. For such networks, especially when spanning wide areas, it is normally the case that the operator is offering a public service. The operator might be a state telecommunications service or a large private company licensed by the government. Until deregulation of telecommunications in several large countries during the 1980s, public network provision was fairly uniform. However, since then, a selection of variants has appeared, with a hint of the *ad hoc* rather than any systematic underpinning. This section reviews the main classes of public network facility now, and for the future. Of course, there are also private networks that fit the various categories also.

7.5.1 PSTN

For completeness, the Public Switched Telephone Network (PSTN) must be included. This offers the Plain Old Telephone Service (POTS) — the standard telephone system that has been evolving since the 19th century. From a computer communications point of view, this can be regarded as a facility for providing a circuit-switched network service, as discussed in Chapter 1. However, the quality is low: a maximum transmission rate of around 30 kbits per second, with the possibility of significantly high error rates such as 1 in 10 000 bits lost. Since the service is designed for transferring speech, modems are needed to represent binary information sent over the circuits.

7.5.2 CSPDN

A Circuit Switched Public Data Network (CSPDN) is operationally similar to a PSTN, except that it is intended for the transmission of binary computer information rather than audible human information. It involves the establishment of a dedicated circuit for each connection established between computers. The interface is similar to the telephone system, except computerized. A computer wishing to establish a connection listens for a digital dialling tone received from the network, then dials its desired partner by transmitting a digital telephone number. The ITU-T protocol used for this is called X.21. A variant called X.21*bis* also exists for use when a historic analogue network is hiding behind the digital interface. The dedicated circuit is turned into a reliable connection-oriented service by the use of the ITU-T LAPB protocol between a DTE and a DCE and, most probably, between network switches also.

7.5.3 PSPDN

A Packet Switched Public Data Network (PSPDN) is normally an X.25 packet switching network. The major distinction from a CSPDN is that packet switching is involved, rather than circuit switching. The X.21 or X.21 *bis* protocols are still used to establish a connection between a DTE and a convenient DCE. This connection is a dedicated circuit, and LAPB is used over the circuit to make it reliable and flow controlled. However, using this circuit, a collection of X.25 connections to different points on the PSPDN can be established by the DTE. These communications are multiplexed on to the DTE-DCE channel. The DCE is responsible for ensuring that messages are routed to their correct destinations. As mentioned in Section 7.4.3, public X.25 networks normally offer a slow transmission rate, typically 64 kbits per second.

7.5.4 N-ISDN

The Integrated Services Digital Network (ISDN) project is introduced in Chapter 1. It was embarked upon in the 1980s, with the aim of building new circuit switched telephone networks to replace PSTNs. As its name suggests, ISDN is a digital technology, rather than analogue, and supports both voice and data in an integrated manner.

A Narrowband ISDN (N-ISDN) service is now available in many parts of the world, but unfortunately the whole concept has been overtaken by changes in user requirements and also technological advances. The basic service is to provide an N-ISDN **bit pipe** as the channel between a user and the N-ISDN network. Communications over various types of channel can then be multiplexed on to this raw bit pipe. The standard combination provided to normal customers provides two 64 kbits per second channels and one 16 kbits per second channel. The two faster channels can be used for digitized voice or for computer data; the slower channel is an extra control channel.

This provision already seems fairly archaic. A completely naïve representation of speech by bits uses just under 64 kbits per second. However, advances in speech compression mean that as low as 8 or 16 kbits per second is quite enough. Meanwhile, this rate is too slow for other applications. In the home, transmission of video material is desirable, but this requires much higher transfer rates. Elsewhere, transmission of computer data proceeds at Mbit per second speeds. Thus, N-ISDN offers a service that seems out of touch with modern computer communications.

Having said that, the N-ISDN data rate is still attractive to those using modems and the POTS for computer communications, since they can get a more than doubled transfer rate from one N-ISDN channel. Indeed, it is possible to obtain a quintupled rate if all three N-ISDN channels are combined into one, and devices are available to do this. In addition, the digital N-ISDN channels are not affected by the analogue noise that afflicts telephone channels.

For backwards compatibility, the interface to N-ISDN offers both the PSTN and PSPDN interfaces mentioned above as well as its native digital telephone-like service. The frame relay service was originally introduced as an extra N-ISDN service. However, it is now also provided by public network operators as a service in its own right. In addition to the basic data-carrying services, N-ISDN can be also used for enhanced teleservices that have been standardized by ITU-T, such as facsimile, teletex and videotex.

7.5.5 SMDS

SMDS (Switched Multimegabit Data Service) was the first switched high-speed digital communication service offered by public network providers. The fact that the service is switched means that it can be used to make calls to different points. This is an advance on the use of leased line services, which can only be used to connect fixed points. The SMDS service was directed at the transmission of computer information, specifically to inter-connect LANs, and the standard speed is 45 Mbits per second. Unlike standard PSPDN services, the SMDS service is a *connectionless* packet switching service.

Messages consist of an eight-byte destination identifier, an eight-byte source identifier, and up to 9188 bytes of information. The identifiers are standard ITU-T E.164 15-digit decimal ISDN numbers. Subscribers agree to an average transmission rate, and are charged accordingly. The network imposes flow control on the user: a 45 Mbits per second peak rate is allowed for short bursts, but policing of message entry to the network ensures that the average rate is restricted appropriately. SMDS also offers a multicasting service, whereby previously agreed identifiers can be used to specify a group of receiving computers. More details of the SMDS service are included in the first case study, which is in Chapter 9.

7.5.6 B-ISDN

Broadband ISDN (B-ISDN) is the high speed version of ISDN that is intended genuinely to become the successor to the POTS, given the severe limitations of the original N-ISDN service. B-ISDN is intended to be appropriate for high speed transmission of audio and video, and also of computer information. In 1997, the basic user service to be offered seemed likely to be that offered by the existing Internet, rather than the more primitive bit transmission service of N-ISDN.

A basic transmission rate of 155 Mbits per second is to be offered, with 622 Mbits per second also possible, and higher rates appearing as technology advances.

After some evolution in the specification, four classes of B-ISDN service emerged, distinguished by different communication behaviour over time:

- Constant Bit Rate (CBR): bits are transmitted at a uniform rate, for example, when emulating a dedicated physical connection.
- Variable Bit Rate (VBR): bits are transmitted at variable rates, with prompt delivery being required despite the burstiness of the traffic. There are two categories of 'prompt' — real time, where promptness is very important, and non-real time.
- Available Bit Rate (ABR): bits can be transmitted at a guaranteed minimum rate and, if network resources are available, at higher rates.
- Unspecified Bit Rate (UBR): bits can be transmitted at some bit rate, but are prime candidates for discarding if the network is congested or other classes of traffic are competing.

The underpinning switched technology for the B-ISDN service is ATM cell relay, which explains its prominence in discussions of the future of networking (ATM can also be used to underpin services such as SMDS). The different classes of B-ISDN service correspond to different types of flow specification when ATM connections are established. DQDB is an alternative underpinning broadcast technology for B-ISDN.

The raw ATM message switching service is not offered directly to a B-ISDN user. Instead, a selection of different **ATM adaptation layers** (AALs) have been defined to make the service more accessible. These were summarized in Chapter 4, but will be enlarged upon here. Essentially, the role of an adaptation layer is to implement the required time behaviour of a user in terms of the appropriate time behaviour for the cell relay network. Unfortunately, views on appropriate different AALs have evolved rapidly, leading to confusion and extinction of some earlier AALs. Roughly speaking, the extant AALs correspond to the above different service types.

AAL1 is used for the CBR service. The continous bit stream is segmented into 46 or 47-byte units, which are transmitted in ATM messages. There is no error-correction mechanism, since delays cannot be introduced. However, the messages carry sequence numbers, so that missing messages can be replaced by dummy messages passed to the receiving computer. Buffering at the receiving end is used to ensure that small variations in message inter-arrival time are ironed out, to give a constant rate bit stream to the receiver. Thus, the communication time behaviour observed by the receiver is the same as the behaviour that applied originally at the sender.

AAL2 was meant to be used for the VBR service. A main difference from AAL1 is stronger error detection, on the basis that AAL2 information streams would be more susceptible to damage from occasional garbled bits than AAL1 information streams. For example, one perceived use would be for compressed video information, which is not as robust to errors as uncompressed video information. Also, the division of an incoming bit stream does not yield same-sized units, given the variable bit rate. However, there were technical problems with defining the AAL2 standard, and so it was not actually released.

AAL3 and AAL4 were intended for the non-time dependent ABR or UBR services, to support connection-oriented and connectionless information transfer, respectively, However, in the absence of any major differences emerging, AAL3 and AAL4 were merged into AAL3/4. The service allows the transmission of messages of up to 65 535 bytes long. It may either be reliable or unreliable, in the sense of whether a guarantee of delivery is given or not, rather than whether the delivered message is error-free or not. It is possible to multiplex comunications over several different channels on to each real ATM connection established. Data is divided up into 44-byte units for transmission in ATM messages.

AAL1, AAL2 and AAL3/4 were seen by the computer industry as being over-complex and inefficient, in other words, they seemed like typical standards emerging from the telecommunications industry. As a result, AAL5 was added. It is an alternative to AAL3/4, and has become the major AAL of interest. Its operation is described in detail in the video conferencing case study presented in Chapter 11.

Like AAL3/4, the AAL5 service allows the transmission of messages up to 65 535 bytes long. Delivery of messages may be guaranteed, or on a best-effort basis. Error detection is carried out, and flawed messages may either be discarded, or delivered with a warning. Multicasts are supported as well as broadcasts. Messages are divided up into 48-byte units for transmission in ATM messages.

This is one of the gains with AAL5, since each ATM message is able to carry four more data bytes than with AAL3/4. There are other efficiency gains in simplified handling of cells, which is very important for high speed operation.

The problems of agreeing on standards B-ISDN and ATM may appear large, but these are nothing compared with the major problem, which is carrying out a programme to replace the worldwide telephone system with a worldwide B-ISDN service supported by a largely untried technology: ATM cell relay. This is the big challenge for message switched networking.

7.6 EXAMPLE OF A HIGH-LEVEL SWITCHING NETWORK IMPLEMENTATION

The preceding two sections, plus Chapter 6, tend to reinforce the idea that networks are formed from physical channels between computers. There is much truth to this, in practice, since networking is an *essential* technique necessary to allow arbitrary communication patterns between computers. However, there is no reason why new networks cannot be built on top of channels provided by other networks. Chapter 8 demonstrates this at length, but it is assuming that the characteristics of the networks, and so the channels, differ.

One good example of a network built upon another network, but with switches not required to act as translators, is the Internet Multicast Backbone (MBONE). This is a network offering multicast channels, and it is implemented using the unicast channels offered by the basic Internet. The MBONE is described in Chapter 8, in the context of a general description of the Internet and its operation.

Another good illustration is given by electronic mail networks, and here one particular example of an electronic mail network is used: the Internet electronic mail service. There are other examples: from the past, UUCP (Unix to Unix Copy) mail networks and, perennially hovering on the future horizon, ITU-T X.400 mail networks. The latter, which has also been adopted by ISO as its **Message Oriented Text Interchange System** (MOTIS), makes use of a general-purpose message switching service to support a specific **inter-personal messaging** service. A problem arising from the existence of multiple standards, which could have been included in Chapter 8 but is not, is how UUCP, Internet and X.400 mail network channels can be harnessed to form a unified electronic mail network.

Internet mail message switching

The purpose of the Internet mail message switching network is to allow any person using a computer attached to the Internet to send messages to other people who use computers attached to the Internet. In principle, there can be a unicast channel between any pair of computers attached to the Internet. Therefore, at first sight, there is no networking required here: one computer can just send electronic mail messages to any other computer directly. In practice, the picture is not so simple

```
Return-Path: <gordonl@pomegranate.kartieri.com>
Delivery-Date: Mon, 6 May 1996 11:38:06 +0100
Received: from relay-2.mail.cheap.net (actually host dispose.cheap.co.uk)
         by grind.arkwright.co.uk with SMTP (PP);
         Mon, 6 May 1996 11:37:27 +0100
Received: from kartieri.cheap.co.uk ([158.152.10.190])
         by relay-2.mail.cheap.net  id aa24841; 6 May 96 10:24 +0100
Received: from localhost by kartieri.co.uk (4.1/SMI-4.1)  id AA01975;
         Thu, 2 May 96 15:02:03 BST
Date: Thu, 2 May 1996 14:24:03 +0100 (BST)
From: Gordon Long <gordonl@pomegranate.kartieri.com>
Reply-To: Gordon Long <gordonl@pomegranate.kartieri.com>
To: william@arkwright.co.uk
Cc: tull@drill.seed.co.uk
Subject: Warning
Message-Id: <Pine.SUN.3.92.960502095301.1928A-100000@pomegranate>
Mime-Version: 1.0
Content-Type: TEXT/PLAIN; charset=US-ASCII

Dear Mr Arkwright,
This is to warn you that there is trouble at the mill.

Respectfully yours,
Gordon Long.
```

Figure 7.7 Example of Internet electronic mail message

as this. Most organizations centralize the processing of electronic mail to some extent. Just as an office building might have a postroom to deal with incoming and outgoing letter mail on behalf of individuals, so certain computers are used as hubs for electronic mail distribution.

Information

The standard format for an Internet electronic mail message was defined in RFC 822. An example of a message is shown in Figure 7.7.

Messages have two components: a header and a body. Messages are represented by lines of text using ASCII characters; the header is separated from the body by a blank line. The characters have a well-known representation in terms of bytes, so a message can be transmitted as a sequence of bytes communicated over an Internet channel.

The header of a message has a sequence of lines giving information about the message. These come into two categories. There are header lines related to the actual delivery of the message — akin to the information on an envelope that is of interest to the postal system. Two essential header lines are the **From:** and **To:** lines, which show the sender and recipient of the message. This feature is standard to all connectionless-style services. Further extra recipients can be given in **Cc:** (carbon copy) lines. There are also header lines related to the message itself —

akin to the information on a letter that is of interest to its reader. The first category of header line is the only one relevant to the electronic mail network.

In RFC 822, the bodies of messages consist of lines of text. With the advent of multimedia computing, this now seems a modest provision. As an improvement, MIME (Multimedia Information Mail Extension) has been defined. This allows computer information, sound, pictures and video to be included in an electronic mail message. Each message can consist of a number of sections, each containing a different type of information. The defined types are encoded in ASCII characters using standard representations, so that the mail message is in the appropriate form for transmission by the electronic mail system.

Time

Human factors condition the rates at which electronic mail messages must be transmitted. At best, most people would be happy with delivery in a time equivalent to making a telephone call. At worst (unless the message is junk mail), most people would want delivery in a time equivalent to that available from the postal system. In computer terms, these time allowances are generous compared with the communication times needed for applications such as distributed computer systems and real-time video transmission.

Within these constraints, the channels used for electronic mail communications can be fairly slow and have high latencies. Also, the switching computers have considerable leeway in terms of introducing delays. Some examples would be computers that only process electronic mail at night, when they are more lightly loaded. More extreme would be a computer that was out of action for one or more days for maintenance or repair work. Alternatively, channels might only be used at off-peak times, when cheap pricing is available.

The electronic mail network implements unicast or multicast communications, corresponding to mail sent to an individual or mail sent to a mailing list respectively. A broadcast service would be distinctly unwelcome. The basic service is a reliable connectionless service. If a message cannot be delivered for any reason, human or computer, then a warning message (usually including the undeliverable message) is transmitted to the sender as a negative acknowledgement. Some elaborations on the basic service allow an optional positive acknowledgement message to be sent back after successful delivery or after the message has been read by a person.

Some mail networks derived from the Internet network have other time-related extensions. One possibility is expiry times carried with messages, so that messages can be discarded by the network if not delivered within some time period. Another possibility is to include a priority with messages, ranging from 'urgent' to 'junk', to ensure that appropriate attention is paid to prompt delivery.

Space

The area covered by the Internet electronic mail network is worldwide. There is a standard hierarchical identifier scheme for Internet electronic mail users. Sender and recipient identifiers are of the general form *person@domain*. The *domain* part is a standard Distributed Name Service (DNS) name for a computer, and has one or more Internet addresses corresponding to it. DNS naming is discussed in more detail in the next chapter. The *person* part is of local significance to the computers that send and receive the message — it has no significance to the mail network.

There is no complete directory of all people who can send and receive Internet electronic mail. Senders have to learn recipients' identifiers from various information sources, and also from incoming electronic mail. The identifiers used by the human sender are what determine the space of electronic mail communications. As long as these are all accessible on the Internet, the communications can be implemented by the mail network.

Channel implementation

A unicast channel for mail sent to a single recipient is directly available just by employing an Internet unicast channel to the computer referred to by the domain part of the recipient's identifier. A multicast channel for mail sent to a mailing list can be implemented simply by using an Internet unicast channel to each of the different computers referred to by the domain parts of the recipients' indentifiers. An Internet standard protocol — the Simple Mail Transfer Protocol (SMTP) — can be used to pass electronic mail messages over Internet channels. SMTP involves the establishment of a connection between the two computers. Within this, there are some preliminary handshakes to check the validity of the sender and recipients of the mail, and then the mail message itself is transferred.

The genuine networking side emerges when the computers of the mail sender and mail recipient(s) do not communicate directly over Internet channels. This arises from the willingness of computers to perform mail relaying. That is, if a computer receives mail that is not addressed to one of its own users, it passes it on to another more appropriate computer. This is normal message switching behaviour. The need for relaying emerges from two sources, both involving a computer transmitting a mail message. This computer looks up the *domain* part of the recipient's identifier to determine which computer the message should be sent to. First, for administrative, security or technical reasons, the computer may not be able to use a direct channel to that computer. Instead, it must use a contactable computer to act as a relay. Second, the computer advertised as the appropriate recipient for the domain may not be the actual computer that the mail is finally delivered on. Instead, it is a computer that is prepared to carry out the final relaying.

In both cases, the relaying computers are most likely to be an organization's specialist mail handling computers. As well as centralizing mail activity, there is another naming advantage. A general domain name can be advertised for the organization, rather than specific names for all of its computers. Then, electronic mail sent to people with the general name in the recipient identi-fier can be delivered to whichever computer is most appropriate at the time of arrival. For example, arkwright.co.uk might be a general name for Ark-wright Mills Ltd, and loom.arkwright.co.uk and dye.arkwright.co.uk might be specific names for two computers. If incoming electronic mail is addressed to william@loom.arkwright.co.uk, it might lie unread if the relevant person uses only dye.arkwright.co.uk for several days. In contrast, mail addressed to wil-liam@arkwright.co.uk, besides sparing the sender from knowing about computer names, can be delivered locally to the correct computer.

Switching of electronic mail messages is carried out using tables stored on each computer. Standard mail handling packages such as Sendmail, MMDF and PP allow human administrators to set up fairly complex rules on how mail messages should be routed. The simplest rule is to use non-selective switching: a computer sends all messages to the same computer. This is appropriate when an organization wants all outgoing mail to be handled by a central computer. The simplest rule at the other extreme is always to use a direct Internet channel to the recipient's computer. This rule might be appropriate for a central mail handling computer itself. There are other possibilities in between. For example, computers might be allowed to use direct channels to computers in their own organization, but have to go through a central computer for external mail.

Before the Internet was so pervasive, source routing of electronic mail mes-sages was necessary, with the details of the route to be followed supplied by the human sender. The mechanism to support this still survives in the standard format allowed for mail identifiers. However, humans are now spared from having to specify a suitable route for their messages. This brings electronic mail into line with the normal expectation of the postal system, where only the recipient's postal address is required.

When an electronic mail message is delivered, the history of its path through the mail network is recorded in header fields. Each relaying computer adds a **Received:** line to the beginning of the message. This information is of little interest to the average mail reader. Indeed, these header lines are often not displayed by mail reading facilities. However, the information can be useful when network problems are being diagnosed.

Communication multiplexing

There is not too much to say on the subject of multiplexing in the Internet electronic mail network. Congestion can occur, for example, if a large amount of electronic traffic is stored at a computer, awaiting forwarding. This may be due to natural causes, or due to accidents in which torrents of electronic mail have been generated. The amount of storage space allocated for mail awaiting forwarding is usually large, and sufficient to cope with everything except emergencies. In emergencies, one possibility is human adjustment to routing tables. If this is not feasible, returning excess mail to its sender is the best option, as long as this does not cause further congestion. Failing this, discarding is necessary. Congestion in general, or discarding in particular, is usually signalled to the human senders by human communication mechanisms. These advise people on what recovery action is best, and relieve the computers of any direct responsibility.

7.7 CHAPTER SUMMARY

Networks in which computers perform selective switching generally offer a connection-oriented message switching service. The networks may be based on physical channels between computers, or on channels implemented using underlying networks. Traditionally, switching networks have occurred as Wide Area Networks, but the technology has now been scaled down even to Desk Area Networks.

One main implementation problem is making available a collection of unicast channels between arbitrary computers, using the channels that are available to the network. This involves computers acting as switches, and forwarding messages in a way that efficiently sends them towards their final destinations. To do this, computers need routing information, and this can be kept up to date in a variety of ways. The other main implementation problem is avoiding congestion caused by excessive multiplexing of communications onto particular network channels. Congestion control can involve avoidance based on careful routing, or a variety of flow control mechanisms between computers and the network.

Message switching networks are a necessity to harness collections of physical unicast channels. Packet switching, frame relay and cell relay are the three main examples of networking technologies used for such networks. These form the basis for various public networking services: PSPDN, ISDN, SMDS and B-ISDN. Message switching networks are also used at higher levels. One example is the Internet electronic mail network, which is founded on channels provided by the Internet.

7.8 EXERCISES AND FURTHER READING

7.1 Give examples of human communications where information is passed by message switching. Comment on the reliability of each communication.

7.2 What are the main reasons why message switching networks usually offer a connection-oriented service, whereas message broadcasting networks usually offer a connectionless service?

7.3 A reliable connection-oriented service may be implemented using an end-to-end protocol between entry and exit switches, or using a point-to-point protocol between each pair of intermediate switches. Discuss the advantages and disadvantages of each approach.

7.4 For any message switching network to which you have access, try to find out how many computers are connected by it, and what geographical area is spanned by it.

7.5 A wide area message switching network typically has many simultaneously active communications. Explain why it is not practicable to multiplex communications at a global level (as in message broadcasting networks).

7.6 A motorist is to drive from a small village on one side of a country to a small village on the other side, for the first time. How might the motorist choose the roads to drive along?

7.7 Write down a summary of the advantages and disadvantages of the five main methods of routing — non-adaptive isolated, broadcast isolated, adaptive isolated, adaptive centrally-controlled and adaptive distributed.

7.8 Explain why, with distance vector routing, news about good routes travels quickly and news about bad routes travels slowly.

7.9 What part would a graph algorithm for finding shortest paths (e.g., Dijkstra's algorithm) play in link state routing?

7.10 Comment on the ways in which congestion control might be carried out in the road system of a city.

7.11 For a connection-oriented service supplied by any message switching network that you have access to, find out what (if any) limits are placed on the information flow within a connection.

7.12 Switched ethernets have proprietary backplanes to inter-connect the switch. Obtain details of the transmission characteristics of some of the available backplanes, and comment on their effectiveness.

7.13 Full duplex ethernet is an available technology that sounds like a contradiction in terms. Explain how switched ethernet can make full duplex transmission possible.

7.14 If the information content of an X.25 message is restricted to a maximum of 128 bytes, what is the maximum duration of a message communication when a 64 kbits per second channel is used?

7.15 The X.25 sliding window size can range between 1 and 7 messages. Comment on what this range suggests about network size and network channel rates.

7.16 Discover more about how IBM's SNA packet switching networks work, in particular, how the routing scheme works.

7.17 Find out whether a frame relay service is available from any of your local common carriers and, for each, note what maximum message sizes and message transfer rates are offered.

7.18 Why do you think that ATM was chosen to underpin the B-ISDN service, rather than STM?

7.19 What is the reason for the basic ATM service being connection-oriented, rather than connectionless?

7.20 If you are interested, find out about the Batcher-Banyan switch, and compare its properties with those of the simple crossbar switch.

7.21 ATM has a header bit with a similar purpose to the FECN header bit in frame relay. Why does ATM not have a bit corresponding to the BECN header bit in frame relay?

7.22 Study the ATM Ring technology, which blends use of ATM messages with an FDDI-style dual ring interconnection, to produce a standard intended for use in areas similar to those targeted by DQDB.

7.23 Obtain details of the ITU-T X.21 protocol, and note the close similarity to what happens when a person makes a telephone call.

7.24 If a local common carrier offers both a leased line service and an SMDS service, compare the prices charged for each.

7.25 List applications that have communications characteristics appropriate to each of the four B-ISDN service classes.

7.26 Look up information on the original ITU-T rationale for choosing different types of ATM adaptation layer. Why have AAL1 and AAL5 emerged as the dominant adaptation layers?

7.27 What is a realistic timescale for B-ISDN services becoming widespread throughout the world?

7.28 Investigate the Fibre Channel standards for high performance (up to 1000 Mbits per second) point-to-point channels between computers and other devices, and assess the part that such channels can play in both message broadcasting and message switching networks.

7.29 If you can receive Internet mail, look at the **Received:** lines at the top of messages to discover the route followed.

7.30 Look at RFC 822 to discover the exact range of header lines that can appear in Internet mail messages.

7.31 If you can send Internet mail, experiment with source routing of messages by using the extended form of destination identifier (details in RFC 822).

7.32 Look at RFC 2046 to discover the types of information that MIME allows to form part of electronic mail messages.

7.33 Read RFC 821, which describes the Simple Mail Transfer Protocol (SMTP). Note the connection-oriented form of this protocol, with handshake communications occurring within the connections.

7.34 Usenet is a facility for electronic newsgroups, which is supported by the Internet (as well as other networks). Comment on the basic similarities and differences between Usenet's requirements and the electronic mail network's requirements.

Further reading

Wide area networks, specifically packet switching networks, have been a main topic in general communication textbooks for a long time. This is because such networks are the longest-established types. For more information on recent high speed networking developments, see *Gigabit Networking* by Partridge (Addison-Wesley 1994). However, most recent editions of general textbooks have coverage of high speed networking, particularly ATM. There is an increasing amount of research literature on the subject of ATM and its properties. The subject of routing in packet switched networks is a much-researched topic, and a large number of papers have been published since the mid 1970s. The subject of congestion control in message switching networks has become a more pressing concern with increasing communication rates, and has a rapidly increasing research literature. For a good general survey of the area, see "A Taxonomy for Congestion Control Algorithms in Packet Switching Networks" by Yang and Reddy, published in IEEE Network Magazine, **9**, July/August 1995.

EIGHT

INTER-NETWORKS

The main topics in this chapter about inter-networks are:

- information, time and space basics of inter-networks
- switching in inter-networks
- bridges between IEEE 802 LANs
- the Internet: the Internet Protocol (IP), IP routing
- ISO standard inter-networking

8.1 INTRODUCTION

Chapters 6 and 7 focused on networks. A characteristic of a network is that there are general agreements on information, time and space matters between all of the computers involved. For example, there are network-wide message formats, time packages and identifier schemes. Importantly in these two chapters, channels did not exhibit major differences, and so switching computers did not need to carry out translations to reconcile channel behaviour. This chapter tackles cases where such translations are necessary. In principle, it might consider a computer connected to two physical channels with very different properties. In general though, major differences between channels arise because they are implemented by different kinds of network. For this reason, the chapter title refers to **inter-networks**.

The function of an inter-network is to allow different networks to cooperate. In general, the networks will be of different kinds, and so extra functions are needed to implement the inter-network. The larger the differences between networks, the more extra functions are needed. Note that the term **internet** is very frequently used as an abbreviation for 'inter-network'. Unfortunately, this invites a degree of special casing, given the existence of 'the Internet', the most famous inter-network of the 1990s, and which is discussed in Section 8.3.2. To avoid confusion, the term 'inter-network' is used throughout this chapter (and the book), despite being rather more of a mouthful.

Sometimes, it is hard to draw exact lines between what is a network and what is an inter-network. A prime instance is a Metropolitan Area Network, using DQDB or FDDI technology for example. Although this is called a 'network', a MAN is usually used as a way of connecting together Local Area Networks. Thus, it might be argued that a MAN which acts in this way is, in fact, an inter-network. The lack of certainty is reflected in the terminology used in such circumstances: constituent LANs are often referred to as 'subnets' of the MAN. To add to the confusion, as mentioned in Chapter 7, some authors use the term 'subnet' to refer to the internal fabric — switches and physical channels — of a message switching network based on physical channels.

The confusion does not arise with the terminology used in this book. Applying the guidelines of the first paragraph, the LAN/MAN problem can be resolved satisfactorily. Each LAN is a network. The MAN, interpreted as its directly connected computers and the physical channel(s), is a network. Taken together, the collection of LANs and the MAN is an inter-network. Where a communication channel is required between a computer on one LAN and a computer on another LAN via the MAN, the function of the inter-network provision is to reconcile differences between the two LANs and the MAN.

Message switching is the implementation method for inter-networks. In simple cases where two or more message broadcasting networks can be connected together by a non-selective switch to form a broadcasting inter-network, the result is more truly regarded as just being a single network. The mere fact that broadcasting still works means that next to nothing needs to be done to reconcile differences between the joined networks. To carry out message switching across an inter-network, certain computers must be connected to channels implemented by more than one constituent network. These provide the basic switching capability to route messages between computers in the inter-network. In this respect, the inter-network is just a message switching network founded on channels implemented by the consituent networks. Thus, implementation of switching involves solving the same problems as discussed in Chapter 7.

However, the major difference is that the switching duty does not involve just forwarding messages, as in a message switching network. The staging computers between networks must take account of the different types of channels offered by the networks. The complexity of this task depends on the type of channel that is offered by the inter-network as a whole. The more modest the service, the easier it is. A lowest common denominator service can be based on the worst service offered by any constituent network, since it is usually far easier to make the good networks' services worse than it is to make the poor networks' services better.

The general problem of harmonizing different network services raises issues concerning information, time and space, and these are discussed in the Sections 8.1.1, 8.1.2 and 8.1.3.

8.1.1 Information basics

Staying consistent with the policy of the book, the **message** will be taken as the unit of information communicated using an inter-network. Different types of network will have different types of message, with variations in format and size. The common features are that each message carries some kind of identifier component and some kind of information content. The identifier component might be the explicit identifiers of the source and destination of the message, or it might be a connection identifier. The information content can be regarded as a sequence of bits.

Further discussion on the identifier component of messages is deferred to the consideration of space-related matters. Apart from this, the essential problem is to reformat a message so that it can progress from one network to another. This may involve the removal, modification or introduction of additional components that are present for control purposes related to time and space issues.

A general physical problem is that the resulting message may be too long or too short to be forwarded, either because of surgery on control components or because the information content has an inappropriate size. Often, this problem can be avoided if there is an agreement on inter-network message sizes that ensures they are acceptable to all constituent networks. However, if this is not the case, the communication of an over-size message must be segmented into sub-communications of smaller messages. An under-size message might be padded to be long enough, or the communications of several under-size messages might be concatenated into one super-communication. If it is important that the structure of the original communication is preserved at the receiving computer, then a mechanism must be provided to recombine or divide up any introduced sub-communications or super-communications.

8.1.2 Time basics

The rate at which messages can be transmitted varies between different networks. The variation might be large, for example, between an X.25 packet switching network at 64 Kbits per second and an FDDI MAN at 100 Mbits per second. It might be relatively small but still problematic, for example, between an IEEE 802.3 Ethernet operating at 10 Mbits per second and an IEEE 802.5 token ring operating at 4 Mbits per second. One solution is for the maximum inter-network transmission rate to be bounded by the smallest transmission rate of any constituent network. However, this is rather wasteful, and something more flexible is desirable.

If messages are sent at the peak rates of networks, there is no problem when messages must hop from a slower network to a faster network. The problem is in the other direction. For a single message of manageable size, buffering in the switch offers a solution to extending the time period used for communicating the message. However, for communications which continue over long time periods, the buffering will become exhausted. If this is the case, a mechanism is needed to control the flow coming from the faster network. This might be localized, in the sense that flow control is used within the faster network. However, in turn, this may mean that flow control is needed on networks involved on the other side of this network. A safer solution is for the inter-network service to include flow control between the source and destination computers, which takes account of the capabilities of intermediate networks.

A combination of differing network rates, and additional delays introduced by longer distances and extra inter-network switching, means that communication time periods both are extended and become less predictable. This has implications for timeouts and expiry periods. These can be finely tuned on single physical channels and on some network types. However, such tuning is much harder for protocols used in inter-networks. A further consideration is that inter-networking procedures must take care not to trigger accidently network timeouts or expiry periods by warping time periods to a larger extent than expected by the network.

In terms of time packages, both connectionless and connection-oriented services may be offered by networks, with some intermediate classes in between also possible. Inter-networking has to deal with such differences in service. The situation is complicated further by the extras that may come with the time packages. These include flow control and correct sequencing of messages sent over connections, and also error detection and correction.

If the inter-network service is chosen to be minimalist, that is, an unreliable connectionless service, the implementation task is easiest. Clearly, any networks within the inter-network that offer such a service can be used directly. Also, it is not hard to implement the service on any more heavyweight network services, such as a reliable connection-oriented service, using the method discussed in Section 4.5.1. Thus, the inter-networking protocol can focus on information and space issues.

In contrast, if the inter-network service is chosen to be more lavish, for example, a reliable connection-oriented service, the implementation task is hard. There are two basic strategies for tackling this. The first is to build an unreliable connectionless service for the inter-network, as described in the previous paragraph. Then, using this as a mechanism for unreliable communication between arbitrary computers on the inter-network, build end to end reliable connections on top of the unreliable end to end service. This can be done using standard methods of the type described in Section 4.5.1.

The second strategy is to build a reliable connection-oriented service on channels implemented by any constituent network that does not offer such a service already. Then, a reliable end to end channel between arbitrary computers on the inter-network can be built by using a series of channels across networks. This approach is analogous to building a connection-oriented service on a message switching network using a series of connection-oriented network channels.

In practice, the first strategy is more popular. This is because it needs just to assume a level of network service that can be expected from all networks. The intermediate connectionless service that it builds may also be useful for inter-network users that do not require a full-blown reliable connection-oriented service. The second strategy means that individual action needs to be taken to integrate each different type of network into the inter-network. The more diverse the range of networks, the more work this involves. Such an approach is only likely to be used when a collection of connection-oriented networks is being linked, that is, when the inter-network provision does not need to enrich the individual network services and so only needs to connect network channels in series.

8.1.3 Space basics

The areas covered by inter-networks differ greatly. The smallest type of inter-network spans a single building, connecting a collection of LANs. The largest inter-networks span the world, connecting varied networks in different countries. The collection of networks involved in an inter-network evolves gradually, in the same way that the collection of computers involved in a network evolves gradually. Combining the two effects, the collection of computers involved in an inter-network usually evolves fairly rapidly if the inter-network is of any non-trivial size.

Given that inter-networks are the implementation mechanism that allows communications between any of the computers involved, some sort of identifier scheme is needed so that computers can be identified uniquely. Each computer already has a unique identifier on its own network (or networks if it is attached to more than one). However, there may be overlaps in the identifier sets used in different networks. There are various ways in which an inter-network identifier scheme can be derived, but most of them involve a two-level hierarchy of identifiers. The top level part of an identifier identifies a particular network, and each network in the inter-network must have a unique identifier. The other part of an identifier specifies a particular computer within the network specified by the top level part. This part of the identifier may or may not be the same as the computer's native network identifier.

The two-level hierarchy allows the allocation of inter-network identifiers to be devolved to the operators of individual networks. A flat inter-network identifier scheme is only practicable in an inter-network that is very centrally managed, and this is not usually the case. Inter-networks are somewhat distributed entities, compared with individual networks. When there are a variety of network types involved in an inter-network, computers are likely to have distinct individual network identifiers and within-network inter-network identifiers. This is because the formats and style of network identifiers will vary, whereas the inter-network identifiers have a common form. This common form is for addresses used by the computers involved in implementing the network — a more flexible naming system is usually provided for human consumption.

The complete inter-network map at any particular point in time is unlikely to be known by any of the computers involved in the network, even specialist network management systems. This is because of the rapid rate of change in the inter-network. As in a WAN, computers must acquire information about other computers, either from human informants or by making enquiries of known computers.

Inter-networks allow unicast communications between pairs of computers. This can be used as the basis for multicasting or broadcasting. However, some inter-networks provide a built-in multicasting service as well as a unicasting service. In principle, broadcasting can be done, but this is not usually appropriate in inter-networks, where there is little likelihood of communicated information being of universal interest to all of the computers involved.

The routing mechanisms for implementing appropriate channels in a message switching inter-network involve the same principles used in message switching networks. It is necessary to establish paths through the inter-network, with switching computers being used to pass messages from one network to another. One hop in an inter-network, which involves a path over one network, is analogous to one hop in a message switching network, which involves a transmission over one network channel. Thus, the principles for channel implementation described in Section 7.2 apply here as well, and so there is no need to repeat them. The difference between inter-networking and networking is that switches have to perform a more complex job: to reconcile different networks, perhaps with different routing mechanisms. This new topic is covered in Section 8.2.

In fact, routing can be simpler sometimes, because inter-networks usually have a physical interconnection pattern that is tree-like to an extent. At the heart of the inter-network, there is a **backbone network**, which forms the root of the tree. Then, the next layer of the tree has a collection of networks, each one connected via a switch to the backbone network. This continues down to individual networks at the leaves of the tree. An inter-network with three tree levels is illustrated in Figure 8.1.

In a simple inter-network, there may be only two levels in the tree, for example, a MAN backbone as the root and a collection of LANs as leaves. The tree gives a basic framework for routing: messages are sent as far up the tree as necessary, and then back down again. In practice, there are other connections in the inter-network besides the ones forming tree connections. These allow short cuts to be made,

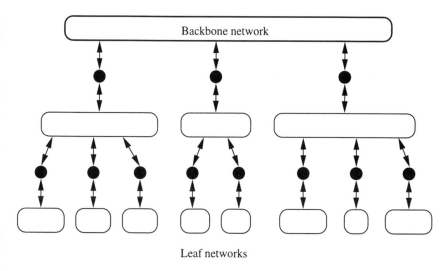

Leaf networks

Figure 8.1 Example of tree-structured inter-network

avoiding the centre of the tree, and this is what adds a potential for subtlety to the routing algorithms.

Communication multiplexing is also a feature of inter-networks. Each computer in the inter-network may have many simultaneous communications in progress. Also, each network and each inter-network switch will have many simultaneous communications in progress. As with routing, the multiplexing mechanisms in an inter-network involve the same principles used in message switching networks. Again, viewing the path through each network as being analogous to a single channel within a network, the multiplexing methods of Section 7.3 apply here, and there is no need to repeat them.

The subject of communication splitting is not mentioned much in Chapters 6 and 7, because splitting is unusual in individual networks. Most message broadcasting networks only have one choice of physical channel. Most message switching networks use fixed paths for all messages sent within connections, if they are connection-oriented, and they do not normally spilt message communications, if they are connectionless. With inter-networks, there is scope for splitting at inter-network switches or where computers interface with the inter-network. Most inter-networks offer more than one possible path between computers, so switches may be able to make use of this dynamically in order to improve performance by the introduction of parallel transmission. As Section 7.2 remarks, a splitting facility is really just an addition to the message routing scheme — it does not introduce extra problems in the way that its opposite, multiplexing, does. In fact, the opposite may be true, since splitting might help to reduce congestion by spreading loading more widely.

8.2 INTER-NETWORK SWITCHING

Inter-networks are distinguished from networks by the fact that there must be some inconsistencies between the networks involved. This is what adds the need for extra functions in the switching task. Simpler special cases will not be regarded as inter-networks here. For example, in some message broadcasting networks, such as ethernet, a switching device called a **repeater** can be used to connect two physical broadcast channels into a single broadcast channel. The effect is to produce an extended network rather than an inter-network. As another example, if two message switching networks of the same type, and with non-overlapping identifier sets, are connected by a switch, again the effect is to produce an extended network. The arrangement in a switched ethernet combines aspects of both these examples: the central switch is just concerned with creating one unified network.

Given that different networks are often under different management, it is not always straightforward to implement switches as single computers attached to more than one network. For this reason, **half-switches** may be used instead, as shown in Figure 8.2. A switch between two networks is implemented as two computers acting as half-switches, one connected to one network, the other connected to the other network, with a direct physical connection between the two computers. This gives a clear separation of responsibility between the two network managements. It does not make any difference to the overall functions of the switch.

The fundamental feature of an inter-network is that it enlarges the collection of computers that can be involved in communications. As already mentioned, this means that there must be a scheme for allocating unique identifiers to the computers in the inter-network. As a minimal requirement, an inter-network must have an agreement on the interpretation of computer identifiers carried in messages. Then, one role of switches is to interpret these inter-network identifiers in terms of network identifiers, in order to ensure that valid routing and delivery of messages is carried out.

When a message is being forwarded by a switch, the choice of network to be used for the next hop is determined by the inter-network destination identifier of the message. However, the destination point on the next network must be specified by an appropriate identifier for that network. In particular, if the final destination of the message is on that network, then the network identifier will be the one that corresponds to the same computer as the inter-network destination identifier.

Unless the inter-network is so uniform and centrally-controlled that each computer has the same network and inter-network identifier, standard network message formats will not have a ready place in which to carry the extra identifier. In general, this is an issue for any inter-network protocols that might be needed: intervening networks are not likely to all have message formats that support these protocols. However, there is no major problem. As a constructed network itself, the inter-network needs its own message formats. Then, inter-network messages are encapsulated inside network messages to be transported across networks. This

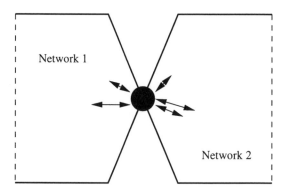

(a) Single switch

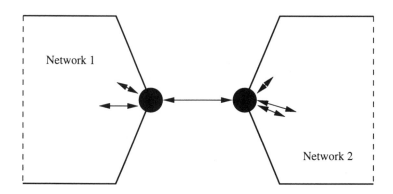

(b) Two half-switches connected by channel

Figure 8.2 A switch implemented as two half-switches

is analogous to the way that network messages may need to be encapsulated for transportation across network channels.

The role of a switch is to deal with receiving and unloading messages from one network, and then appropriately loading and transmitting messages for the next network en route. Terminology is used in various ways by different people, but a switch of this sort can be called a **router** (putting the emphasis more on the routing side of things) or a **gateway** (putting the emphasis more on the preparation of messages for the new network ahead).

If networks in an inter-network are sufficiently similar that a separate inter-network protocol and its associated message types are not necessary, then life is easier for a switch. Dealing with small protocol differences between two networks is not troublesome. Also, loading and unloading of inter-network messages into and out of network messages can be avoided. Instead, messages can be directly forwarded from network to network, with appropriate simple reformatting if necessary. This is most common when a collection of message broadcasting LANs is linked together into an inter-network. A switch of this sort is called a **bridge**, although as with the terms 'router' and 'gateway', there is no precise agreement on a definition of what constitutes a bridge. The only safe statement to make is that bridges carry out simpler jobs than gateways do.

A further technique has the flavour of both methods described in the previous two paragraphs. Suppose that communication is required between two computers that are in two networks of similar type. However, these networks are not adjacent in the inter-network, so direct bridging is not possible. The protocols used in the source and destination networks can be treated as though they were inter-network protocols, and then these are implemented over intervening networks using these networks' native protocols. Essentially, the effect of a single bridge is mimicked by a pair of half-bridges connected by a channel over an inter-network path. This technique of transparently connecting two networks is an instance of tunnelling, a term introduced in Chapter 7. Here, the channel between the two half-bridges is the tunnel. If one views networks as being mountains, then bridging connects mountains by spanning open space, whereas tunnelling connects mountains by drilling through intervening mountains.

8.3 EXAMPLES OF INTER-NETWORK IMPLEMENTATIONS

In this section, three well-known examples of inter-networking technology are described. The level of service provided by each type is examined, in terms of information, time and space properties, as well as the type of switching used between networks. One example excluded is the ITU-T X.75 'inter-networking' protocol. This is because X.75 is only used to connect together X.25 packet switching networks, the end result being an inter-network that is barely distinguishable from a single larger X.25 network. In practice, X.75 is often used as the internal protocol within the black box that is an X.25 network, which serves to confirm that it is not worth considering here.

8.3.1 IEEE 802 LAN bridging

As mentioned in Chapter 6, the IEEE 802 series of standards is a dominating feature of the local area network scene. The 802.3, 802.4 and 802.5 standards for ethernet, token bus and token ring LANs respectively were designed by different groups with different aims in terms of the LAN service provided and the implementation mechanisms used. A further IEEE 802 standardization effort was concerned with the problem of bridging 802 LANs of the same type, and the problem of bridging

LANs of the above three different types. In the first case, the central problem is one of how to route messages between LANs. In the second case, there are extra problems caused by the various inconsistencies between the various types of LAN. Conceptually, each bridge can be considered as a single computer that is connected to two different LANs. However, the true nature of a bridge might be as two half-bridges connected by some direct communication channel.

Information

There are two information-related problems to be overcome. First, there are small differences in message format between the three types of IEEE 802 guided LAN. Second, there are differences in the allowable minimum and maximum message sizes between the three types. The common components of IEEE 802 messages are the destination and source identifiers, the information content and the cyclic redundancy code. A bridge between two different types of LAN has to reformat messages, retaining the identifiers and information, while inserting and removing other components as appropriate. The reformatting necessitates the recomputation of the CRC. A further detail is that the bits of 802.5 token ring messages are transmitted in the opposite order to those of the other types.

The differences in allowable message size stem from temporal considerations, since they reflect bounds on allowable communication time periods. However, it is convenient to treat this as an information-related issue because there is no time-related solution available to a bridge. The problem is that IEEE 802 bridges are not permitted to segment message communications or to concatenate message communications. The maximum 802.3 message size is 1518 bytes, and the maximum 802.4 message size is 8191 bytes. There is no maximum size for 802.5 but, given standard limits on token holding times, there is an effective maximum size of 5000 bytes. Thus, there is a potential problem when bridging to an 802.3 LAN from a LAN of the other types, or when bridging to an 802.5 LAN from an 802.4 LAN. Given that communications cannot be segmented, over-large messages must be discarded by the bridge, which is a serious shortcoming. There is a minimum message size on 802.3 LANs, but this does not pose a problem for bridging.

Time

There are absolute rate differences between members of the IEEE 802 family of guided LANs. The standard transfer rate for 802.3 and 802.4 is 10 Mbits per second, with newer 802.3 LANs being capable of 100 Mbits per second. The standard rates for 802.5 LANs are 4 or 16 Mbits per second. Thus, a bridge between LANs of different types has to include a buffering capability to cope with messages moving from a faster LAN to a slower LAN. If this buffering runs out, then messages have to be discarded. Buffering may also be needed to cope with different network latencies. The 802.3 latency is affected by the length of contention periods, whereas 802.4 and 802.5 latency is affected by token circulation time.

There are other time-related differences. The 802.5 token ring has an automatic acknowledgement feature, using message bits that are altered as a message circulates round the ring. Although not mentioned in Chapter 6, the 802.4 token bus also has an acknowledgement feature, whereby the token is temporarily lent to an information recipient to allow the transmission of an acknowledgement. These acknowledgement mechanisms cannot be faithfully maintained if bridging to an 802.3 LAN is involved. Even when no 802.3 LANs are involved, there is a problem with acknowledgements being delayed when they are coming from a distant LAN rather than the original LAN.

A further difference is that 802.4 and 802.5 both include message priority mechanisms, although these are not directly compatible with one another and so conversion is necessary. When messages are bridged from a LAN with priorities to one without, their priority indication is lost. When messages are bridged the other way, priorities have to be invented.

Space

The redeeming feature of the IEEE 802 standards is that the same identifier scheme is used for all three, so bridges are not faced with any conversion problems. The destination identifiers on messages can be for single computers, multiple computers or all computers, allowing the use of unicast, multicast or broadcast channels for communication. However, the implementation of these through bridges requires appropriate routing, and this issue dominates the next section on inter-network switching.

Inter-network switching

The various conversions on messages carried out by bridges are covered above, as is the need for bridges to respect different IEEE 802 LAN protocols. Just as the original standards committees managed to devise three variously different types of LAN, leading to these bridging problems, so the bridging standards committees contrived to devise two different types of routing for bridges. One type forms part of a standard applicable to bridging all three types of guided LAN. The other type is only used for bridging 802.5 token ring LANs.

Transparent bridge

The first type of bridge is called a **transparent bridge**, reflecting the fact that its presence should essentially be invisible to the inter-connected LANs and their users. An alternative name, reflecting an important feature of its routing scheme, is **spanning tree bridge**. The bridge receives every message broadcast on the LANs that it is attached to. It then decides whether to forward each message to any other LAN. This is done using a routing table indexed (in a hashed manner) by destination identifier. If the identifier is present in the table, then its entry reveals which one of the bridge's attached LANs should be used for forwarding. Clearly, if this LAN is the same as the one the message arrived on, the bridge need take no further action.

Bridges use an adaptive isolated method for routing. The routing tables in a bridge are constructed dynamically. An entry for a destination computer is only added when a message is seen with a source identifier corresponding to that computer. The LAN from which this message arrived is taken as the one to be used for message forwarding in the other direction. Entries only survive for a few minutes after the last time a message was seen with the appropriate source identifier, to ensure adaptation to changes in configuration of the LANs.

When there is no entry for a particular destination in the routing table, flooding is used. The bridge transmits a copy of the message on all of its attached LANs, except the one the message arrived on. All being well, the bridge will see some kind of response message coming back in the other direction soon after, and will then be able to add a routing table entry.

The above scheme relies on one key property of the connections used between bridges: there is a unique path from any source to any destination. This ensures that when flooding is done, there is no possibility of a message returning to a flooder by following an alternate path back, and then being broadcast once again. The fact that there is only one path to any destination also means that a bridge only needs to have one LAN entry for each destination in its routing tables and, further, that this can be learned from messages travelling in the reverse direction.

The necessary property of the inter-network is achieved by forming a spanning tree of the graph that represents the inter-network. This tree is composed of a subset of the connections between LANs: each LAN in the inter-network is a node or a leaf in the tree, and bridges only allow communication between LANs that are adjacent to each other in the tree. Thus, a unique path exists between any two LANs, and it involves going up the tree towards the root, and then going back down again. The tree is constructed dynamically using a protocol between the bridges. It is reconstructed periodically and also when bridges fail, in order to react to changes in the inter-network. The fact that only the tree is followed means that some potential paths through the inter-network are never used. Thus, there is no benefit in deliberately introducing redundant bridging in the inter-network, except as a precaution against complete failures.

Source routing bridge

The second type of IEEE 802 bridge is called the **source routing bridge**. As its name suggests, it relies on the sources of messages to specify the route that the message is to follow through the inter-network. This requires some extra identifiers to be carried in the message, so that the route can be specified. Each LAN is allocated a unique 12-bit identifier by the inter-network management, and also each bridge is allocated a four-bit identifier that is unique to the LANs to which it is attached.

The source route carried by a message is a sequence of alternating LAN and bridge identifiers. The sequence begins with an identifier for the LAN on which the messages originates, and ends with an identifier for the LAN on which the message terminates. As a message progresses through the inter-network, each bridge that receives it, scans the source route sequence. If it finds the identifer of the LAN on which the message arrived, immediately followed by its own bridge identifier, it forwards the message to the LAN identified by the next element in the source route sequence. This source routing method is distinctly simpler to implement in bridges, compared with the transparent bridge method.

The general problem with source routing is how the routes are constructed by the message sender, to ensure efficient message delivery. In the IEEE standard, special discovery messages are used. These are transmitted by a sender, and then flooded through the inter-network by the bridges. The flooding is restricted to follow a spanning tree, of the sort used in the transparent bridge system. However, note that the spanning tree is only used for this purpose, not for transmission of information-carrying messages. The flooding ensures that at least one copy of the discovery message will reach the required destination.

The destination computer then sends back a reply, which is completely flooded back to the original sender. Bridges attach their identifier when such a reply message arrives, and then forward a copy of the extended message on all of their attached LANs (not just those in the spanning tree). The original sender eventually receives one or more copies of the reply message, each with a different suggested route attached, and it can then choose the best one. To avoid the overheads of this discovery process each time a message is sent, the computer stores the route details in a source routing table.

The main advantage of the source routing scheme is that it is able to find optimal routes through the inter-network. However, this is at the cost of introducing manual configuration of LAN and bridge identifiers, making communicating computers responsible for routing matters, and loading the network with flooded discovery and response messages.

8.3.2 The Internet

The Internet has become famous, as the first piece of computer communications technology to emerge from the technical backstage into the spotlight of widespread publicity. Very often, it is not recognized as the inter-network that it is, but is confused with the services that it supports for end users. This is not surprising, since the rise of the World Wide Web as the first friendly face of computer communications shown to non-specialists has been largely responsible for general interest in the Internet. Here, however, there will be no confusion. The Internet is an example of a worldwide inter-network, and is no more than that. The pervasiveness and importance of the Internet means that its features are worthy of a deeper treatment than most of the other examples in this book.

The Internet is designed to connect together any types of network. Thus, it has its own inter-network protocol: the Internet Protocol (IP). Messages for this protocol are transported using the services provided by networks. IP is in transition from one version to another. The old IP has been in use since the late 1970s, and has underpinned the explosive growth of the Internet. However, this unanticipated growth, together with unanticipated applications and unanticipated technologies, led to technical difficulties with this IP. Therefore, in the mid 1990s, a new 'next generation' IP was developed, and its specifications were standardized at the end of 1995.

The new IP retains the overall ethos of its predecessor, but has some notable diferences. The description below covers both versions of IP, and indicates why changes were necessary. For historic reasons, the old IP is regarded as 'version 4' and the new version is regarded as 'version 6' (version number 5 had already been allocated to a rather different non-IP protocol). Therefore, to distinguish the two, old IP is referred to as IPv4 below, and new IP is referred to as IPv6. The transition from IPv4 to IPv6 is expected to take at least a decade. However, IPv6 has been designed so that a gradual migration is possible. The Internet will evolve from one protocol to another, as component parts move from IPv4 to IPv6. This is consistent with the very distributed philosophy of the Internet.

The Internet Protocol is used to provide the basic service of the Internet: transferring information in messages. Another protocol is a required part of the IP mechanism. This is the Internet Control Message Protocol (ICMP), and it is used to allow the transmission of control or error messages relevant to the operation of the Internet itself. ICMP messages are sent using IP as the transport mechanism, rather than using a separate mechanism. Thus, IP is indeed the heart of the Internet provision. With the advent of IPv6, ICMP has also been changed in step, both to reflect changes in IP and also to rationalize ICMP itself. Thus, it is referred to as ICMPv4 or ICMPv6 below, corresponding to its version 4 and version 6 variants.

As concern has risen about security hazards in having the globally open communication mechanism that is the Internet, the term **intranet** has started to be used very frequently. This does not refer to a single network, or something inside a network. Rather, it is used to refer to a private inter-network that is restricted to a single organization. It supplies all the facilities associated with the Internet, *except* that it only involves a smallish collection of computers, rather than the global collection of computers available to participants in Internet communications. An intranet may be connected to the Internet. However, this is via a switch with extra security features, to restrict the types of communications to only those approved by the intranet operator. Such a switch is called a **firewall** — a term that reflects the aim of protecting the intranet from damage.

Information

The IPv4 and IPv6 message formats have already been presented as two of the classics introduced in Chapter 2. In the language of IP, messages are usually

Version number	4 bits
Header length	3 bits
Service type	1 byte
Total length	2 bytes
Identification number	2 bytes
Flags	3 bits
Fragment offset	13 bits
Time to live	1 byte
Protocol	1 byte
Header checksum	2 bytes
Source identifier	4 bytes
Destination identifier	4 bytes
Optional header extension(s)	
Information content	

Figure 8.3 Format of IP version 4 message

referred to as 'datagrams' (or sometimes just as 'packets'). Figure 8.3 shows the basic IPv4 message format again, in a slightly less bit-dependent form, and with names attached to the various fields so that they can be explained here. Figure 8.4 shows the basic IPv6 message format, and allows an immediate visual comparison of the two types to be made. The IPv4 message has a 20-byte basic header, whereas the IPv6 message has a 40-byte basic header. The only consistent feature between the two formats is the four-bit field at the beginning: the version number. This is equal to 4 for IPv4 and 6 for IPv6, and so allows messages of the two types to be differentiated when both types are present in the Internet at the same time.

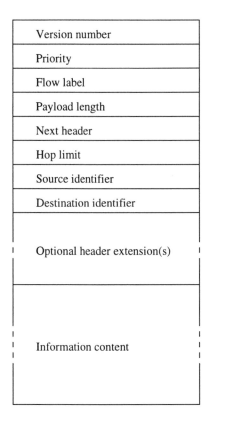

Version number	4 bits
Priority	4 bits
Flow label	3 bytes
Payload length	2 bytes
Next header	1 byte
Hop limit	1 byte
Source identifier	16 bytes
Destination identifier	16 bytes
Optional header extension(s)	
Information content	

Figure 8.4 Format of IP version 6 message

Two components of both IPv4 and IPv6 messages are a source identifier and a destination identifier; indeed, this is a common feature of all messages used in a connectionless environment. The difference is that IPv4 identifiers are four bytes long, whereas IPv6 identifiers are 16 bytes long. This was one of the major technical changes, and the reasons for it are discussed below as a space issue. Another component that the two versions have in common is a length field. However, again this differs between IPv4 and IPv6. In IPv4, there is a 'total length' field, which contains the total length of the message including the header. In IPv6 there is a 'payload length' field, which contains the length of the message, minus the 40 bytes of its header.

So far, only the always-present IP message headers have been illustrated. In both IPv4 and IPv6, there are optional extensions to the header, which can contain other control information before the actual information content of the message. In IPv6, some of the old fixed IPv4 header components have become optional components in IPv6. The way in which header extensions are added to the fixed header differs between IPv4 and IPv6.

In IPv4, there are six types of component that can be added as extensions to message headers, only four of which are actually in use in the global Internet:

- strict source routing;
- loose source routing;
- record route; and
- Internet timestamp.

The different types of extension have different lengths, and particular types may have variable lengths as well.

In all cases, the first byte of the component indicates which type of extension it is. For the four types of interest, the second byte gives the total length of the component. When more than one extension component is present, they are contiguous in the message, following the fixed header and before the information content. In the fixed part of the message header, there is a 'header length' field, which gives the total length (in 32-bit units) of the header, including any extension components. Given that this forces the header length to be a multiple of four bytes, the final extension component (if there is one) is padded with zeros if necessary.

In IPv6, there are also six types of allowable extra components, one of which (routing) overlaps with two of the IPv4 types (strict source routing and loose source routing), the others having different functions:

- hop-by-hop options;
- routing;
- fragmentation;
- authentication;
- security; and
- destination options (unused in the initial IPv6 specification).

The different types of extension have different lengths, and particular types may have variable lengths. However, all lengths must be a multiple of eight bytes.

Unlike IPv4, there is not a distinctive byte at the beginning of each extension, to indicate its type. Nor is there a header length field in the fixed part of the header. Instead, a one-byte 'next header' field appears both in the fixed header and in each extension component. This field gives the message data structure the flavour of a linked list of components, since it indicates what comes next. The next item can either be a header extension component or the beginning of the information component. The exact mechanism involves a unification of IPv4's use of header extension identifiers with its use of a 'protocol' field in the fixed part of the message header.

In IPv4, it was assumed that the information content of an IP message would be a message belonging to some protocol that was using IP as its inter-network transport mechanism. The protocol field of each IPv4 message indicates which user protocol was responsible for the message. Possible values are drawn from a large list maintained centrally, the two most frequent being the Internet's own

Transmission Control Protocol (TCP), which has value 6, and its Universal Datagram Protocol (UDP), which has value 17.

In IPv6, each of the possible header extensions is deemed to belong to a different little sub-protocol of IPv6, and each has a centrally allocated protocol number. For example, the authentication extension is allocated value 51. Then, the idea is that an IP message consists of the fixed length header, followed by zero or more messages for these IPv6 sub-protocols, followed by a message for a user protocol. The fixed header's next header field then has the same role as IPv4's protocol field, as does each extension component's next header field. This uniformity allows simplification of message handling by switches.

The fixed part of the IPv4 header contains a 16-bit header checksum, which is the standard Internet summation code described on page 58 in Chapter 2. It is computed by treating the header as a sequence of 16-integer integers, summing them using one's complement arithmetic, and taking the one's complement of the result. IPv6 does not have such a checksum, since it was felt that detection of errors at the IP message level was not necessary, and so the computation and checking of header checksums was a waste of valuable message handling time.

In IPv4, the maximum size of a message is 65 535 bytes, which is forced by the fact that the length field in the header occupies only 16 bits. This is almost the case in IPv6 too, where the maximum size is 65 574 bytes, since the fixed-length header is not included in the payload length field. However, IPv6 also has a mechanism to allow larger messages, thus offering a service of interest to high speed computers transmitting information over high speed networks. One use of the hop-by-hop options header extension (in fact, the only use in the initial specification of IPv6) is to carry a 32-bit 'jumbo payload length'. A message making use of this feature is termed a **jumbogram**.

In practice, the IP message sizes actually used are much smaller than the maximum allowed, since it is desirable for each message to be directly transportable by the networks it must cross. A requirement for both IPv4 and IPv6 is that all inter-network switches and computers must be able to deal with messages of size up to 576 bytes long. This may still be larger than the limit for particular networks, for example, the typical limit of 128 bytes in X.25 packet switching networks. Where IP messages are too large to be transported as single units, their communication must be segmented. The mechanism for doing this is discussed further in the next section, on page 288.

An important use of IP messages is for carrying ICMP messages. This is denoted by the protocol field having the value 1 for ICMPv4 and the value 58 for ICMPv6. Figure 8.5 shows the general format that applies to both ICMPv4 and ICMPv6 messages. The first byte indicates which type of message it is, the values used being different for ICMPv4 and ICMPv6. The second byte is a code value that gives extra information about the message type. The third and fourth bytes contain a 16-bit checksum over the ICMP message. This checksum is a standard Internet checksum, as used in IPv4. In ICMPv4, the checksum is computed only over the ICMP message itself whereas, in ICMPv6, the checksum is computed over the

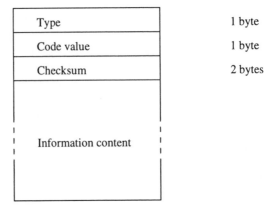

Type	1 byte
Code value	1 byte
Checksum	2 bytes
Information content	

Figure 8.5 General format of ICMP message

IPv6 header as well. This compensates for the fact that the IPv6 header is not checksummed separately. The various types of ICMP message will be described in context, when discussing time and space issues in later sections.

IP and security

Two issues that were not addressed in the original Internet, and therefore not addressed in IPv4, were authentication and privacy. The source identifier field of each IPv4 message was assumed to be a reliable indication of which computer actually sent the message. Further, the contents of any message were open to inspection by any point on the Internet that it passed through. These features were not regarded as desirable as the Internet evolved into an infrastructure to underpin sensitive activities, such as banking and commerce (the military had already made their own arrangements in this area). IPv6 has additions to address these two problems. Both are implemented using header extension components, so that they are not an unnecessary inefficiency in handling messages that do not require security precautions.

The authentication header extension component of IPv6 messages contains a 32-bit Security Association Identifier (SAID) together with variable-length authentication data. The destination computer of a message uses the SAID, together with the message source computer identifier, to identify a pre-established **security association** with the source computer. This includes finding out which authentication method is being used for the security association. In the default authentication provision, the 32-bit SAID is used as an index to look up a secret key at the receiver. Using this key, a computation is carried out using the MD5 (Message Digest 5) algorithm, mentioned in Chapter 2, applied to the contents of the message. The 128-bit result is compared with the authentication data supplied along with the SAID. If they are the same, the message is deemed authentic because it is reasonable to assume that no imposter could have computed the appropriate 128-bit value without knowledge of the secret key.

The security header extension component of IPv6 messges begins with a 32-bit security association identifer, which plays a similar role to the one in the authentication header. In the default security provision, the SAID is used as an index to look up a secret key, which is used for Data Encryption Standard (DES) Chain Block Cipher (CBC) decryption. The SAID is optionally followed by a variable-length initialization vector for the decryption process. In the case of DES CBC, this is a 64-bit vector. After the SAID and the initialization vector, all the rest of the security header, and the rest of the IPv6 message, is encrypted. If it is necessary to send IPv6 messages that are completely encrypted, i.e., including *all* of the header fields, then they can be encapsulated as messages inside IPv6 messages that have security headers.

The use made of other IP message header components not mentioned so far, both in the fixed header and in optional extensions, will be covered in the following discussions of time or space issues, as appropriate.

Time

No absolute time bounds are associated with the service offered by the Internet. Two communicating computers might be located on the same desk, connected by a high speed LAN. Alternatively, two communicating computers might be located in different continents, connected via tens of switches and dubious physical channels such as telephone lines. For any particular communication, the transmission rate and latency are of interest, but they can only be determined by experimentation or by detailed knowledge of how the communication is implemented. Thus, the only service guarantee associated with the Internet is one of 'best effort' in delivering messages promptly.

Priorities and flows in IP

The Internet Protocol incorporates a notion of priority for messages. In IPv4, the 'service type' byte in the fixed header includes a three-bit 'precedence' field. This has eight possible values, ranging from the lowest value 0, which is normal precedence, to the highest value 7, which is network control. In practice, however, the setting of this field is ignored by inter-network switches, and so it has no effect.

In IPv6, there is a 4-bit 'priority' field in the fixed header. Values in the range 0 to 7 are used for traffic which can be controlled by its sender when congestion occurs, and values in the range 8 to 15 are used for traffic which cannot, for example, continuous real-time traffic. In the former range, the increasing priorities reflect the need for better responses times. Thus, priority 2 is recommended for background electronic mail, priority 4 for attended file transfers and priority 6 for interactive traffic. In the latter range, the increasing priorities reflect the importance of messages not being discarded. For example, priority 8 is recommended for high-fidelity video traffic and priority 15 for low-fidelity audio traffic. This seems counter-intuitive at first sight, but is because far less information is communicated in the low-fidelity case, and so this communication is far more susceptible to the loss of any information.

The two different priority schemes in IPv4 and IPv6 are related to other features affecting message flow. In IPv4, the service type byte also contains three one-bit fields: D, T and R. When set, these indicate that the message requires low delay, high throughput and high reliability respectively. It is recommended that no more than two of these bits should be set simultaneously. However, this is largely academic, since these bits, like the precedence field, are ignored by most inter-network switches.

In IPv6, there is a 24-bit 'flow label' field. This is related to the idea of flow specifications being given when establishing connections over high speed networks. The field is used as a sort of connection identifer although, strictly speaking, IP only supports a connectionless service. The idea is that a sequence of messages flowing from a source to a destination all carry the same flow identifier, and it can be recognized by the switches *en route*. In the initial specification of IPv6, no mechanism for how this field is utilized had been decided upon, but a suggestion is that facilities will be provided to allow communicating computers to specify resources required for a particular flow. Resources include such things as low delay, high throughput or high reliability.

The notion of flows in IPv6 adds support for a connection-oriented style of service. However, the standard service offered by the Internet Protocol is a 'best effort' connectionless datagram delivery service. Thus, communication time periods consist of IP message communication times. The duration of these periods depends on the message length, as well as the (unknown) characteristics of the communication path followed by the message. As already mentioned above, there is a maximum message length forced by the 16-bit size of the IP message length field, with the optional exception of IPv6 jumbograms.

IP fragmentation and reassembly

If a message is too large to be forwarded through the Internet, it is necessary for its communication to be segmented into several sub-communications of shorter messages. For IP, this process is called **fragmentation** and **reassembly**. Fragmentation is the dividing up of one message into several fragments. It may be carried out by a source computer or, in IPv4 only, by an intermediate switch. Reassembly is the recombining of fragments into a message. It is performed only by destination computers, not by any intermediate switches. An absolute time constraint is placed on the reassembly process: all of the sub-communications must be completed within 60 seconds of the first sub-communication being observed at the destination computer.

The fragmentation and reassembly processes of IPv4 and IPv6 are similar, but there are detail differences. The fixed header of IPv4 messages has a 16-bit 'identification' field, a three-bit 'flags' field and a 13-bit 'fragmentation offset' field. These are used when a message is fragmented, in order to ensure that it is reconstructed correctly later. Each message travelling between a particular source and destination carries a unique value in its identification field. If the source computer just uses a 16-bit counter to generate the identifier, there is no problem with clashes assuming that one message is not still in transit between the two computers when its identifer is reused 65 536 messages later — a reasonable assumption in the IPv4 Internet. When a message is fragmented, its identifier is copied into all of the fragments.

In each fragment, the fragment offset field specifies the offset, measured in units of eight bytes, of the fragment message's information within the original message's information. Thus, the first fragment's offset field is set to zero, for example. The 'more fragments' bit in the flags field is used to indicate the final fragment of the original message. This bit is set to zero in the final fragment, and to one in all the preceding fragments. This extra bit is necessary, since the destination computer does not know the length of the original message, because each fragment only carries its own length in the total length field. Figure 8.6 shows an example of a message broken into three fragments. There is also a 'do not fragment' bit in the flags field — if this is set by a message sender, intermediate switches are not allowed to fragment the message. In such cases, if a switch receives a message that is too large to forward, it sends back an ICMP 'destination unreachable' type error message, with the further information that this was because fragmentation was needed.

In IPv6, fragmentation and reassembly is catered for using a header extension, rather than fields in the fixed header. Further, as already stated, only the source computer is allowed to fragment messages, so the effect is similar to always setting the IPv4 'do not fragment' bit. If an intermediate switch receives a message that is too large, it sends an ICMP 'message too big' type error message back to the source computer. This message contains the maximum message length that can be accommodated by the next hop needed in the message's journey.

Advised by this message, the source computer can perform fragmentation of over-large messages. This mechanism can also be used as part of a discovery process by computers, to establish the maximum message size that can be sent to a destination, prior to attempting to send real information. This type of discovery process in not new with IPv6 — it was pioneered with IPv4, making use of its 'do not fragment' facility. However, the IPv4 version is more complex, since the ICMPv4 error report does not indicate what the message size limitation was.

The fragmentation header extension has the same components as the IPv4 fixed header has for fragmentation use. The only difference is that the identification field is 32 bits long, rather than 16 bits long. This allows scope for there to be many more messages in transit between two points simultaneously, which is possible when high speed networking is used over long distances. Unlike IPv4, IPv6 sends ICMP error messages to a message's source if there are problems during reassembly: if all fragments do not arrive within the 60-second time limit,

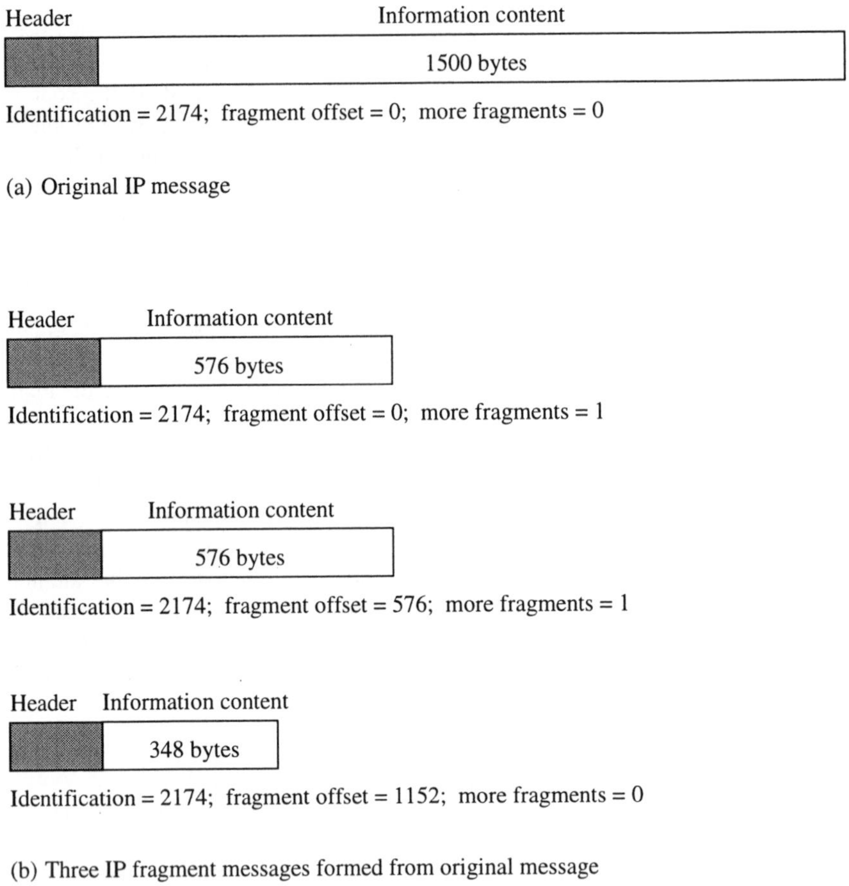

Header   Information content

1500 bytes

Identification = 2174; fragment offset = 0; more fragments = 0

(a) Original IP message

Header   Information content

576 bytes

Identification = 2174; fragment offset = 0; more fragments = 1

Header   Information content

576 bytes

Identification = 2174; fragment offset = 576; more fragments = 1

Header   Information content

348 bytes

Identification = 2174; fragment offset = 1152; more fragments = 0

(b) Three IP fragment messages formed from original message

Figure 8.6 Example of IP message fragmentation

if fragments have invalid lengths or if the reassembled message has a payload length longer than 65 535 bytes.

IP expiry times

There is one more IPv4 feature that might be regarded as time related. This is the eight-bit 'time to live' field in the fixed header. This specifies how long, in seconds, the message is allowed to survive in the Internet, and so places an absolute bound on the communication time period. In fact, it is hard to implement this feature accurately, since a switch receiving a message does not know how long a message has spent being transmitted from the previous switch. Thus, the rule is that the field in decremented by one each time it passes through a switch — this makes it more a space issue than a time issue — but it is also decremented for each second it is queued in a switch awaiting onward transmission. If the time to live field becomes zero at a switch, the message is discarded, and an ICMPv4 'time exceeded' message is sent back to the message source.

IPv6 has a similar mechanism, using an eight-bit field in its messages. However, it calls the field the 'hop limit' and uses it solely to count hops between switches rather than attempt to measure an expiry time. If the hop limit field reaches zero, an ICMPv6 'time exceeded' message is sent back, so the time-related theme is maintained, in name at least.

IP and flow control

IP supports a connectionless service, so there is no context in which flow control can be performed. Flow control between communicating computers is the province of the connection-oriented Transport Control Protocol (TCP) that is very frequently used on top of IP. However, with IPv4, there is a mechanism for reporting, and reacting to, congestion. If an IPv4 message is discarded by a switch because of congestion, an ICMPv4 'source quench' message is sent back to the message source. As the name suggests, when a source receives such a message, it is meant to reduce the rate at which it is transmitting messages to the destination of the message that was discarded. There is no ICMPv4 mechanism to reverse a source quench. Instead, computers gradually start increasing the message rate again once source quench messages stop arriving.

There is no such mechanism in IPv6. When messages are discarded, an ICMPv6 message must not be sent back. This reflects the move to higher speed networks where, during the time taken to send back a source quench message, large numbers of new messages will still be heading towards the congested location. A flow specification agreed in advance is necessary to avoid this problem.

Space

The Internet allows both unicast and multicast communications to take place. IPv6 also adds the capability to perform anycast communications: communication between a computer and just one of a set of computers. The type of communication is denoted by the destination identifier field carried by a message. For unicasts, the destination identifier is just the Internet identifier of the receiving computer. For multicasts, the destination identifier is the unique identifer allocated to the group of computers involved in multicasting or anycasting. There are some group identifers permanently defined, but most groups are created and maintained dynamically. For IPv4, this is done using a separate protocol, the Internet Group Management Protocol (IGMP). Like ICMP, IGMP is an integral part of the IP system, and IGMP messages are transported within IP messages. For IPv6, given that multicasting had become well established on the Internet, IGMP was removed and, instead, ICMPv6 includes the functions of IGMP in addition to its other functions. For anycasting, normal unicast identifiers are used to specify anycast groups of computers.

IP *identifiers*

The computer identifier scheme used in the global IPv4 Internet is introduced in Section 5.2.1. 'Dotted decimal' notation is used when writing down 32-bit IPv4 identifiers for human reading: this consists of the four bytes of the identifier, in decimal, written from most significant to least significant, left to right, separated by dots. For example, the identifier **11010100 00010011 10111010 00001111** in binary is written as '212.19.186.15'.

In Section 5.2.1, Class A, Class B and Class C identifiers, used for unicasting, are described. These identifier types consist of a network number, followed by a computer within network number. This two-level hierarchy in the identifier scheme reflects the physical fact that the Internet is composed of a collection of inter-connected computer networks. Class D identifiers, which begin with the bit pattern **1110**, are used for multicasts, the remaining 28 bits being a multicast group identifier. Class E identifiers, which begin with **1111**, have not been used in service.

A third level of hierarchy can be imposed within the Class A, B and C identifier scheme if required, to split a network's identifier space into a set of sub-network identifier spaces. From the point of view of routing IP messages, sub-networks can be treated as separate entities, rather than being viewed as all part of a single point on the Internet. The splitting up of networks is done using a **subnet mask**, that specifies which bits in identifiers constitute a network plus sub-network identifier. In practice, subnet masks are used to specify a contiguous group of bits at the beginning of the 32-bit identifiers, so the effect is to indicate that some of the most significant bits of each computer within network number are to be treated as a sub-network within network number.

For example, if a Class C network has identifier

$$195.22.132.0$$

and a subnet mask

$$255.255.255.192$$

then, noting that 255 is **11111111** in binary and 192 is **11000000** in binary, the subnet mask means that the first 26 bits constitute a network plus sub-network number. Since Class C addresses have a 24-bit network number, this means that there is a two-bit sub-network within network number. This leaves six bits for the computer within sub-network number.

In the global Internet, network identifiers are allocated centrally. Once a network identifier has been allocated uniquely, individual network operators can allocate the within-network identifiers to specific computers. Certain Class A, B or C identifiers have a special significance, which allows useful communication spaces, some of them multicast, to be used without knowledge of specific Internet identifiers. These are shown in Table 8.1. From this, it can be seen that eight-bit numbers equal to 0 or 255 have a special significance. In summary, 0 means 'this'

Table 8.1 IPv4 identifiers with special significance

Identifier(s)	Meaning
0.0.0.0	this computer
$0.0.0.c_1$	computer c_1 on this Class A network
$0.0.c_1.c_2$	computer $c_1.c_2$ on this Class B network
$0.c_1.c_2.c_3$	computer $c_1.c_2.c_3$ on this Class C network
127.*.*.*	loopback within this computer
255.255.255.255	all computers on this network
n_1.255.255.255	all computers on Class A network n_1
$n_1.n_2$.255.255	all computers on Class B network $n_1.n_2$
$n_1.n_2.n_3$.255	all computers on Class C network $n_1.n_2.n_3$

and 255 means 'all'. There are also pre-defined multicast identifiers for various purposes. One example is 224.0.0.1, which is used to mean 'all computers attached to this LAN' — not necessarily the same set as all those with this Internet network number.

The fundamental problem with the IPv4 identifier scheme is that it was designed without prior knowledge of the actual growth pattern of the Internet. This resulted in the available Class B network numbers being rapidly used up. This was because many network operators were concerned that the 256 computer numbers available with a Class C network number (actually 254, excluding 0 and 255) would not be enough. Therefore, they requested Class B numbers, although their networks were never likely to have anything like 65 536 computers attached.

One emergency solution adopted to the problem was to issue appropriately sized blocks of adjacent Class C numbers to network operators, rather than one over-large Class B identifier range. This was part of the CIDR (pronounced 'cider', and short for Classless Inter-Domain Routing) standard, which is mentioned further in the discussion of IP routing below.

However, this was no more than a temporary fix for a problem that was going to grow in the future, as the Internet continued to expand. Thus, IPv6 was designed with a certain amount of urgency, so that its vastly increased identifier range could be introduced before the IPv4 identifier scheme collapsed completely. IPv6 identifiers are 128 bits long — four times the length of IPv4 identifiers. The actual choice was a compromise between various factions. Other suggestions were more modest 64-bit identifiers, more lavish 160-bit identifiers consistent with ISO standard NSAP identifiers, and also variable-length identifiers. IPv6 identifiers have a fixed length meant to be bullet-proof against future expansion (in theory, each atom on the surface of the Earth could be given a unique identifier) but which does not impose an excessive overhead on message size or message handling time.

Table 8.2 IPv6 identifier prefixes and their significance

Identifier prefix	Meaning
0:0:0:0:0:0	IPv4-compatible IPv6 identifiers (in remaining 32 bits)
0:0:0:0:0:FFFF	existing IPv4 identifiers (in remaining 32 bits)
0:2 and 0:3	ISO standard NSAP identifiers
0:4 and 0:5	Novell Netware IPX identifiers
4 and 5	global provider-based identifiers
8 and 9	geographic-based identifiers
F:E:8 to F:E:F	local use identifiers
F:F	multicast identifiers

There is also one other subtle change to the identifier scheme for IPv6. Identifiers no longer refer to computers and switches, but instead refer to *network interfaces* on computers and switches. Thus, if a computer or switch has more than one interface to the Internet, it has more than one IPv6 identifier. In fact, it is normal practice to do this with IPv4 also, but the IPv6 specification made the practice explicit.

A new written notation was devised for IPv6 identifiers, since extension of the IPv4 dotted decimal notation would be rather clumsy. IPv6 identifiers are written as eight groups of 16 bits. Each group is written as a four-digit hexadecimal number, with a colon between each number. To make this more manageable, it is possible to omit leading zeros in hex numbers, and also to omit one sequence of four-zero hex numbers completely. As an example,

$$1080:0000:0000:0000:0008:0800:200C:417A$$

can be written as

$$1080::8:800:200C:417A.$$

The loopback identifier 0:0:0:0:0:0:0:1 can be written as just ::1, and the 'unspecified' identifier 0:0:0:0:0:0:0:0 can be written as just ::.

Initially, much of the available IPv6 identifier space (85%, in fact) was unassigned, reserved for future developments. Table 8.2 shows the identifier ranges that were assigned, listed by their coloned hexadecimal prefixes.: The first four categories shown incorporate existing identifier schemes within the IPv6 scheme. Inclusion of the IPv4 identifier scheme is particularly necessary, to allow comfortable migration from IPv4 to IPv6 in the Internet. IPv6 equipment interacting with IPv4 equipment can be identified by the first type of identifier, and IPv4 equipment without IPv6 capabilities can be identified with the second type of identifier.

Provider-based identifiers allow for an Internet future dominated by various companies acting as global Internet providers to customers. This type of identifier has a hierarchical identifier scheme, with a few registries for geographical regions at the top, then providers within these regions, then customers, etc.

Geographic-based identifiers allow for an Internet future dominated by geography and nationality. The use of this type of identifier was not specified initially, but this type of identifier could have a hierarchy identifier scheme based on countries, regions of countries, towns, etc.

The local use identifiers are for use on intranets. Since the identifiers are not used in the Internet itself, there is no need for uniqueness outside individual intranets, and so no need for any form of central allocation.

All of the above types are unicast identifiers. Anycast identifiers are also allocated from this range, and are indistinguishable from unicast identifiers. An anycast message is always sent to the nearest computer in the anycast group, according to routing distance. The idea is that, within the smallest Internet region containing all of the group, all members advertise routes to the anycast identifier. Outside this region, only one route to the identifier is advertised. A main perceived use for anycasts is to assist computers in finding the nearest server of some type. All available servers would be part of an anycast group, and an anycast communication will go to the nearest.

Multicast identifiers are an extension to the IPv4 scheme. There is a 112-bit group identifer, qualified by a one-bit flag to indicate whether the group is permanent with a centrally allocated number, or is transient. There is also a scope field used to restrict the range of multicasts: these can be local to a computer, a network, a site, an organization or the whole planet. There are no separate broadcast identifiers in IPv6, since this is just a special case of multicasting.

Mobility and IP identifiers

The Internet is truly worldwide, spanning all continents, including Antarctica. The fact that the unicast identifier scheme in both IPv4 and IPv6 has a hierarchical structure based on identifiying networks, then computers within networks, presents a problem with the rise of mobile computing. It is no longer the case that all computers are quasi-permanently connected to one network, perhaps moving to another network occasionally, at which point they can be given a new Internet identifier. Mobile computers should be connected to the nearest available network in the Internet as they are moved around from place to place. It is desirable for computers' identifiers not to change frequently, otherwise it is hard for other computers to communicate with them, given the lack of continually up-to-date maps of the Internet. However, this is exactly what would happen if a mobile computer acquires a new identifier each time it changes network.

To deal with this problem, an alternative scheme has been devised. The idea is that a mobile computer has a **home network**, and its identifier is within the range allocated to that network. When the computer is attached to a different network, it registers itself as being present. This **foreign network** then informs the home network of the computer's presence and, from then on, the home network forwards messages for the mobile computer to the foreign network.

For this to work, both the home and foreign networks must have computers available that are prepared to help out with the process. The foreign network computer finds out about new arrivals by periodically broadcasting an advertisement of its services. When a visiting computer responds, the foreign network computer tells the visitor's home network computer. The home network computer then arranges to receive all messages addressed to the mobile computer. Any such message is then sent to the mobile computer over an IP tunnel to the foreign network computer. The foreign network computer finally forwards the message to its rightful owner. The overall scheme allows support for mobile computing without making any changes to the basic IP fabric of the Internet.

Names corresponding to IP identifiers

IPv4 identifiers were pretty unmanageable for humans to use, as well as being rather specific about which network a computer is on. IPv6 identifiers are even more cumbersome for humans. Luckily, people are rescued from having to know numerical identifiers. In addition, higher level applications can be isolated from the harsh reality of a network-plus-computer based identifier scheme.

The saviour is the Domain Name System (DNS), which supports a directory service. DNS provides a distributed mechanism for mapping textual names, organized in a hierarchical identifier scheme, to and from Internet identifiers, as well as doing some other types of name services. DNS is implemented using the basic IP service to transmit queries to, and responses from, name servers. Note that DNS is not a necessary component of the IP service, unlike ICMP.

The style of names used is familiar to anyone who has ever used the Internet. Two examples are `lundy.dcs.pop.ac.uk` and `gorilla.ibm.com`. The hierarchy applies top to bottom, from right to left in the dot-separated string forming the name. Each country has a top-level entry in the hierarchy, denoted by its international standard country code (UK is a small exception, used for historical reasons instead of GB, the code for Great Britain). In addition, there are a few other non-country generic top-level entries: COM (commercial), EDU (educational), GOV (US governmental), INT (international), MIL (military), NET (network) and ORG (non-profit organizations). These are mainly used to name computers in the USA, in preference to using the US country code, largely for historical reasons predating the internationalization of the Internet. Below the top level of the hierarchy, the next levels can be defined and populated by the organization responsible for the top-level domain. Delegation can then continue down to lower levels.

Inter-network switching

The Internet Protocol was deliberately designed to make minimal demands on the networks forming the Internet. This was to make it easy to harmonize the widely varying services available. As has been seen from the detailed description of IP, the only service asked of a network is for it to transfer a message from one point to another on a best effort basis. There are no specific requirements for timing or for reliability. As a result, Internet switches are not involved in any major conversion work, even of the type necessary for IEEE 802 bridges which, in principle, one would expect to have an easier task.

One issue is that Internet-wide identifiers have to be mapped to network identifiers in order to transmit the network messages to the right place. The opposite operation may also be necessary, so that computers that know their network identifier can discover their Internet identifier. Switches maintain tables for identifier mapping. In a fairly static network, these tables can be maintained by hand. However, there are standard protocols to assist. The Address Resolution Protocol (ARP) allows a computer or switch to broadcast an IP identifier, and await a response giving the correct network identifier to use. The Reverse Address Resolution Protocol (RARP) allows a switch to broadcast its network identifier, and await a response giving the IP identifier it should use.

Apart from dealing with identifier mapping, the only real network inconsistency problem that switches have to watch out for is messages that are too large to travel over particular networks. Given this, the main function of Internet switches is then the routing of messages towards their destination. For this reason, the switches are usually called 'routers' in the Internet context. It is fair to say that Internet switches probably have the most challenging routing task in worldwide computer communications, given the size and diversity of the Internet. There is no single standard method of routing used in the Internet. However, there are a few prevalent standards, which will be described briefly here.

Source routing in the Internet

A first routing method, already hinted at in the description of IP messages, is source routing. It is possible for a source computer to attach a route to a transmitted message, to specify the route to be followed through the Internet. The route is specified as a list of IP identifiers of the switches to be reached at each hop.

Two types of source routing are supported: strict source routing, where only the switches with identifiers on the route list can be visited, and loose source routing, where it is allowable also to visit other switches in between those with identifiers on the route list. Source routing is not of much use for normal message transmissions, since it demands that the source computer has a detailed knowledge of the physical structure of the Internet. However, it is useful for some network management functions, since it allows specific paths through the Internet to be checked for correct functionality.

In IPv4, strict source routing and loose source routing are two of the types of optional header fields allowed. A message begins its journey with its destination identifier set as the identifier of the first switch it should visit. The source routing header field contains a list of identifiers of the computers the message should visit after the first switch. The final identifier in this list is the ultimate destination of the message. As the message progresses, each identifier in the source routing list is used in turn to determine the next hop for the message. When a switch uses an identifier from the list, it replaces it by its own identifier. Thus, when the message arrives at its destination, the original source routing list has been completely replaced by a list recording the intermediate switches actually visited by the message.

IPv4 also has a recorded route optional header field, which causes a message to accumulate a similar list of switches visited. This option does not involve source routing, however. The timestamp optional header field is a further elaboration of this, which causes a message to accumulate a list of both switch identifiers and timestamps of when the switches were visited. These two options are useful for network investigation and management.

IPv6 has an optional extended header component for routing. This has the capability to support different types of routing, allowing different formats for each type. Initially, there was only Type 0, which is very similar to IPv4 source routing. One obvious difference is that the identifiers in the route list are 128 bits long, rather than 32 bits long. The idea of strict and loose source routing is generalized from IPv4, where the whole route must either be strict or loose. In IPv6, a bit map is carried with the route list. This indicates, for each hop of the route, whether it is strict or loose. That is, whether or not the computer specified by the next identifier on the list must be a direct neighbour of the current computer.

Routing tables in the Internet

While IP has a source routing capability, using it is the exception rather than the rule. Most messages are routed on the basis of routing tables stored in Internet switches. The first point to note is that, with the massive growth of the Internet, routing tables can potentially become huge. It is certainly not practicable to have an entry for each computer on the Internet. It is desirable to have an entry for each network or, better, for each sub-network when subnet identifiers are used within networks. However, this is still not feasible because of the scale of the Internet, where the estimated number of networks had reached six figures by 1996.

With the IPv4 identifier scheme, a belated attempt was made to introduce some hierarchy into network identifiers, in order to avoid massive flat-identifier routing tables. Four ranges of available Class C identifiers were allocated to different regions: Europe, North America, Central and South America, and Asia and the Pacific. This allowed a top-level geographical routing hierarchy.

Further hierarchy was also introduced, to allow blocks of adjacent network identifiers allocated to the same organization to be treated as a whole for routing purposes. This was done in tandem with the scheme to allocate blocks of contiguous Class C network identifiers in place of single Class B network identifiers, already mentioned in the earlier discussion of IPv4 identifiers.

The overall scheme was called CIDR, standing for 'classless inter-domain routing' — classless, because it worked for not just Class C network identifiers, but also for classes A and B. Essentially, CIDR was an emergency provision for IPv4, pending the arrival of IPv6. The design of the IPv6 identifier scheme incorporates the ideas of CIDR, and the future allocation of IPv6 identifiers in a hierarchical manner is designed to protect switches against the need for massive routing tables.

Adaptive routing in the Internet

The size of routing tables is only one problem. The most significant problem is how the contents of routing tables are maintained, in order to allow efficient routing of IP messages through the Internet. The routing of messages within individual message switching networks is not of concern from an Internet point of view. Internet routing concerns the forwarding of messages from one network to another. This is not just a technical problem, but may also be an administrative problem when networks have different owners or are in different countries. Because of this, there are two levels of routing within the Internet.

An **Autonomous System** (AS) is a collection of networks and switches controlled by a single administrative authority. Each AS is centrally registered, and is allocated a unique AS number. When a request is made for a new network identifier to be allocated in the Internet, the AS containing the network must be specified.

Adaptive distributed routing is the norm within the Internet, although not the only type, and two kinds of protocol are used for keeping routing tables up to date. Within a single AS, **interior gateway protocols** (IGPs) are used and, among the ASs, **exterior gateway protocols** (EGPs) are used. The main difference is that an IGP has complete technical freedom, subject to the needs of the AS owner, whereas an EGP has to incorporate political compromise into its operation.

Thus, the main goals of an IGP are finding optimal routes and responding quickly to changes, whereas the main goals of an EGP are finding reasonable routes and providing stability. Note the use of the word 'gateway' in the terms IGP and EGP. This reflects the fact that 'gateway' used to be the preferred Internet term for what is now usually called a 'router'. There is little consistency in such matters.

If an AS is small and changes infrequently, then it is possible for routing tables to be updated manually by a human administrator of the AS. However, this is not practicable for most AS configurations. A variety of IGPs are in use around the Internet. Since each IGP is confined to a single AS, there has never been a need for uniformity. Three examples of IGPs in widespread use are:

- RIP (Routing Information Protocol);
- OSPF (Open Shortest Path First); and
- IGRP (Inter-Gateway Routing Protocol).

These three protocols illustrate the different ways that things propagate in the computer world. RIP originated at Xerox, but gained wide usage because its implementation was distributed free with the Berkeley BSD Unix system, rather than because it is a particularly good IGP. OSPF was designed in the Internet community explicitly as an IGP and made available as an open protocol. Finally, IGRP is a proprietary protocol designed by Cisco, a company which is a major producer of Internet switches.

Both RIP and IGRP use distance vector routing. Periodically, each switch broadcasts its entire routing table to neighbouring switches. On receipt of new information from a neighbour, a switch uses the Bellman-Ford graph algorithm (mentioned on page 157 in Chapter 5) to determine whether any new 'shorter' routes are available via this neighbour. If so, it updates its routing table appropriately. RIP just uses a simple hop count as its measure of 'distance' in determining whether one route is shorter than another. IGRP uses a more complicated metric, that takes delay, capacity, reliability and load into consideration.

In contrast, OSPF uses link state routing, a technique which scales better than distance vector routing. OSPF is now the recommended Internet IGP. Each switch broadcasts information about the reachability of its neighbouring switches periodically. On receipt of such information, a switch uses it to update its own map of the current state of connectivity. This map contains distances between neighbouring switches. Dijkstra's graph algorithm (mentioned on page 157 in Chapter 5) is then used on the map in order to compute shortest paths between non-neighbouring switches.

OSPF has various features reflecting the evolution of the Internet. First, an authentication mechanism is used for OSPF messages, to avoid bogus routing information being promulgated. Second, a large AS can be partitioned into **areas**, with OSPF broadcasts being confined to individual areas; this adds a level of hierarchy to the intra-AS routing. Third, three different types of best routes can be stored: lowest delay, highest throughput and highest reliability. These types correspond to the service types that can be specified in an IPv4 message.

An earlier EGP was the eponymous EGP. This protocol was only appropriate for a tree-structured Internet with a core AS at its root, which was indeed the initial form of the Internet, as it evolved from the ARPAnet network. The protocol that is now recommended is BGP (Border Gateway Protocol), named because it involves communication between switches on the borders of autonomous systems.

In BGP's model, there are three categories of AS, depending on their ability or willingness to assist with Internet routing. A **stub AS** has only a single channel to one other AS, and so can only carry its own local traffic. A **multihomed AS** has channels to more than one other AS, but refuses to carry through traffic between two other ASs. A **transit AS** has channels to more than one AS, and is designed (under certain policy restrictions) to carry both local and through traffic.

BGP uses distance vector routing, like RIP and IGRP, but with a key difference. The information exchanged between neighbours does not just contain the costs associated with the best paths to destinations. It also contains the actual paths, i.e., the sequence of ASs that is traversed. This extra information allows AS routing policies to be enforced. For example, a path is not valid if it involves a multihomed AS somewhere. Alternatively, a path is not valid if the starting AS is not allowed, or does not wish, to route messages through a particular AS that appears on the path. The fact that explicit paths are exchanged also allows switches to respond very quickly to failures. When a switch fails, all paths involving it can be removed immediately. With a basic distance vector scheme, this is not possible and it usually takes some time for news of failures to propagate fully.

Multicast routing in the Internet

For multicast routing, the crucial difference from unicast routing is that switches must forward copies of messages to more than one network at appropriate points. This is to ensure that all members of the multicast receive a copy of the message. The key issue is to decide when the 'appropriate points' are.

In extremis, flooding would be a solution: send copies of messages everywhere and, as in a message broadcasting network, rely on only the members of the multicast group to read the messages. This is a very wasteful solution, for example, a multicast might just involve three computers in the same room. What is actually required is to compute a spanning tree of the multicast group, rooted at the message's source. The leaves of this tree are the switches that give entry into the networks containing computers in the multicast group. The nodes of the tree represent the appropriate points for message duplication.

In the global IPv4 Internet of the mid 1990s, few of the installed switches supported multicasting. Most of the routing and forwarding required was actually done by general-purpose computers that run multicast routing software. These computers and the multicast switches that do exist, together with tunnelled channels between them, constitute the **Multicast Backbone** (MBONE). This has been in existence since 1992. As mentioned at the beginning of Section 7.6, the MBONE is an example of a network implemented using the services of another network: it uses channels provided by the Internet. As switches are upgraded, and as IPv6 spreads, the general-purpose computers will no longer be required to help with multicasting, and the tunnels between MBONE switches will be eliminated in favour of direct channels between Internet switches.

An experimental Distance Vector Multicast Routing Protocol (DVMRP), an extension of RIP, has been used in most of the initial MBONE to find trees for multicast routing. Although described as 'experimental', DVMRP has actually been around since 1988. Other protocols co-exist with DVMRP in the MBONE. In particular, MOSPF (Multicast OSPF) is an extension to OSPF that allows multicast routing trees to be found.

In addition to the OSPF feature of storing complete information about the AS connectivity at each switch, the MOSPF extension involves complete information about membership of multicast groups also being stored at each switch. When a multicast message is received by a switch, it constructs a shortest path spanning tree, rooted at the message's source, for the whole AS, using the stored connectivity information. If the message's source is not in the same AS, or it is in a different area of the same AS, an approximation to a shortest path tree is generated, based on summary information available about the Internet outside the area of the AS.

The switch then prunes all tree branches that do not contain multicast group members, and saves the resulting tree for use with any subsequent messages from the same multicast source. Finally, based on its own position in the tree, the switch forwards the message appropriately.

MOSPF is designed for routing multicasts within one AS. An architecture designed for routing at all levels was developed in the mid 1990s. This is called PIM (Protocol Independent Multicast), and it has two modes. One version is called 'dense mode', and it is similar to DVMRP. It is suitable for use when flooding is a possible routing method *and* when a multicast group is large. It essentially involves broadcasting initially, and then switches that have no multicast group members attached requesting that they be pruned out of the broadcast tree. This results in a final multicasting tree. The other version is called 'sparse mode', for use in other cases. It essentially involves setting up one or more master computers for the group. These then handle requests from switches that wish to be included in the multicast tree.

8.3.3 ISO inter-networking

After the lengthy presentation of the Internet in Section 8.3.2, a reader could be forgiven for thinking that it represents the whole of worldwide inter-networking. However, as well as there being various proprietary inter-networking products, there is also an ISO international inter-networking standard. Like most ISO communications standards, this one has been tremendously eclipsed by the rise of the Internet and its protocols. However, unlike most ISO communications standards, this one is very strongly based on the Internet, rather than being designed by committee or modelled on practice in the telecommunications industry. Because of this, it is possible to present a short summary here, in order to point out the main similarities and differences from the Internet standard.

The ISO inter-networking standard defines a connectionless inter-network service and a protocol to support it, as well as an inter-network identifier scheme and inter-network routing protocols. The inter-network protocol is usually known as **CLNP** (ConnectionLess Network Protocol) or, sometimes, as ISO-IP to reflect its ancestry.

Information

The format of a CLNP message is shown in Figure 8.7. Most of the components of the fixed header are analogous to components of the IPv4 message header. There are destination and source identifiers for the message. CLNP computer identifiers are variable length, so there is a length field for each of the identifiers. There is

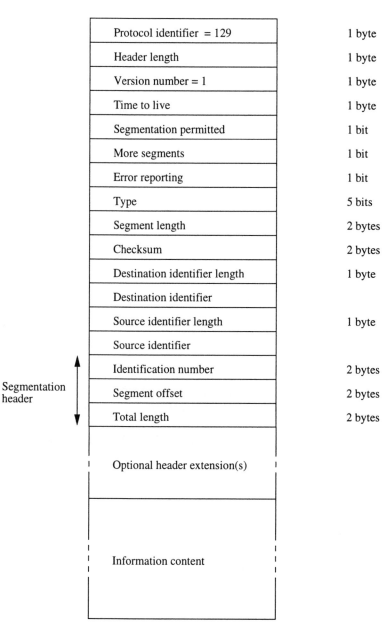

Figure 8.7 Format of ISO CLNP message

an eight-bit header length and also a 16-bit message length (including header), both measured in bytes. At the very beginning, there is a protocol identifier byte with the value 129, to distinguish the message as belonging to CLNP rather than any other protocol. There is also a one-byte protocol version number, which is set equal to one for standard CLNP. There is a 16-bit header checksum, which is computed over the bytes of the header. This does not use the same algorithm as IP. Rather, it uses the ISO standard summation code described on page 59 in Chapter 2.

Each message has a five-bit message type field. There are only two different types of CLNP message. One is the **DT** (data) message (type **11100**), which is equivalent to an IP message. The other is the **ER** (error report) message (type **00001**), which is used to report discarding of data messages, and so performs a similar function to the correspoding ICMP message type.

In addition to the components mentioned so far, there are other IP-style components that will be outlined in following sections. These include segmentation fields (which need not be present if not required) and five different kinds of optional extensions to the standard header. One of these options is a security level that applies to the message. There is a similar sort of option in IPv4; however, it was not mentioned earlier because it has only been used for military applications.

Time

As for IP, the only service offered by CLNP is a 'best effort' message delivery service. One of the optional components of the CLNP header is a priority value, ranging from 0 (lowest) to 14 (highest). Another optional component is for specifying a preferred quality of service. This is a little more elaborate than IPv4's D, T and R service type bits. It allows sequencing to be traded off against delay, and a three-way trade-off between delay, reliability and cost. It can also be used to report to a destination that a message experienced congestion *en route*.

Fragmentation and reassembly in CLNP is similar to that in IP. Messages carry a 'segmentation permitted' bit in the main header. If this is zero, the six-byte segmentation part of the message header is omitted. The segmentation header contains a 16-bit unique message identifer, a 16-bit segment offset in bytes and a total original message length in bytes. The first two are as in IP, but the third is an extra. Although this total length field is present, there is still a 'more segments' bit in the main header, as in IP.

The final feature analogous to IP is that the main header contains a message lifetime value, measured in 500 millisecond units, which is counted down as the message travels through the inter-network.

Given that a connectionless service is offered, there is no context in which flow control can be performed. However, it is possible for a message source to monitor the performance of the inter-network, and react if necessary. Messages carry an 'error control' bit. If this is set in a data message, and the message is discarded, an error control message is sent back to the data message's source with

16-bit code explaining the reason for the discarding. When this indicates switch congestion or expiry of message lifetime as the cause, the message source gets a warning of trouble in the inter-network.

Space

The original ISO connectionless network service was a unicast service. Since then, provision for a multicast service has been added. As already seen, and as normal in connectionless services, CLNP messages carry a destination identifier and a source identifier. These identifiers are ISO NSAP (Network Service Access Point) format identifiers, as described on page 145 in Chapter 5.

The Domain Specific Part (DSP) of an NSAP identifier usually has a hierarchical structure. This can reflect a structure akin to that of the Internet: the inter-network being divided into autonomous systems, which can be divided into areas, which contain networks, which can contain sub-networks, which contain computers. For example, in implementation terms, the structure might correspond to five levels of channel implementation using networking, where each of the four higher-level structural units is a network implemented using channels provided by other networks. By convention, there is one lowest level of structuring in NSAP identifiers: the final byte is used to differentiate between different processes within a single computer.

NSAP identifiers are even more unwieldy than IPv6 identifiers. Thus, names are needed to protect everyone and everything, apart from the inter-networking system, from these identifiers. It is possible to use an *ad hoc* naming mechanism, but it is preferable to make use of an established naming facility. One option is just to use Internet-style naming, since there is an experimental extension to the Domain Name System (DNS) that supports NSAP identifiers, in addition to the normal IP identifiers. This provision was suggested in anticipation of the use of CLNP within parts of the Internet. However, this pre-dated the decision to migrate towards IPv6.

Another option is to make use of the very general-purpose ISO standard directory service. This provides a general way of organizing **objects** in a distributed tree structure: inserting objects, removing objects and searching for objects. Hierarchical names for computers can then just be regarded as one particular type of object, and so can be managed using the directory service.

Inter-network switching

The ISO CLNP requires switches to perform much the same role as the Internet IP does. That is, apart from dealing with over-large messages, switches are mainly concerned with routing. Like IP, CLNP has provision for source routing, using an optional header field that contains a list of the switches to be visited. The only difference is in the names: strict and loose source routing are referred to as 'complete' and 'partial' source routing respectively.

In the absence of source routing information, switches need to select appropriate paths for forwarding messages to their destinations. Before outlining the standard routing protocols, it is useful to know some ISO inter-networking jargon. An **Intermediate System** (IS) is a switch connecting networks in the inter-network. An **End System** (ES) is a computer that is using the inter-network communication service. A **domain** is roughly equivalent to an Internet autonomous system. There is an ES-IS protocol, an intra-domain IS-IS protocol and an inter-domain IS-IS protocol.

The ES-IS protocol can be used to allow a computer to discover the network identifier of its local switch, and also for a computer to make itself known to that switch. This allows CLNP messages to be routed between the computer and its local switch. The protocol involves broadcasting by a computer to find its switch, and also broadcasting by a switch to advertise itself to potential computer clients. Because of this broadcasting, the ES-IS protocol is only intended for use on message broadcasting networks. For other networks, it is assumed that appropriate configuration of computers is done manually.

The distinction between the two types of IS-IS protocol is similar to that drawn between IGPs and EGPs in the Internet. That is, the inter-domain IS-IS protocol has to take account of legal, contractual and administrative concerns. Both of the IS-IS protocols are complex (and their additional jargon makes them seem even more complex). In summary, the intra-domain IS-IS protocol is used to implement link state routing. Each switch broadcasts link state information, and this is used as input to Dijkstra's shortest path algorithm at each switch. In order to cut down on the extent of broadcasting, and on routing table size, a domain can be split into areas. Within each area, 'Level 1' switches exchange routing information. Between each area, 'Level 2' switches act as a backbone for the domain, inter-connecting the areas. Thus, the overall flavour of this protocol is like that of OSPF. In a similar vein, the inter-domain IS-IS protocol has an overall flavour like that of BGP. The switches exchange distance vector information that also contains exact information about the paths used.

8.4 CHAPTER SUMMARY

An inter-network is a network that is implemented using channels that are implemented by a range of different networks. Such channels can have widely varying characteristics, depending on the networks used. When an inter-network is constructed using channels of different types, switching computers have to do more work. In addition to carrying out selective switching of messages, they must perform translations to reconcile different channel properties. In order to keep this as simple as possible, it is normal for the inter-network to offer a modest level of service that makes minimal demands on the component networks. Unicast channels with 'best effort' connectionless message transmission are a typical simple level of service.

The switching computers that support channels to different networks have a variety of names. A 'bridge' is a switching computer that connects two or more Local Area Networks together. There is an IEEE standard for bridges that connect IEEE standard LANs together. A 'router' is a switching computer that predominantly concentrates on switching. A 'gateway' is a switching computer that also has to carry out non-trivial translation work to reconcile differences. However, the three terms 'bridge', 'router' and 'gateway' are used in different ways by different people, so these should not be regarded as standard definitions.

The Internet is the most famous inter-network. The core of its service is provided by the Internet Protocol (IP). This gives a 'best effort' connectionless service between any two computers connected to the Internet. A multicasting service is also available. IP began a transition from version 4 to version 6 in 1996. One of the main reasons for this change is the need to increase the available identifier space as the number of computers has grown exponentially. IP is supported by various other standard protocols, including protocols for exchanging routing information. There is an ISO standard inter-networking protocol also, and it is very strongly based on IP.

8.5 EXERCISES AND FURTHER READING

8.1 Draw up a list of as many different network service characteristics as you can think of, and comment on how much variability between networks there might be for each characteristic.

8.2 A common arrangement for ethernet-based LANs is to have one backbone ethernet, with other local ethernets linked to the backbone via computers with two ethernet interfaces: one to the backbone, the other to a local ethernet. How much work will switches in this sort of inter-network have to do?

8.3 What problems might be introduced if inter-network routing allows splitting of communications?

8.4 Investigate the history of the IEEE 802 standardization effort, to discover why there are inconsistent LAN standards, and also two different bridge standards.

8.5 Design a communication service specification for an inter-network formed from IEEE 802 LANs. It should require a fairly precise 'lowest common denominator' of the different 802 LAN services, and so make life simple for bridges between LANs.

8.6 Find out more about both the IEEE standard transparent bridge, and the IEEE standard source routing bridge, and compare their advantages and disadvantages.

8.7 Read RFC 791, which specifies the Internet Protocol (IP), version 4, and also RFC 792, which describes ICMPv4.

8.8 Survey RFC 1550, then optionally RFC 1667–1680, 1682/3, 1686–88 and 1705/19, then RFC 1726, then RFC 1752, and finally the IPv6 specifications in

RFC 1883–87. This sequence records the process by which IP version 6 was designed.

8.9 Consult the most up-to-date version of the 'Assigned Numbers' RFC. In particular, look at the assigned version numbers to see why IP uses the numbers 4 and 6, and at the assigned protocol numbers to see the range of protocols (and sub-protocols) that can be carried in IP messages.

8.10 Investigate the IPv6 authentication and privacy mechanisms, by writing programs that implement the mechanisms for simulated IPv6 messages. If you cannot implement MD5 for authentication, or DES-CBC for privacy, invent simpler algorithms as alternatives.

8.11 In the description of message priorities in IPv6, some examples of priority choices for different types of traffic were given. Justify why these choices were suggested.

8.12 Why do you think that fragmentation of IPv6 messages may not be carried out by inter-network switches, whereas this was allowed in IPv4?

8.13 Read RFC 1191, which describes the mechanism that can be used to discover the maximum message size that can be sent over an Internet path without fragmentation being necessary.

8.14 What range of values is typically used in the time to live field of IP messages in the global Internet?

8.15 In what ways were the facilities of IGMP, used for multicasting with IPv4, changed when it was incorporated into ICMPv6, used with IPv6?

8.16 How many IPv4 Class A network identifiers have been allocated?

8.17 Look up the IPv4 Class D multicast addresses that have a special pre-defined significance.

8.18 Suggest ways in which the mobile IP routing mechanism, *as described in outline in the text,* could be made more efficient. (Try to do this without consulting the RFCs that describe how mobile IP can be implemented efficiently!)

8.19 IPv6 identifiers beginning with 0:2 and 0:3 carry ISO NSAP identifiers. Discover how 20-byte NSAP identifiers can be carried by the 12 bytes available after the four-byte prefix.

8.20 Revisit Exercise 5.4 on DNS, in the light of the material contained in this chapter.

8.21 Obtain the DNS names for any Internet-linked computers that you use. What organizations are responsible for allocating identifiers at each hierarchical level in the names?

8.22 If you have access to the `nslookup` command on a Unix computer system (or equivalent on another system), use it to look up the Internet identifiers for different DNS names, and vice versa.

8.23 Read RFC 826, which describes the ethernet Address Resolution Protocol (ARP). Also, read RFC 903, which describes the Reverse Address Resolution Protocol (RARP).

8.24 The IEEE 802 source routing bridge involves switches scanning the route list carried by messages, and not altering it. In contrast, the IP source routing involves switches modifying the route list. Why do you think this difference exists?

8.25 Consult RFC 1519, to see how the Classless Inter-Domain Routing (CIDR) strategy tackled the problems of both shortages of Class B IP identifiers and routing in the rapidly expanding Internet.

8.26 Which Autonomous System(s) contains any Internet-linked computer(s) that you have access too?

8.27 List the reasons why OSPF is regarded as a better interior gateway protocol than RIP. Consult appropriate RFCs to assist in this.

8.28 List the reasons why BGP is regarded as a better exterior gateway protocol than EGP. Consult appropriate RFCs to assist in this.

8.29 If you have access to the `traceroute` command on a Unix computer system (or equivalent on another system), use it to investigate the routes that IP messages take to a range of computers in various parts of the world.

8.30 Discover the status of MBONE access to any computers that you use. If you can directly use the MBONE, find out what multicast groups are in operation, and what types of information are being multicast.

8.31 Consult RFC 994, which contains the draft version (from 1986) of the ISO CLNP. Note the general style of an ISO standard document, and use it to compare the details of CLNP with the details of IP.

8.32 Investigate whether there is any significant use of the ISO CLNP inter-networking in any organizations in your area or country.

8.33 If you have access to information about the ISO ES-IS and IS-IS routing protocols, compare them to the corresponding Internet routing protocols. Note that, although ISO standards documents are not usually readily available freely, draft versions of these two standards are usually available via the World Wide Web.

8.34 Read RFC 1637, which describes how ISO NSAP identifiers could be incorporated into the Domain Name System.

8.35 Track the growth of intranets that use standard Internet protocols, but are not part of the global Internet, within commercial organizations.

8.36 Explain the main problems that a firewall switch should deal with, in order to allow an organization to make general use of the global Internet facilities, but to prevent the global Internet making general use of the organization's facilities.

Further reading

All relatively recent general textbooks contain material on the Internet, and its working. A good specialist textbook is *Internetworking with TCP/IP* Volume 1 by Comer (Prentice-Hall 1995). A book specifically on routing in the Internet is *Routing in the Internet* by Huitema (Prentice-Hall 1996). Detailed material on LAN bridges can be found in *Local and Metropolitan Area Networks* by Stallings (Macmillan 1997). The Internet Requests for Comments (RFCs) are a valuable source for detailed material on the internal workings of the Internet. A great many academic papers have been concerned with Internet-related experiments, or suggested improvements to the protocols used in the Internet.

NINE

CASE STUDY 1: ACCESSING THE WORLD WIDE WEB

The main topics in this case study about accessing the WWW are:

- The World Wide Web
- HTTP protocol and HTML markup language
- The Internet Transmission Control Protocol (TCP)
- IP transmission over a telephone channel using PPP
- Structure of the global Internet
- IP transmission using an SMDS service
- IP transmission over an ethernet

9.1 INTRODUCTION

The World Wide Web (WWW), or just 'The Web' for short, was the revolution that brought the Internet to the attention of the general public. The ideas behind the WWW originated at CERN — the European Centre for Nuclear Research — in 1989, as a way of allowing physics researchers based in different countries to exchange information. The first prototype implementation appeared at CERN in 1990, and the first example of a friendly user interface appeared in 1993. Since then, interest in, and use of, the WWW has grown explosively. The influence of the WWW is so great that, in many people's minds, the WWW *is* The Internet. In fact, the WWW is just one of many applications supported by the basic inter-network message transfer service of the Internet.

The WWW is a mechanism for allowing access to documents stored on computers around the world. A major attraction for users is that it can be used without needing to know any details of how computer communications work. Just as user-friendly applications such as spreadsheets and word processors liberate people from having to know how computers work, so the WWW liberates people from having to know how communications work. In particular, the availability of colour WIMP (Window, Icon, Menu and Pointer) interfaces for using the WWW assists in this process. Of course, what is experienced by the end user may be affected by the implementation, that is, by the way in which computers or their communications work.

The World Wide Web, as its name may suggest, is essentially a giant data structure, with components stored on computers in many different places, and evolving continuously. The data is the collection of documents, and the structure comes from links between documents. Thus, the WWW can be regarded as a directed graph, with documents as vertices, and links between documents as directed edges. Normally, WWW documents are just referred to as **pages**. Links appear within pages: where it is appropriate to refer to some other document in context, the page includes a link. The difference from references in conventional documents is that the links are *active*. A reader can explore the directed graph via a computer screen, and follow links immediately, rather than having to note a reference and then seek out the referenced document.

If all of the pages just contained textual information, then the facility provided by the WWW would be called **hypertext** (on a worldwide scale). However, the further attraction of the WWW is that it supports multimedia information: not just text, but also pictures, sound and video clips. This is called **hypermedia** technology. Figure 9.1 illustrates a WWW page produced by a fictitious company. The screen display was rendered by the Netscape Navigator **WWW browser** — a tool for browsing the WWW. For obvious reasons, this does not show off the audio or video capabilities. However, it does show a picture included among textual information. The representation here does not include the colours that existed in the original screen display. It also does not indicate that the mysterious 'Crawl Crawl is Fantastic!!!' line was, in fact, a snapshot of a continuously animated display of 'Web Crawl is Cool!!! Web Crawl is Fantastic!!!'. The hypertext aspect comes from the items that are underlined (and also the picture, in fact). By clicking on these with a mouse, a reader can immediately see another WWW page that is being referred to.

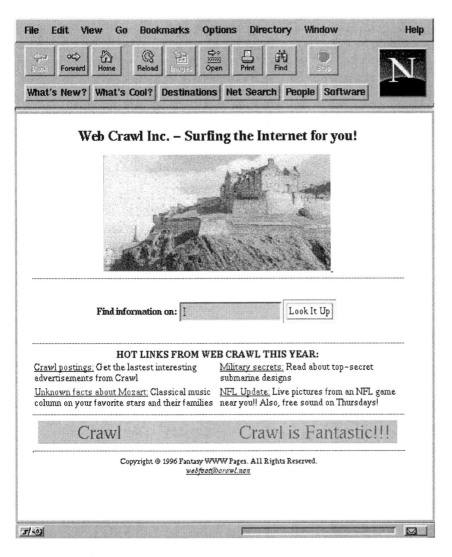

Figure 9.1 Example WWW page displayed by Netscape Navigator

The links between pages allow a user to explore the WWW by hopping from page to page. Obviously, before doing this, some starting point has to be found. WWW browsing tools allow a user to specify a page of interest directly, using a unique identifier for it. The page is then displayed directly. Identifiers can be learnt from newspaper or magazine articles, or from many advertisements. An alternative is to make use of a WWW **search engine** — a computer that offers a searching service. Search engines explore the WWW systematically, by following all links that they can find, and accumulate information about all of the pages found. A user can then give one or more search words to the search engine, and it supplies a dynamically produced document containing links to pages that seem to have the most relevance to the search words. The example WWW page given in Figure 9.1 is a modest example of an interface to an imaginary search engine. Search words can be typed in the area following 'Find information on:' and then a search initiated by pressing the 'Look It Up' button using a mouse.

The above description gives a basic outline of how the World Wide Web works. Any further details are best gained from practical experience. From the point of view of computer communications, the basic problem can be stated simply. Given that a computer has an identifier for a required WWW page, how is a representation of that page obtained from a computer on which it is stored? All other issues concerned with supplying the look and feel of the WWW to a user are matters for WWW page designers and WWW browser designers. Note, in particular, that although the WWW can be regarded as a graph, this has no connection with the fact that computer networks can be regarded as graphs. In the WWW case, the edges reflect connections of human interest; in the network case, the edges reflect ways of transferring information.

9.2 THE PROBLEM: INFORMATION, TIME AND SPACE ISSUES

The problem in this case study is obtaining a WWW page for a WWW user. This involves a communication in which information is shared between a server computer where the page is stored and the client computer of the user. To make it reasonably interesting, the client computer will be in the user's home somewhere in Europe, and the server computer will be in a large organization in the USA.

The obvious piece of information sharing is that the contents of the page are sent from the server computer to the client. Another piece of information sharing, but in the opposite direction, is that the server computer learns that the client computer wanted the WWW page. The communication occurs during a time period starting when the client computer requests the page, and finishing when the client computer has fully received the page. The space of the communication consists of the client and server computers, and a channel between them. In this case study, the channel will be provided by the Internet.

Note that the particular communication will be part of a gigantic super-communication that represents all information sharing occurring over the World Wide Web. From the local perspective of the client computer, the communication is one in a sequence, corresponding to the pages accessed in turn by the user. From the local perspective of the server computer, the communication is one of perhaps many, corresponding to page accesses by different computers.

Information

There are two types of information being shared. First, there is a type that describes the WWW page required. This information is given to the server computer by the client computer. Second, there is a type that describes the contents of the WWW page required. This information is given to the client computer by the server computer.

Uniform resource identifiers

The first type of information is an instance of a **Uniform Resource Identifer** (URI). A URI is used to identify an accessible resource, using a name, a location or some other characteristic. The type of URI used in the World Wide Web is a **Uniform Resource Locator** (URL), which identifies a resource by specifying its location: a computer, and a position within that computer. This makes implementation straightforward, since it is easy to know which server computer to contact, and easy for the server to find the required page. The disadvantage of URLs is that a location, rather than the resource itself, is identified. This causes problems if a resource is moved between locations, ceases to exist at some location or is replicated at several locations. One bane of a WWW user's life in the mid-1990s was the link to a page that no longer resides at the location implied by its identifier. Another nuisance is search engines that report the same page at numerous different locations. In some cases, replication is deliberate, by the use of **mirror sites**, which are servers that store copies of resources, to make them more easily accessible to clients. For example, a resource created and maintained in one continent might be mirrored in several other continents.

In due course, a **Universal Resource Name** (URN) system should prevail, in which resources are allocated unique names by naming authorities. This might encompass existing schemes, such as International Standard Book Numbers (ISBN) and Universal Product Codes (UPC). A prerequisite of any such scheme is that there should be efficient mechanisms for finding one or more locations for a resource with a given URN.

Although the general format of identifiers used for WWW resources allows for different URI systems, the formats actually used are exclusively for URLs. A full URL has three components. The first specifies the protocol to be used to access the resource. The second specifies the computer that supports the resource, using its Internet Domain Name System (DNS) name. The third specifies a location at that computer, using a scheme based on Unix file naming. Given a URL, a client computer uses the specified protocol for a communication with the specified server computer, in order to obtain information from the server's specified location. As an example, a URL for the WWW page in Figure 9.1 might be:

```
http://www.crawl.non/searcher/front-page.html
```

In this, `http` is the name of the protocol (the Hypertext Transfer Protocol, to be described later), `www.crawl.non` is the (fictitious) DNS name of the server and `searcher/front-page.html` can be regarded as the name of the file containing a representation of the page.

HTTP is the standard protocol for accessing pages that were designed for the World Wide Web. The example used in this chapter will assume the use of HTTP. More precisely, it will assume the original version of HTTP — HTTP/1.0 — which was still in general use in early 1997. An improved version — HTTP/1.1 — was agreed upon during 1996, and then appeared as an Internet RFC (RFC 2026) in January 1997. This new version dealt with a major communications problem that is described in this case study. However, it is instructive to use HTTP/1.0 here, to allow this problem to be exposed.

HTTP is not the only protocol that can be used. There are other collections of information created prior to the WWW, or created to allow access via a non-WIMP interface, and these can also be accessed using their own protocols via the same URL mechanism. The protocol names include FTP for the standard, and long-standing, Internet file transfer protocol; GOPHER, PROSPERO and WAIS for three other information access systems; and NEWS or NNTP to access Usenet newsgroup articles.

In addition to accessing objects with these various protocols, it is possible to obtain access to services, with the protocol names MAILTO for sending electronic mail and TELNET for conducting an interactive terminal session. For some of the non-HTTP protocols, the computer name in the URL must be embellished with a computer user name and/or password, since open access may not be possible. The idea is that a URL contains all of the information that a direct human user of a particular resource-accessing service would have to provide. Everything is in the identifier.

WWW page representation

The overall format of a WWW page consists of a page of text. Unlike printed pages, there is no length restriction on a WWW page. The textual characters may be displayed in a variety of sizes, fonts and colours, and be positioned at various points on the page. Pictures can be included among the text. Also, hyperlinks to other pages, and to other WWW objects such as audio and video clips, can be inserted as required.

All of this is represented using a description composed of lines of text, expressed in the **Hypertext Markup Language** (HTML). A markup language is a way of placing instructions about the required layout of a document within the document itself. These instructions are interpreted by the tools used to display the document on screen or on paper. This approach is in contrast to the WYSIWYG (What You See Is What You Get) style, where documents are described by their final display format, taking no account of the abilities of the display tools, or of the screens and printers available to them.

Figure 9.2 shows the HTML representation of the WWW page shown in Figure 9.1. A tutorial on the use of HTML will not be given here, since it is not a matter of direct relevance to computer communications. However, a casual reader should be able to make connections between the different components of the HTML description and the different components of the resulting page as displayed using Netscape Navigator. Certain aspects of the HTML description deserve further comment, in order to explain the informational properties of WWW pages.

The most interesting HTML construct is the Anchor facility. This begins with with `<a href="`*URL*`">` and ends with `</a>`. An anchor represents a hyperlink, and the URL identifies the page that the hyperlink is pointing to. The material that appears between the `<a>` and the `</a>` is displayed with any text underlined or differently coloured, to indicate that it can be clicked on with a mouse to hop to the referenced page.

HTML is only used to describe the textual content of WWW pages. The inclusion of pictures has to be indicated using a specific HTML construct. The ninth line of Figure 9.2 shows how a picture can be included using the Image facility `</img...>`. The `src="..."` argument gives a URL for the image. In most cases, as here, the URL does not contain protocol name and computer name components, in which case the image location is taken relative to where the HTML document was located. Images can be represented in various standard formats. GIF (Graphical Interchange Format) is supported by virtually all browsers, and most also support the more economical JPEG format. Here, the image is in JPEG format.

So far, the use of WWW pages has been described as a completely passive activity. Users can obtain pages, and read them. This was all that was supported by the earliest versions of HTML. However, the desirability of a more interactive approach rapidly became obvious. In the first instance, limited interaction with the server computer was added via the Forms facility. This allows a user to input one or more items of information, which are then sent to a handler identified by a URL. The input strings are appended to the end of the URL, to be read by the handler, which is usually located in the same place as the WWW page.

Most handlers make use of an existing standard, the **Common Gateway Interface** (CGI). This involves writing a script or program to massage the input received, pass it to other programs for processing if required, and then produce output in HTML form. The output is sent back, to be displayed for the user. An example of a simple one-item form occurs in the example of Figure 9.2, and its display is shown in Figure 9.1.

Forms allows modest interaction, because processing is done by the distant server. For better quality interaction, it is preferable for processing to be done by the client computer, under the direction of the server computer. In essence, the server computer must program the client to tell it how to behave. This means that, as well as standard representations for information such as text and pictures, a standard representation for programs is needed. As mentioned on page 41 in Chapter 2, the language that has become a standard for the WWW is called **Java** — an object-

```
<html>
<head><title>Web Crawl Inc. Home Page</title></head>
<body bgcolor="#ffffff">
<center>

<h1>Web Crawl Inc. - Surfing the Internet for you!</h1>
<p>
<a href="/info/plans/crawl.html">
<img border=0 src="/info/logos/crawl.jpg" width=300 height=150>
</a>

<hr>
<form action="/cgi-bin/crawler" method=GET>
<b>Find information on:</b>
<input name="query"><input type=submit value="Look It Up">
</form>

<hr>
<b>HOT LINKS FROM WEB CRAWL THIS YEAR:</b>
<table width="100%" cols="2" cellspacing="0" cellpadding="2" border="0">
<tr><td width=50%>
<a href="http://www.crawl.non/junk/">Crawl postings:</a>
Get the lastest interesting advertisements from Crawl
</td><td width=50%>
<a href="http://www.submarine.nav/torpedo.html">Military secrets:</a>
Read about top-secret submarine designs
</td></tr><tr><td width=50%>
<a href="http://www.rock.music.bus/Mozart/">Unknown facts about Mozart:</a>
Classical music column on your favorite stars and their families
</td><td width=50%>
<a href="http://www.stats.nfl">NFL Update:</a>
Live pictures from an NFL game near you!!  Also, free sound on Thursdays!
</tr>
</table>

<hr>
<applet code="Blink.class" width=475 height=30>
<param name=lbl value="Web Crawl is Cool!!!  Web Crawl is Fantastic!!!">
<param name=speed value="4">
</applet>

<hr>
<font size="-1">
Copyright &copy; 1996 Fantasy WWW Pages. All Rights Reserved.<br>
<address><a href="mailto:webfoot@crawl.non">webfoot@crawl.non</a></address>
</font>

</center>
</body>
</html>
```

Figure 9.2 Example WWW page in HTML representation

oriented programming language developed by Sun Microsystems. The appearance of Java is somewhat similar to that of C or C++, although conceptually it is closer to the object-oriented language Smalltalk. Java is compiled to a standard machine-independent byte code, which is the representation communicated. Browsers must supply an interpreter than can obey the byte-coded instructions correctly.

A piece of Java code included in a WWW page is called an **applet** (meaning a small application). Applets have uses far beyond just processing user inputs and generating outputs. They can also be used to extend the capabilities of a browser, by adding the ability to interpret new types of representations for information, or to understand new protocols for information transmission. The general capability is a simple example of **network computing**, where a computer has little built-in software, but acquires what it needs from other server computers. This approach in general, and the use of Java in particular, has to be adopted with care. Imported code is able to access resources on the user's computer, and care is necessary to protect against the introduction of malicious, or just erroneous, software that might cause damage or perform undesirable acts.

The example in Figure 9.2 includes a a simple applet that is capable of generating a modest animated display. The Java program for this is not included, being of little relevance from a communications point of view, and having a product that cannot be shown in one static snapshot. In addition, although this program was actually obtained from a public WWW page, its author might not wish his work to be displayed in a printed book.

Time

There is a desirable absolute time period for the communication of a WWW page. This is determined by research into cognition, which shows that it is best for human users to get a response to requests within two to four seconds. In practice, this is wishful thinking for most WWW pages, due to various technical factors:

- delays at the server computer;
- delays on the communication channel; and
- delays at the client computer.

Examples of server delays are information retrieval times or congestion caused by multiple requests. Delays may occur on the communication channel because of latency (distance dependent), transmission time (information quantity dependent) and possible congestion or error recovery. Finally, there may be delays at the client while received information is being rendered for display. Because of these potential delays, it is not usually possible to give any absolute time guarantees.

The basic time package for the communication is a simple two-stage hand-shake. In the first stage, a request containing a URL is sent from the client computer to the server computer. In the second stage, a response containing the required WWW page is sent from the server computer to the client computer. The HTTP

```
GET http://www.crawl.non/searcher/front-page.html HTTP/1.0
If-Modified-Since: Friday, 10-May-96 14:27:43 GMT
User-Agent: Tapestry/11.96
```

Figure 9.3 Example of HTTP request for WWW page

protocol has provision for this handshake as its most central operation. Its handshake involves an exchange of textual messages. The format of the messages is very strongly based on the format of Internet electronic mail messages. That is, messages have a textual header consisting of a number of lines and each line begins with a standard header name.

Figure 9.3 shows an example of the HTTP request that might be sent for the example WWW page. The important part of this is the first line. It specifies that the operation to be carried out is getting an object, gives an identifier for the object, and states the version of HTTP that is being used. Note that the term 'HTTP' is used throughout the description here to mean the original version of the protocol: HTTP/1.0. The second line is also important from a temporal point if view. It specifies that the object need only be sent if it has been changed since the absolute time given. The presence of such a line indicates that the client computer already has a stored copy of the required object. A new copy is only needed if the object has been modified since the stored copy was made. The third line is only for information — it names the browser that was used to generate the page request. There are several other informational header lines that can be used, in addition to the three types shown in this example.

The response to an HTTP request consists of lines of text, followed by a representation of the requested object if required. Figure 9.4 shows an example of the HTTP response that might be sent back after the request shown in Figure 9.3. There are three components. The first line is a status report. This is followed by response information lines. Finally, after a blank line, the object representation is included. In this case, the representation is just the text already seen in Figure 9.2. The status report line has three features. The first is an indication of the version of HTTP being used (HTTP/1.0). The second and third are status indications for computer and human use respectively: a number for the computer and a text string for the human. A range of standard numerical status codes is defined; there is no standard definition of corresponding text strings. One particular example is the code 304, which means 'not modified'. This would be returned if a page had not been modified since the date given in an If-Modified-Since: line in the request.

The other header lines in the response convey various information, and most of those in the example are self-explanatory. The Date: (of responding) and Last-modified: (referring to the page) header lines are directly time-related. The content-type line specifies the format used to represent the object being transferred. That is, it concludes a dynamic agreement on the implementation of the information to be transferred. It uses formats drawn from the standard

```
HTTP/1.0 200 Document follows
Date: Mon, 20 May 1996 07:47:41 GMT
Server: Webwide/7.2.18
Content-type: text/html
Last-modified: Tue, 14 May 1996 23:32:16 GMT
Content-length: 1523

<html>
<head><title>Web Crawl Inc. Home Page</title></head>
<body bgcolor="#ffffff">
<center>

<h1>Web Crawl Inc. - Surfing the Internet for you!</h1>

    etc. etc. etc.

</center>
</body>
</html>
```

Figure 9.4 Example of HTTP response to request for WWW page

range of types defined for MIME — the multimedia mail extension mentioned in Section 7.6. The names used have two parts, separated by a slash: a main type and a subtype. Here, the name `text/html` means line-structured text in the HTML format. Other common possibilities include:

- `text/plain`: line-structured text with no implied syntax;
- `application/octet-stream`: sequence of raw bytes;
- `image/gif`: picture in GIF format;
- `image/jpeg`: picture in JPEG format;
- `audio/basic`: sound sample in simple PCM format;
- `video/mpeg`: video sample in MPEG format.

In HTTP/1.0, the `content-type:` line must be present if an object representation is transferred. The `content-length:` is optional; if it is absent the length of the content is just defined to be the remaining length of the HTTP message.

The temporal structure of the communication is more complex if the WWW page contains embedded pictures, as given by `<IMG>` statements in the HTML representation of the page. For each of these, there must be an extra two-stage HTTP handshake to request and receive the images identified by the URLs in the `<IMG>` statements (assuming that the URLs are not non-HTTP). There is no requirement for these extra communications to take place in any particular order, or to take place after the main page communication is complete. The earliest time that each can start is when the client computer has received and interpreted the associated `<IMG>` statements.

Space

In principle, the communication space is straighforward: the client computer, the server computer and a channel between them. However, an HTTP communication may involve a small amount of HTTP message switching to implement the communication space. This is because **proxies** or **gateways** may be involved. These are intermediate computers, a proxy being at the client computer end and a gateway being at the server computer end. Note that the terms 'proxy' and 'gateway' also have more general usage away from the WWW scene.

A proxy is a computer that makes requests on behalf of other client computers. One reason for the existence of a proxy is that an organization may not wish to give Internet access to all of its computers. Instead, these computers issue requests to the proxy, which then forwards them to servers, possibly applying some filtering to enforce an access policy. Another reason is that proxies may assist client computers that do not speak the full range of protocols possible for WWW access. For example, a client might only understand HTTP, and a proxy could conduct FTP or Gopher conversations on its behalf. A final reason is for efficiency reasons. A proxy can maintain a cache of WWW pages that have already been fetched by its various clients. Then, any further requests from any clients can be immediately serviced by the proxy.

A gateway is a computer with a symmetric role to that of the proxy, except at the server computer end. It can carry out security filtering, protocol conversion or, where a number of servers are behind the gateway, cacheing of pages.

In the example here, it will be assumed that there is direct communication between the client computer and the server computer. That is, the channel is not implemented with an intermediate proxy and/or gateway. The properties of this assumed communication would apply equally to three other cases: proxy/server, client/gateway and proxy/gateway communication. The only additional fleshing-out would be to examine any localized client/proxy and/or gateway/server communication. Such communication is much more straightforward, in fact.

Summary of the problem

The abstract problem of fetching a WWW page has been reduced to a computer communications problem that requires detailed implementation. There is one master pair of handshake sub-communications: in the first, a textual HTTP message is sent from client to server; in the second, a textual HTTP message plus, usually, a page representation is sent from server to client. The time period for the overall communication begins when the first sub-communication begins and ends when the second sub-communication ends. If the WWW page contains pictures, then one or more further pairs of handshake sub-communications are required. Here, it is assumed that these further handshakes occur between the same pair of computers. However, because of the way HTTP/1.0 works, they use different channels, a matter which is discussed later. The effect is that splitting occurs. Figure 9.5

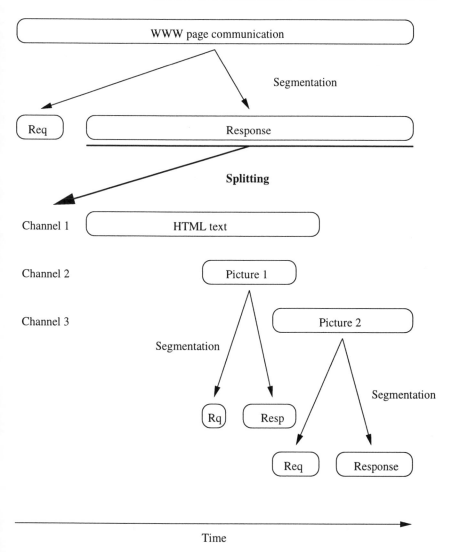

Figure 9.5 Implementation of WWW page communication

illustrates the implementation of the overall communication, for an example where the WWW page contains two pictures. A major difference with HTTP/1.1 is that there is normally no splitting — the same channel is used for each handshake. The segmentation and splitting involved in the implementation here results in six component sub-communications: three requests and three responses.

The next section considers how the basic requirement of these communications can be implemented. That is, how bytes of information can be reliably and promptly sent in both directions between the two computers via a channel. For the handshake style required here, half duplex transmission would be sufficient. However, because of its general availability, a full duplex channel is actually used.

9.3 RELIABLE END TO END COMMUNICATION USING TCP/IP

The case study involves World Wide Web access over the Internet. Thus, the standard Transmission Control Protocol (TCP) is used to supply the necessary reliable end to end communication service. TCP provides a reliable connection-oriented service, and its overall operation was outlined as one of the examples in Section 4.5.2. TCP makes use of the connectionless service provided by the Internet Protocol (IP), which was extensively described in Section 8.3.2. TCP is intimately entwined with IP, and cannot be directly used on top of different connectionless services. Thus, it is normal to refer to the combined pair of protocols **TCP/IP**, rather than just TCP on its own. One particular feature of the service provided by TCP, of relevance in this example, is that it allows multiple unicast channels to exist in parallel between pairs of computers. This is in contrast to IP, which creates the illusion of a single channel, using appropriate channels through networks.

Information issues

The TCP message was shown in outline in Chapter 2 as one of the classic message types. Figure 9.6 shows the contents of the TCP message header in more detail. This shows the traditional format, where TCP/IP uses IPv4. The use of IPv6 does not force any changes on the TCP message format. However, IPv6 forces a few changes to the operation of TCP, which means that implementations have to be changed, and so there has been discussion of whether this should be used as an opportunity to improve the TCP message format as well. At the time of writing in early 1997, no concrete proposals had been finalized for a dramatic change to TCP.

The sequence number, acknowledgement number and window size fields of the TCP header contain items already discussed in Section 4.5.2, in an explanation of TCP's sliding window mechanism. They contain the information serial number, the piggy-backed acknowledgement serial number and the piggy-backed receive window size, respectively. One of the six code bits, the ACK bit, indicates whether or not the acknowledgement number is valid.

The checksum was also mentioned in Section 4.5.2. It is computed not just over the TCP message, but over a 12-byte 'pseudo header' prepended to the message. The format of this header is shown in Figure 9.7. These items are included in the checksum, to ensure that the message has arrived at the correct computer, and is a genuine TCP message. The contents of the pseudo-header are

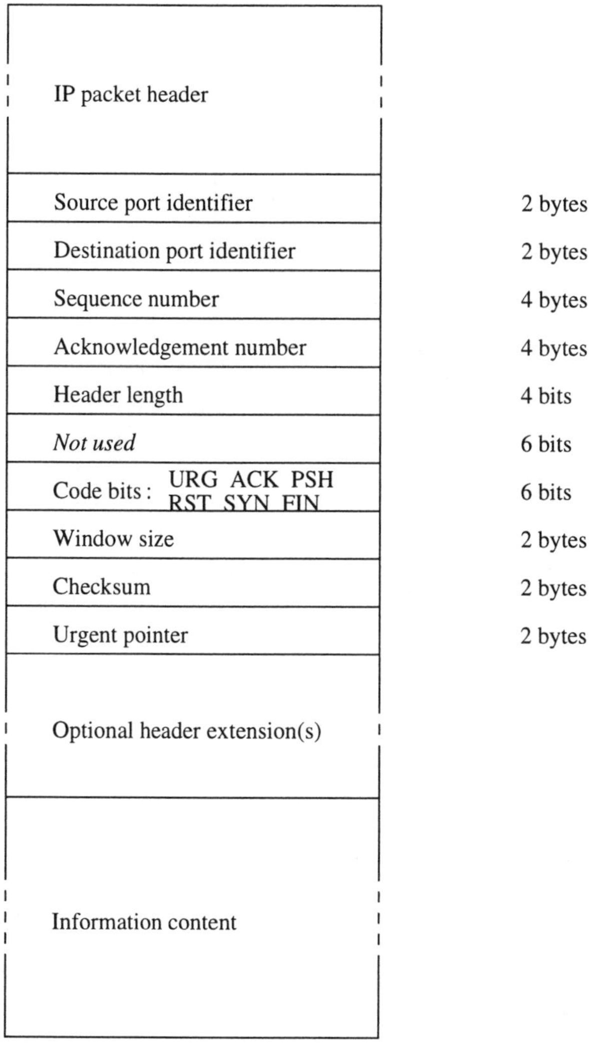

IP packet header	
Source port identifier	2 bytes
Destination port identifier	2 bytes
Sequence number	4 bytes
Acknowledgement number	4 bytes
Header length	4 bits
Not used	6 bits
Code bits : URG ACK PSH RST SYN FIN	6 bits
Window size	2 bytes
Checksum	2 bytes
Urgent pointer	2 bytes
Optional header extension(s)	
Information content	

Figure 9.6 TCP message format

one aspect that ties TCP closely to IP. The fact that IP identifiers are included forces a change in the pseudo-header when IPv6 is used instead of IPv4.

Other TCP header fields will be discussed in the next sections on time and space issues. The information content of the TCP/IP message consists of a sequence of bytes. There is a maximum length restriction on the size of this sequence. The default maximum length is 536 bytes. This corresponds to the default maximum size of 576 bytes for an IP message, minus 20 bytes for a standard IP header, and minus another 20 bytes for the standard TCP header. Thus, in a TCP/IP

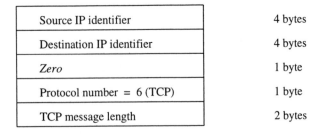

Source IP identifier	4 bytes
Destination IP identifier	4 bytes
Zero	1 byte
Protocol number = 6 (TCP)	1 byte
TCP message length	2 bytes

Figure 9.7 TCP pseudo-header format

message with the default maximum size and with standard headers, 93% is used
for carrying information and 7% is an overhead for TCP/IP headers. However, it
is possible for a different value to be negotiated. With a larger maximum size, the
header overhead can be reduced.

When a TCP connection is established, the TCP messages exchanged can
carry an optional header field containing a 16-bit maximum length to be used
within the connection. The notion of an agreed maximum length is unnecessary
when using IPv6, since communicating computers need to know the maximum
allowed IPv6 message size because *en route* fragmentation is not done. This
known size can just be used as a basis for determining the maximum allowed TCP
message length.

Since TCP messages are to be used to transfer HTTP requests and responses
in this case study, the maximum message length is almost certain to force segmentation of communications. A typical HTTP request might fit into one 576-byte
information field — for example, the request shown in Figure 9.3 is 139 bytes long
(for 133 characters, plus three CR-LF sequences at the end of lines). However,
while the header of an HTTP response is likely to fit into one message, the object
representation will be too long. For example, the response header in Figure 9.4 is
187 bytes long, but is followed by 1523 bytes for the HTML file. Other examples
are the byte code for the Java applet in the example, which was 2315 bytes long,
and the GIF image data for the picture, which was 36 539 bytes long. The exact
usage of TCP messages will be considered in the next subsection, on time, as part
of an explanation of the entire time package involved.

Time issues

As already described in Section 4.5.2, TCP implements a connection-oriented
service, and this is used by HTTP. The implementation of HTTP/1.0 involves
establishing a TCP connection for each request-response handshake. The client
sets up the connection, then sends its request. The server sends its response, and
then shuts down the connection. The behaviour of HTTP/1.1 is different: a single
TCP connection can be used for a sequence of handshakes. Further, pipelining
is possible, in that it is allowable to send a further request before a preceding

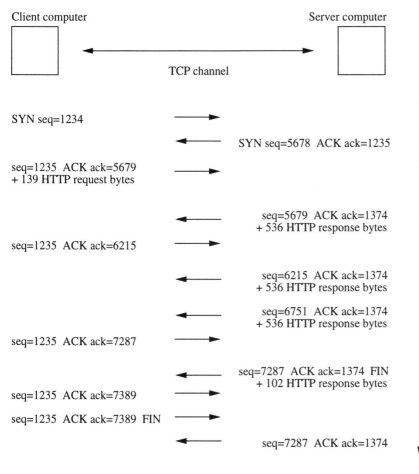

Client computer

Server computer

TCP channel

SYN seq=1234

SYN seq=5678 ACK ack=1235

seq=1235 ACK ack=5679
+ 139 HTTP request bytes

seq=5679 ACK ack=1374
+ 536 HTTP response bytes

seq=1235 ACK ack=6215

seq=6215 ACK ack=1374
+ 536 HTTP response bytes

seq=6751 ACK ack=1374
+ 536 HTTP response bytes

seq=1235 ACK ack=7287

seq=7287 ACK ack=1374 FIN
+ 102 HTTP response bytes

seq=1235 ACK ack=7389

seq=1235 ACK ack=7389 FIN

seq=7287 ACK ack=1374

Time

Figure 9.8 TCP dialogue for example HTTP handshake

response has finished. Note that this does not extend to overlapping of requests,
or overlapping of responses. These are separate in time, and in sequence.

The overall TCP message exchange for the HTTP/1.0 handshake shown in
Figures 9.3 and 9.4, for fetching the HTML document of Figure 9.2, is shown in
Figure 9.8. This assumes that no TCP messages are lost in transit. The details of
the message exchange are explained over the next few paragraphs. Each of the
TCP messages shown is transmitted from one computer to the other encapsulated
inside an IP message. The messages may travel by different routes, but this is a
matter for the IP service, not the TCP service.

A connection is set up using a three-stage handshake, as described in Section 4.4. The three messages exchanged use two of the code bits in the TCP header:

the SYN bit and the ACK bit. The first message has the SYN bit set, the second message has the SYN and ACK bits set and the third message has the ACK bit set. The first two messages can carry a header option specifying a maximum TCP information field length, in order to negotiate this parameter for the duration of the connection.

A connection is shut down gracefully using a four-stage handshake, which consists of two successive two-stage handshakes. These involve the FIN code bit and the ACK bit. The computer shutting down the connection sends a message with the FIN bit set, and the other responds by sending a message with the ACK bit set. Then, after informing the user of the connection-oriented service about the demise of the connection, the other sends a message with the FIN and ACK bits set. Finally, the first computer responds by sending a message with the ACK bit set. In cases of abnormal problems, a less graceful mechanism is available. This uses the RST code bit. If either computer sends a message with the RST bit set, the connection is terminated immediately, with no further messages being sent.

Within a connection set up for an HTTP handshake, the required information flow is implemented by the exchange of TCP messages. First, the bytes representing the HTTP request are transferred in one or more TCP messages from the client computer to the server. Then, the bytes representing the HTTP response are transferred in one or more TCP messages from the server to the client. The number of messages needed depends on the agreed maximum information field length.

In the absence of message losses, acknowledgements flow back in the opposite direction at a rate sufficient to prevent timeout and retransmission. As shown in the example of Figure 9.8, it is possible for an acknowledgement for the last one or more HTTP request messages to be piggybacked on the first HTTP response message. If any messages are lost, then retransmission is necessary, using Karn's algorithm, described in Section 4.5.2. In summary, the timeout for the first retransmission is based on a combination of the estimated round trip time and the estimated variance of the round trip time; timeouts for subsequent retransmisions are chosen using exponential backoff, the timeout period typically being doubled for each retransmission.

The rate at which HTTP information can be transmitted depends on the transmission window size, which can vary throughout the duration of the connection. The window size field gives the number of extra information bytes that the receiver is currently prepared to accept, and so controls the amount of information that can be sent ahead before an acknowledgement is received. In the example of Figure 9.8, the window size field of messages is not shown but, as can be seen, at the point where the HTTP response starts being sent, it allows at least 1072 bytes (two full messages) to be sent ahead.

The window size allows a receiver to control the rate of transmission to a manageable level from its point of view. However, the capacity of the TCP channel connecting the computers must be taken into account. A transmission rate agreeable to both computers may be too high for congested networking facilities to supply. In such a case, messages will be discarded. In the modern Internet, congestion rather than transmission errors is the main reason for message losses. To deal with this, transmitters maintain two window sizes. One is the size notified to the receiver. The other is an estimate of the capability of the channel. The minimum of these two sizes is used to control the rate of transmission. Two related techniques are used to maintain the channel capacity window: **slow start** and **multiplicative decrease**.

Slow start applies when connections begin, and after periods of congestion have been experienced. The channel window size is set to one maximum length message size, so only one message is sent ahead. If this is acknowledged in time, the window size is doubled, and two messages are both sent ahead. If these are both acknowledged in time, the window size is doubled again. This continues until the channel window size reaches the receiver's window size. If slow start is being used after a period of congestion, the doubling stops when the size reaches half of its previous value. It is incremented by the maximum message length thereafter, rather than being doubled. Multiplicative decrease applies after a message has been lost, i.e., is not acknowledged in time. The window is reduced by half for every loss, until it reaches the size of one maximum length message.

The use of slow start and multiplicative decrease allows TCP to deliver good performance over Internet paths that suffer from varying amounts of congestion. However, while this statement applies to 'normal' connections, which last for a relatively long time, it does not apply to HTTP/1.0 connections, which have a short duration. Slow start adversely affects both the sending of the HTTP request and the sending of the HTTP response. In both cases, the channel window must grow from its minimum size, and it is quite likely (especially for the HTTP request) that information transmission is complete before the window size has grown to its natural maximum. Thus, extra round trips times are needed while the transmitter waits for acknowledgements to arrive. One way for the client and server to ameliorate the effect of slow start is for them to negotiate a large maximum message length value, which allows more information to be sent for each slow start induced acknowledgement handshake.

In addition to this inefficiency, setting up and shutting down overheads are incurred for HTTP/1.0 handshakes. As shown in the example of Figure 9.8, some concatenation of the initial and final handshake communications with information carrying communications is possible. Request data bytes can be sent with the third message of the setting up handshake. Response data bytes can be sent with the first message of the shutting down handshake, and they can be acknowledged by the second message. However, this still means that four extra TCP messages must be sent: two for setting up and two for shutting down. So, two extra round trip times are added to the time period needed for the HTTP/1.0 handshake. The situation is worse than this, because TCP demands that, after the server computer has sent the final message of the four-stage shutting down handshake, it waits for 240 seconds before finally killing the connection. This is in case any TCP messages from within the connection are still floating around the Internet. The period is chosen to be twice the maximum time to live of the IP messages containing the encapsulated TCP messages. For some busy servers, this can lead to thousands of connections being in the wait state, which wastes the computer memory used to store details of connections.

The problem with HTTP/1.0 and TCP mentioned in the previous two paragraphs stem from a single time-related factor: a fresh TCP connection is established for each HTTP handshake. If it was the case that a typical WWW session only involved a single HTTP handshake between pairs of computers, then the overheads are unavoidable. However, this is not the case. Fetching a WWW page is likely to involve several HTTP handshakes, because of pictures embedded within pages. Further, overall usage patterns tend to involve accessing a sequence of WWW pages from the same location. In this case study, the absolute time penalty of extra message round trip times depends on exactly how the channel is implemented. The use of one TCP connection for each HTTP/1.0 handshake stems from a desire for simplicity of implementation, and dates from the earliest days of the World Wide Web. However, there is distinct scope for improvement.

A modest suggestion for improvement was made by Jeffrey Mogul: Persistent HTTP (P-HTTP). As its name suggests, connections are made persistent. The server does not shut down the connection after sending back an HTTP response. Instead it leaves it open for a time period, and the client is able to reuse it if necessary. Experiments have shown this leads to significant performance improvements, by eliminating unnecessary round trip time delays. This idea has been incorporated into HTTP/1.0.

The idea can be taken further, and is one of several proposed new features for Next Generation HTTP (HTTP-NG), which was being considered at the time of writing in 1997, and is largely based on the work of Simon Spero. HTTP-NG provides a mechanism that allows several HTTP communication channels to be multiplexed on to a single TCP channel. Thus, parallel handshakes are possible, in addition to HTTP/1.1's sequential handshakes within a single connection. One further feature is that responses may be sent before requests, when the server can predict what a client will need next.

In more detail, several parallel channels are implemented by HTTP-NG. One of these is used for control information: requests sent from client to server, and meta-information responses from server to client. The other channels are used for sending information responses from server to client. Another change proposed for HTTP-NG is that lines of text with MIME conventions are not used to represent requests and responses. Instead, a simplified version of ASN.1 is used, to facilitate computer processing. In addition to these performance-related enhancements, HTTP-NG also has enhancements reflecting the fact that numerous commercial applications make use of the WWW. These include security features, means for identifying client users, and means for conveying information about charging and restrictions.

Finally, returning to TCP briefly, the reader might have observed that a TCP message carries six code bits in its header, but that only four (ACK, FIN, SYN and

RST) have been mentioned so far. For completeness, the remaining two — PSH and URG — will be mentioned here, since they have time-related significance, although are not relevant to the case study example. The PSH bit is used to turn the basic stream-oriented service of TCP into a unit-oriented service. The information in a message with the PSH bit set is regarded as the end of a unit. The practical significance of this is that a receiver must forward the information to its user immediately, rather than waiting for any more information to arrive in subsequent messages. The main reason for using the PSH bit is to avoid transmitters trying to perform concatenation by waiting for a short period to see if further information appears from the user.

The URG bit is meant to place the receiver into an 'urgent' mode, that gives high priority to handling incoming information. The urgent pointer field of the TCP message header is also involved. It gives an offset from the message sequence number at which the urgent information ends. All unprocessed information with sequence numbers up to, but not including this number, should be processed urgently at the receiver. It is up to the user at the receiver end to work out which information is really urgent. The main reason for having the URG bit was to allow users to interrupt normal communication, for example, if a terminal user presses an interrupt key to abort some operation in progress. In this case, a single character (e.g., control-C) is likely to be in the urgent message, and will be treated as the only real urgent item, with earlier information being discarded. In practice, the urgent feature has been little used, and an alternative use has been similar to that of the PSH bit. That is, units can be identified within the stream of information. In this way, PSH and URG allow two different types of units to be identified, if this is useful to an application.

Space issues

In this case study, the two communicating computers remain the same, but a different TCP channel is used for each HTTP/1.0 handshake. The different channels are multiplexed on to a single IP channel between the two computers. Note that IP provides a connectionless service, so communications over the IP channel are unsegmented, that is, each communication conveys one IP message. In turn, the IP channel is implemented using some combination of filtering, switching, multiplexing and splitting over real physical channels, as described in Section8.3.2.

As well as its various time-related features, TCP also provides the support for the multiplexing of different TCP channels on to communications over the single IP channel. The central notion needed is that a computer has a collection of **ports**, each with a unique identification number. TCP channels are between pairs of ports, rather than computers as a whole.

A TCP channel can be uniquely identified in the Internet by combining one endpoint's computer identifier and its port number with the other endpoint's computer identifier and its port number. This gives a channel identifier which, together with the unique computer identifiers, gives an exact definition of the communication space for each communication. Conceptually, it is reasonable to regard each such channel as existing permanently, and being used by a sequence of successive TCP connections over time. In practice, the actual number of active channels is limited by the resources available at computers.

Each TCP message header has a source port field and a destination port field. These identify the ports used at the sending computer and receiving computer respectively. For TCP/IP, these port numbers, together with the source and destination identifiers carried earlier in the IP header, mean that each message carries the unique identifier of its channel. Given that port numbers are 16 bits long, up to 65 536 ports could be provided, in principle. When a computer receives a TCP message, it uses the destination port number to decide which process should be given the message for further processing.

When a connection is established, the channel to be used has to be chosen, that is, the port numbers have to be chosen. One possibility is to have a prior agreement between two applications about which port numbers will be used on their respective computers. Then, these numbers are included in the TCP messages. However, it is more usual not to have such a pre-agreement, and this is the case in client-server relationships of the type occurring in this case study. Servers are not aware of the identity of all of their potential clients, never mind pre-negotiating port number choices with all of them.

To deal with this, the Internet has a notion of **well-known port numbers**. Each well-known Internet application has a port number allocated to it. HTTP has the number 80 allocated to it. Other examples include FTP (file transfer protocol), which is 21; TELNET (terminal protocol), which is 23; and SMTP (electronic mail), which is 25. Then, when setting up a connection, a client chooses a unique spare port number at its end, and uses the well-known number at the server end. Port numbers in the range 0 to 1023 are centrally allocated as well-known port numbers. Port numbers higher than this range are available for general use. For information, a central list of well-known uses of these higher port numbers is published. However, any such port numbers may also be freely used by other applications.

The same notions of ports and well-known ports occurs in the User Datagram Protocol (UDP). This supplies a connectionless service between ports on computers which is analogous to TCP's connection-oriented service between ports. TCP and UDP share the set of well-known port numbers and, where a particular application may use both TCP and UDP, the same well-known number is used. Apart from using port numbers, UDP does not involve extra implementation work, since IP supplies an adequate underlying connectionless service. UDP messages may carry a TCP-like checksum in the UDP header, to allow error detection, but this is optional.

For a sequence of HTTP/1.0 handshakes to fetch a WWW page, the identifiers of the channels used will have three components in common: the client computer identifier, the server computer identifier and the server computer port number (80); only the client computer port numbers will vary. The number of channels in active use by connections may be limited by the client and/or the server computers. One popular WWW browser never uses more than four channels for HTTP/1.0 at once, by default, in order to reduce loading on the client and also to reduce the risks of these multiplexed channels slowing each other down. For HTTP/1.1, a browser normally uses only one channel at a time, and at most two. Most servers have a limit on the total number of channels being serviced at one time, and this impacts on the number of channels allowed for a particular client. In addition, the compulsory 240 second delay after closing each connection may cause the apparent number of simultaneously active channels to be artificially high.

Summary of the TCP/IP issues

The problem is to implement appropriate HTTP request-response handshakes to fetch a WWW page. This is solved using an exchange of IP messages over a best-effort channel between the client computer and the server computer. Each IP message carries a TCP message, belonging to a particular TCP channel between the two computers. The source and destination port fields of the TCP message header are used to identify which of the TCP channels is being used. Each HTTP/1.0 handshake involves TCP being used to set up, use, then shut down, a reliable connection over one of the TCP channels. This involves time overheads. Also, the normal TCP congestion avoidance algorithm is applied independently to each channel, and is likely to lead to further time overheads. The information exchanged for the HTTP handshake is encoded as a sequence of bytes, and these are carried in the information fields of TCP messages. The number of TCP messages needed depends on the maximum per-message information length agreed by the computers when each connection is set up.

A summary of the implementation can be gleaned from Figures 9.5 and 9.8. Each request-response communication pair shown in Figure 9.5 is implemented using a sequence of IP message communications of the type shown in Figure 9.8.

Now, the main remaining problem is the implementation of the IP channel between the client and server computers. Here, this will take place via two **Internet providers**. An Internet provider is an organization, sometimes commercial, sometimes governmental, that allows individuals or organizations to connect to the Internet. Together, the Internet providers, and the channels between them, form the backbone of the Internet. Computers are then connected to this via their provider. There is a rough correspondence between Internet providers and the administrators of Autonomous Systems, described in Section 8.3.2.

Sections 9.4, 9.5 and 9.6 deal with the implementation of IP channels for three parts of the journey involved:
- between the European home computer and its local Internet provider;
- between two Internet providers, one European and the other American;
- between the American organization computer and its local Internet provider.

After this, the implementation of the overall communication channel, using these sub-channels, will be considered in Section 9.7.

9.4 HOME COMPUTER AND THE INTERNET

In this case study, the client computer is located in its user's home. The means of connection to the outside world will be the most common one: the normal telephone system. In some countries, an ISDN channel might be a practical and affordable alternative. Other possibilities are radio, if the user is also an amateur radio enthusiast, or cable, if the user has an enlightened cable television company operating in the area.

When the normal telephone system is used, it is necessary to insert a modem between the computer and the telephone line, in order to represent binary computer data using analogue telephone signals intended for representing speech. The outside world is represented by an Internet provider. By paying a subscription and/or connection time costs, the user is able to telephone a modem belonging to the provider, and then exchange information between the home computer and one of the provider's computers.

From a space point of view, the two modems and the telephone system can be regarded as implementing a direct channel between the two computers. The normal interface between a computer and a modem involves the exchange of bytes of information. In detail, the actual interface is likely to involve each byte being transmitted synchronously as a sequence of bits, with start and stop bits, as described in Section 3.2. However, this level of implementation detail can be ignored here. At the byte level, there is an asynchronous interface, in that bytes need not be transferred at regular time intervals. Thus, the raw service provided by the channel is a full duplex, asynchronous byte transfer service. This is not an error-free service, since telephone line noise might damage bytes. In an all-digital telephone system, the error rate is likely to be very low; however, in older analogue systems, the error rate might be fairly significant.

The requirement is to allow the exchange of IP messages between the two computers using the telephonic channel. In principle, it is possible to envisage using the channel directly, just by transmitting each IP message as a sequence of bytes, and by regarding channel errors as being within IP's guarantee of a best effort only service. In practice, there are a number of issues that are better addressed by an extra protocol to provide an extra layer of implementation between IP and the telephone line.

Here, the Point-to-Point Protocol (PPP) is used for the implementation work. This connection-oriented protocol is designed for use in a variety of situations where two computers are directly connected by a simple full duplex channel that does not reorder information. Because of this generality, there is quite a lot of detail associated with PPP. Here, only the features relevant to the case study are covered. An older alternative, which is rather simpler but less powerful, would be the Serial Link IP protocol (SLIP). PPP is discussed under the usual headings of information, time and space.

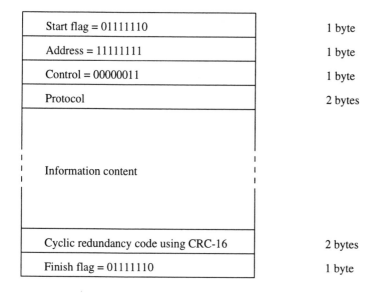

Start flag = 01111110	1 byte
Address = 11111111	1 byte
Control = 00000011	1 byte
Protocol	2 bytes
Information content	
Cyclic redundancy code using CRC-16	2 bytes
Finish flag = 01111110	1 byte

Figure 9.9 Format of PPP message

PPP information issues

PPP information is transmitted in the information field of an HDLC-style message. The format is illustrated in Figure 9.9. Note that the illustration includes the standard HDLC-style framing bytes, used to mark the beginning and end of the message. Within the message, **byte stuffing** — a byte version of bit stuffing, as described on page 90 in Chapter 3 — is used to prevent the framing byte value occurring and, optionally, to prevent other values agreed by the two computers from occurring. When a stuffable byte occurs, it is preceded by a byte with the value **01111101**, and is itself exclusive OR-ed with the value **00100000**. For obvious reasons, the byte value **01111101** is treated as stuffable.

The address and control fields are standard in HDLC-style message headers, but here always have the same value: the address field **11111111** means 'all stations', and the control field **00000011** means that the message is an unnumbered information frame, in HDLC parlance. Given that these values are constant, messages can be compressed by omitting the two bytes entirely. The protocol field that follows the address and control fields is not permitted to begin with a **11111111** byte, so the compression cannot cause ambiguity. This compression is not done by default, but can be agreed by the two computers when a PPP connection is established. At the end of the message, a standard CRC-16 cyclic redundancy code is carried, as normal for HDLC-style messages.

The protocol field specifies the protocol that owns the message encapsulated in the PPP message. In the case of IP, the value of this 16-bit field is 0021 (hexadecimal), and the bytes of the IP message follow in the information content.

This is not the only value of interest, because two other sub-protocols are needed to establish a PPP channel, and to establish an IP over PPP channel, respectively. The first sub-protocol is the Link Control Protocol (LCP), which has the protocol field value C021 (hexadecimal), and the IP Control Protocol (IPCP), which has the protocol field value 8021 (hexadecimal).

Between establishing a PPP connection and using it to implement an IP channel, an optional authentication handshake is possible. The most secure of the protocols supported for this is the Challenge Handshake Authentication Protocol (CHAP), which has the protocol field value C223 (hexadecimal). The 16-bit values used for protocols always have an even top byte and an odd bottom byte. This allows a simple compression technique for values with a zero top byte: just omit the top byte. If a receiver sees an odd top byte, it can deduce that compression has occurred. To harness this facility, the values used for networking protocols, such as IP, have a zero top byte. The compression is not done by default, but can be agreed when the PPP connection is established.

The default maximum size of the information carried by a PPP message is 1500 bytes. However, this size can be negotiated upwards during the establishment of the PPP connection if both computers wish. For sub-protocols defined as part of PPP, such as LCP and IPCP, the format of the information carried has a standard sub-protocol header format. This has a one-byte code number, a one-byte identifier and a two-byte length. The sub-protocols use request-response handshakes, and the code number indicates the type of request or response. Each request is sent with a unique identifer, and its response is sent back with the same identifier. The length gives the total length of this header plus the information following. For some of the requests and responses, the information following consists of a list of options to be negotiated. Each such option has a standard format, with a header containing a one-byte option type and a one-byte total length of the option. Figure 9.10 shows an example of an LCP request message with two options present (the HDLC-style header and trailer are omitted).

As has already been mentioned, when the PPP connection is established using LCP, compression can be agreed, to eliminate the address and control bytes, and one of the two protocol bytes. This can reduce the per-message overhead from eight bytes to five bytes. When the IP channel is activated using IPCP, compression of TCP/IP headers using Jacobson's algorithm can be agreed. This compression scheme (mentioned in Section 2.3.1) can reduce the header size from 40 bytes to five bytes for messages sent after establishment, and before termination, of a TCP connection. A further optimization, applicable when information is being sent in one direction only within a TCP connection, can reduce the header size to only three bytes. Note that two of these three bytes are for the TCP checksum, which cannot be compressed without reducing its effectiveness. So, in the best circumstances, a 584-byte PPP message carrying a 576-byte TCP/IP message, of the default maximum length and carrying 536 information bytes, can be compressed to a 544-byte message carrying 536 information bytes. The percentage of information

Protocol = **11000000 00100001**	2 bytes
Code	1 byte
Identifier	1 byte
Length	2 bytes
Option 1 type	1 byte
Option 1 length	1 byte
Option 1 information	
Option 2 type	1 byte
Option 2 length	1 byte
Option 2 information	

Figure 9.10 Example LCP request message with two options

per message increases from 92% to 98.5%. This is significant on a slow-speed telephone link.

PPP time issues

The rate at which information can be sent over the telephone line depends on the quality of the modems used and the quality of the telephone line. The fastest modems are capable of a 33.6 kbits per second transfer rate over a non-noisy analogue telephone link. Some 'intelligent' or 'smart' modems also offer compression which can, in theory, double or even quadruple the transmission rate. However, this improvement is not often obtainable in practice, and only a small rate increase is likely on average. Given that 10 bits are transmitted per byte (start bit, eight byte bits and stop bit), the fastest byte transmission rates are no better than about 3000 bytes per second, i.e., only about two maximum size PPP messages per second.

In the presence of errors, smart modems reduce the transmission rate in an attempt to compensate. The cyclic redundancy code in PPP messages is used to detect errors that do occur. Any messages received with CRC errors are discarded silently. Where necessary, sub-protocols such as LCP and IPCP make use of timeouts to recover from loss of messages. No timeouts are used for PPP messages carrying IP messages. Any losses are treated as a feature of the best effort IP service guarantee.

A basic absolute time period involved in communication between the home computer and the Internet provider is the duration of the telephone call made to supply the physical connectivity. This time period normally corresponds very closely to the time period for the PPP connection. It is possible that there may be a succession of PPP connections during the same telephone call, but this is unlikely unless there is some human management intervention.

During the time period of the PPP connection, there may be one or more activations of an IP channel. In favourable circumstances, there will be a single IP channel that is available throughout a telephone call. However, this may not be the case if a telephone channel becomes error-prone. If PPP detects this, it may deactivate the IP channel, and reactivate it later when the errors have lessened.

The time period of the telephone call should be determined by the usage of the IP channel service on the home computer, assuming the software for Internet access is of a reasonable level of sophistication. When the need to transmit an IP message arises, and no telephone call is in progress, a three-step process takes place. First, a telephone connection is set up, then a PPP connection is set up, and finally an IP channel is made available.

The PPP connection set-up involves an LCP handshake of the type described in Section 4.4. This includes negotiation on such matters as PPP header compression and maximum PPP message size. Another negotiated feature is the use of the Link Quality Report sub-protocol. With this, the computers exchange reports on messages received, discarded and transmitted, at agreed time intervals. From these reports, the quality of the telephone channel can be ascertained, and action such as temporary suspension or termination can be taken. A further matter of negotiation is whether an authentication protocol is to be used to ensure that no imposters are involved in the communication. If it is, authentication takes place immediately after PPP connection establishment, and optionally may be repeated again during the course of the connection.

The CHAP authentication protocol involves a three-stage handshake, and makes use of the MD5 message digest algorithm. The messages exchanged have the same format as LCP or IPCP options: a code byte, an identifier byte, two length bytes and then information. The authentication depends on a secret value previously shared between the two computers. First, the authenticator sends a **Challenge** message to its peer, containing a unique and unpredictable challenge value. The peer then sends back a **Response** message containing a value obtained by applying MD5 to the Challenge message identifier, the secret value and the challenge value. Finally, the authenticator checks the response value against its own computation of the correct value, and either sends back a **Success** message or a **Failure** message.

After PPP connection establishment and successful authentication, there is an IPCP handshake to activate the IP channel. This is exactly the same as the LCP handshake, except that there are different (and fewer) options to negotiate.

One option is whether Jacobson header compression is to be used. The only other option concerns IP identifiers, and it is discussed in the next section on space issues. After the IPCP handshake is complete, the home computer is able to transmit IP messages, as required. Of course, the Internet provider's computer can also transmit IP messages.

It is not desirable to leave the telephone connection set up indefinitely once it is established, for cost reasons or because of inconvenience. However, there is no clear indication of when the IP channel service is no longer required, given that it is a connectionless service. To deal with this, a timeout mechanism is used. A timer is used to measure the time for which the IP channel has been idle — it is reset every time an IP message is sent or received. If the timer reaches a threshold value, a few minutes being typical, the IP connection is deactivated, then the PPP connection is shut down and, finally, the telephone connection is shut down. Simple IPCP and LCP handshakes are used to deactivate the IP channel, and shut down the PPP connection, respectively.

It may be that a user is still navigating the WWW when the IP channel is timed out for inactivity, for example, if someone spends a few minutes reading a page just received. If so, the next request made will create the need to transmit an IP message. Then, the telephone connection will be established again. This causes a non-trivial delay to be introduced. However, such a delay must be traded off against savings made by temporarily terminating the telephone call.

PPP space issues

The communication space is not complicated, since it just involves two computers and a physical telephone line channel between them. One issue is the location of the Internet provider's computer. Ideally it should be reasonably close to the home computer, in order that the telephone channel has good quality and minimum cost. Most major Internet providers make telephone facilities available that can be called at local telephone rates, to minimize the cost to users. The home computer must know the telephone number of the Internet provider, so that it can establish telephone calls to one of the provider's computers. Thus, telephone numbering is the identifier scheme used for computers in communications over the telephone channel.

The other issue concerns Internet identifiers. The home computer needs an Internet identifier so that it can be identified correctly for IP purposes, in particular so that messages are correctly routed to it. Normally, the identifier is assigned by the Internet provider, so that the computer appears to be part of a network administered by the provider. The fact that Internet identifiers are allocated in blocks of at least 256 means that it is not practicable for personal computers to be allocated identifiers independently.

With provider-allocated identifiers, one possibility is for each subscriber to be allocated an identifier permanently. Then, subject to appropriate authentication checks, this can be used each time an PPP connection is established. A more flexible alternative is for the provider to allocate a temporary IP identifier each time a PPP connection is established. This allows the provider to use a range of identifiers related to the number of incoming modems rather than to the total number of subscribers.

IPCP connection establishment includes negotiation of the IP identifier to be used by the home computer. It is possible for the home computer to state which identifier it wants to use and, if this is acceptable, the provider's computer would accept it. Otherwise, the provider's computer can specify an IP identifier that should be used by the home computer for the duration of the IP connection.

Summary of the PPP issues

PPP messages are transmitted over a telephone channel between the home computer and the Internet provider's computer. These messages carry IP messages. A PPP connection is established for the duration of the telephone call, and then an IP channel is made available during the time period of the connection. Then, even with compression of message headers, the fastest IP message transmission rate is around 3000 bytes per second, rather less on a noisy telephone line. The home computer is identified by a unique IP identifier, and the Internet provider must ensure that any messages sent to the computer with this identifier are sent over the telephone channel.

9.5 WITHIN THE INTERNET

In Section 9.4, transmission of IP messages over a single physical channel was considered in some detail. For IP message routing, this channel is just one hop of many in a message's journey through the Internet. The detailed description was possible because the channel is known to be of a specific type: a telephone channel. However, in this case study, most of the hops of the journey occur between the WWW client's Internet provider and the WWW server's Internet provider. In real life, the full details of the physical channels used are not likely to be known to the user and, moreover, may vary from message to message. Therefore, in this section, only a general picture is given, covering typical characteristics of routes between Internet providers in Europe and the USA at the time of writing in early 1997. The main complicating factor is one of sharing: while the capacity of the telephone channel can be dedicated to one computer's IP traffic, other Internet channels will be shared between many users' communications. This makes absolute time behaviour rather more unpredictable.

Inter-networking issues

The Internet began as a resource to support scientific researchers working in different locations, allowing sharing of scientific data and computational resources. In this guise, the Internet providers were research funding agencies in different countries. In particular, the National Science Foundation (NSF) in the USA provided the main Internet backbone. The physical connectivity required was provided by leased lines belonging to national telecommunications organizations. As use of the Internet percolated beyond the scientific research community, commercial Internet providers began to appear in many countries. These were specialized companies making use of leased lines, in some cases, and were the telecommunications companies themselves in other cases.

Given that the whole point of the Internet is to provide connectivity between as many computers as possible, it is not desirable for each Internet provider to be an independent entity, supplying an IP service only for those computers that subscribe to its service. Thus, routing of messages between providers is necessary. This is where the notions of Autonomous Systems (ASs) and exterior gateway protocols (EGPs), described in Section 8.3.2, are important. Each Internet Provider administrates an AS, and EGPs are used to police the routing of messages between providers. Initially, gateways between providers developed on an *ad hoc* basis. However, with an increasing number of providers, and a desire for efficiency, the idea of an **Internet Exchange** (IX) began to appear in the early 1990s. This is a location to which different providers supply channels, and the IX provides a high speed switching service between these channels.

In 1995, commercial Internet provision was sufficiently advanced that the National Science Foundation stopped providing the main Internet backbone, and handed this task over to commercial providers operating in different parts of the USA. Its continued support for the Internet came through the establishment of four Internet exchanges, known as **Network Access Points** (NAPs):

- New York (actually in Pennsauken, NJ), operated by Sprint;
- Chicago, operated by Ameritech Advanced Data Services;
- Washington DC, operated by MFS Datanet; and
- San Francisco, operated by Pacific Bell.

The Washington DC NAP is called MAE-EAST (Metropolitan Area Exchange, East), and it is pleasant to know that MFS Datanet also operate a non-NAP Internet exchange in San Jose, CA called MAE-WEST (as well as five other MAEs in the USA, in early 1997).

The NSF also funds the Routing Arbiter (RA) project, which supplies a routing server at each NAP. The servers do not actually route messages, but rather supply routing information to the switches. Thus, they act as central repositories of information about Internet connectivity. One benefit is that Internet providers only need to supply routing information to the RA, rather than to all other providers meeting at the NAP. The NSF also funds the very high speed Backbone Network Service (vBNS) project, which supplies a next-generation backbone to connect together various NSF high performance computing and communications centres. This five-year project began in 1995, with the vBNS operating at 155 Mbits per second, and the intention that technological advances would allow a 2200 Mbit per second service by 2000. The vBNS is not part of the normal Internet provision.

As well as the NSF Network Access Points, there are many other Internet exchanges. In Santa Clara, CA, the Commercial Internet Exchange (CIX) supplies a switch to connect its members. In Europe, there are exchanges in most of the European Union countries. For example, there is the London Internet Exchange (LINX) in the United Kingdom, the Deutsch Commercial Internet Exchange (deCIX) in Germany, and the Amsterdam Internet Exchange (AMS-IX) in the Netherlands. These commercial IXs (usually non-profit making, in fact) connect together all of the major commerical Internet providers in the country, as well as any government provision.

Against this background of how the Internet is organized, there are many detail possibilities for this case study, depending on which European country is involved, which Internet providers are involved, and which area of the USA is involved. The only common feature that is guaranteed to affect performance adversely is the Atlantic Ocean. This necessarily interposes a distance of around 6000 km, and moreover a distance that is spanned by a limited amount of cabling. Thus, it acts as a lengthy bottleneck to communications.

In addition to the journey under the Atlantic Ocean, other long distance communications will dominate the quality achieved for the WWW page retrieval. These communications may be necessary in the European country and/or in the USA. Any further communications involved in implementing each IP message's journey should only add very small overheads, being short trips around LANs or MANs belonging to Internet providers. Extra significant overheads will only occur if the provider is under-resourced to cope with its total IP message load, something not unknown among smaller and newer providers.

For long distance channels, many providers use leased lines from telecommunication companies. This includes leased lines under the Atlantic. The typical basic speed for such lines in early 1997 was the T1 rate: 1.544 Mbits per second. Some, more expensive, lines operate at the T3 rate: 44.736 Mbits per second. In 1996, a trans-Atlantic line using SONET technology became available between the USA and Sweden, operating at 155 Mbits per second. Of course, much slower lines, down to 64 Kbits per second still exist in some places, either as part of public X.25 networks or making use of ISDN channels.

As an example of the trans-Atlantic bottleneck, it is interesting to look at an example: the UK Joint Academic Network (JANET) which, among other things, is a government-funded Internet provider for academic institutions. In 1995, it had the so-called 'fat pipe' — a 4 Mbits per second link under the Atlantic — shared by all of its trans-Atlantic IP messages. In early 1996, the capacity was doubled to 8 Mbits per second and then, within another few months, six new T1 links were added, raising the total capacity to around 17 Mbits per second. After each upgrade, a WWW user could notice some speed-up, but for only for the first few days before increased usage soaked up the new capacity.

When communicating over leased lines, PPP or something similar is normally used. Thus, the leased line is treated as a distinctly high speed telephone channel. Of course, this is exactly what is being offered by the telecommunications company. The only difference is that the 'telephone call' exists permanently between two fixed end points. To give a different example here, there is a brief look at the information, time and space issues involved in a different level of provision. This is where a Switched Multimegabit Data Service (SMDS) is supplied to the Internet provider by the telecommunications company, rather than just a raw bit transmission capability. A very short summary of SMDS was given earlier in Section 7.5.5. It is interesting to look at SMDS in more detail, because it involves one approach towards using cell-based (i.e., fixed-length message based) networking.

SMDS information issues

SMDS allows the transmission of messages that are up to 9188 bytes long. The maximum was chosen to accommodate the largest type of LAN message. Each message is encapsulated in an SMDS Interface Protocol (SIP) level 3 message for transmission. The format of an SIP level 3 message is shown in Figure 9.11. The header contains an eight-bit sequence number, which can be used to detect missing messages, and a 16-bit message length. These are followed by 64-bit fields for the destination and source identifiers of the message. The other sub-fields following give information about the later components of the message, such as padding and the CRC. The information content is padded so that it is a multiple of four bytes long, to facilitate message handling. The trailer optionally includes a 32-bit CRC using the CRC-32 generator to allow error detection. This is followed by a repeat of the message sequence number and length that is in the header. The maximum possible message length is 9240 bytes.

In turn, SIP level 3 messages are encapsulated into SIP level 2 messages. For an SMDS network using the DQDB MAN interface, these are 48 bytes long, and are then preceded by a five-byte header to form a 53-byte message (a 'cell'). Each level 2 message can carry 44 bytes of information, so each level 3 message communication must be segmented into 44-byte sub-communications. Note that the maximum length of 9240 bytes is an exact multiple of 44. The format of an SIP level 2 message is shown in Figure 9.12. The first two bits of the header identify what kind of level 3 information is being carried:

- complete level 3 message;
- first part of a level 3 message;
- one of the middle parts of a level 3 message; or
- last part of a level 3 message.

This is followed by a four-bit level 2 message sequence number, and a 10-bit level 3 message identifier, which are used to guide reassembly of the level 3 message by the receiver of the level 2 messages. The trailer contains a six-bit message length and a 10-bit CRC using the CRC-10 generator to allow error detection. Note that this message format is not just used for SIP level 2. It is also used for ATM Adaption Layer 3/4 (AAL 3/4).

Given that the interface to the SMDS network uses the DQDB MAN standard, the 44-byte SIP level 2 messages are then encapsulated into 53-byte DQDB

Reserved	1 byte
Sequence number	1 byte
Message length	2 bytes
Destination identifier	8 bytes
Source identifier	8 bytes
Other sub-fields	4 bytes
Optional header extension	0 to 20 bytes
Information content	
Padding	0 to 3 bytes
Cyclic redundancy code	0 or 4 bytes
Reserved	1 byte
Sequence number	1 byte
Message length	2 bytes

Figure 9.11 Format of SIP level 3 message

Message type	2 bits
Sequence number	4 bits
Message identifier	10 bits
Information content	44 bytes
Information length	6 bits
Cyclic redundancy code	10 bits

Figure 9.12 Format of SIP level 2 message

messages of the type described in Section 6.4.5. Note that other interfaces to an SMDS network are possible. For example, a frame relay style interface would use a different form of encapsulation from the cell-style encapsulation described here.

SMDS time issues

The name SMDS incorporates a statement about its speed: the service is intended to operate at multimegabit speeds. A normal interface to the SMDS service uses a T1 or T3 speed link. With the former, a 1.17 Mbits per second service is offered to subscribers. With the latter, 4, 10, 16, 25 or 34 Mbits per second services are offered. The subscribed-to rate is policed by the SMDS provider, to ensure that messages are not sent at too fast a rate, although occasional bursts at the full T1 or T3 rate may be permitted. The SMDS service has also been made available at sub-multimegabit speeds — 56 and 64 Kbits per second and multiples thereof — to give an alternative to the connection-oriented X.25 and frame relay services.

SMDS offers a connectionless service. As just seen in the previous section, the communication of each message is implemented by segmenting it into sub-communications of 44-byte messages. The CRC carried by each SIP level 2 message allows discarding of messages that are affected by errors when in transit. The other control information carried allows fragmented SIP level 3 messages to be reassembled. The reassembly process must incorporate a timeout, to guard against cases where the final piece(s) of an IP level 3 message are lost or discarded. That is, an absolute time bound is placed on the duration of the communication of each message, as seen by the receiver. The complexity of the reassembly process depends on how the SMDS network is implemented. For example, if it is based on ATM, then the 53-byte ATM messages are guaranteed to be delivered in the order of transmission, so reassembly just involves accumulating successive SIP level 2 messages received. The first one contains the SIP level 3 message length, so an appropriate amount of buffer space can be pre-allocated for the accumulation.

SMDS space issues

In a public SMDS service, the identifier scheme used in SIP level 3 messages is the ITU-T standard E.164 15-digit decimal numbering scheme, as defined for ISDN networks. These identifiers can be used to specify MANs that are being inter-connected by the SMDS network. There is also the possibility of SMDS messages being routed to other types of service provided by the telecommunications community. To cater for different identifier schemes in other services, the first four bits of the 64-bit identifier specify the identifier scheme used in the remaining 60 bits. Thus, one of these four-bit prefixes denotes the E.164 scheme. One possible alternative is the use of the IEEE 16-bit or 48-bit LAN identifier scheme, if LANs are connected together by the SMDS service.

The internal structure of an SMDS network is not specified. Indeed, a main feature of SMDS is that it has a technology-free service specification. Thus, for example, it might be implemented using frame relay networking or ATM networking, or just directly using T1 or T3 lines. The implementation might mean that further encapsulation of messages occurs before physical transmission. There is also no requirement that all messages follow the same path, as would be the case for a frame relay service, for example. Therefore, SMDS can be seen as an offering of telecommunications providers in the traditional black box sense. Its major advance is in providing a connectionless service, as opposed to the more usual connection-oriented services.

Multiplexing may affect the use of an SMDS channel in two ways. First, the channel to the SMDS service may be shared by messages that are being sent to different destinations. Second, IP messages corresponding to many different IP channels may be passing along the same SMDS channel. Thus, it is fairly unlikely that one WWW page retrieval would be able to acquire the full transmission capacity available from the service. The second type of multiplexing particularly afflicts most long distance channels that are likely to be involved in implementing the WWW communication.

Summary of within-Internet journey

Each IP message will make several, possibly many, hops from switch to switch during its trip through the Internet. The journey will involve the two Internet providers at each end, and possibly other intermediate Internet providers. Messages pass between Internet providers at Internet Exchanges, or at *ad hoc* gateways between providers. When passing through each provider, there will be one or more long distance hops. These may involve PPP (or similar) being used over leased lines or the use of SMDS, ATM, frame relay or X.25 services. One of the long distance hops will be across the Atlantic ocean — usually underneath, but there are satellite links above as well. Apart from the long distance hops, there are likely to be hops over MANs or LANs belonging to each Internet provider. On each hop, the IP message may be straightforwardly encapsulated, for example, as with PPP, or its communication may be segmented (or possibly concatenated with others), for example, as with SMDS.

The result is that IP messages are delivered over a channel implemented between the points at which the WWW client computer and the WWW server computer link to their respective Internet providers. The delivery time depends on the physical distance spanned (which may be as long as 10 000 km), the speed of physical channels and the level of multiplexing on physical channels. In some cases, congestion on physical channels will lead to the complete loss of messages that have to be discarded.

9.6 WWW SERVER AND THE INTERNET

The final component of the IP messages' journey is between the WWW server and its Internet provider. In the simplest case, where the organization involved does not have sophisticated networking, this might be the same as for the home computer. That is, the WWW server has a dial-up channel using the normal telephone system or ISDN. At the other extreme, the organization might have a DQDB or FDDI backbone network connecting together many local area networks. If so, messages may have to travel a few hops between the WWW server computer and the switch connected to the Internet provider. For example, perhaps a hop over one LAN, then a hop over the backbone and finally another hop over a different LAN. Here, to keep the example rather simpler, an intermediate scenario will be chosen. This is that the WWW server and the Internet switch are both on the same LAN, specifically on an ethernet LAN. Thus, each IP message has to be transported across the ethernet.

The principles and operation of ethernet message broadcasting networks were covered in Section 6.4.1, so this section just summarizes the information, time and space issues involved in using an IEEE 802.3 standard ethernet particularly for the transport of IP messages. This involves the use of the IEEE 802.2 Link Level Control (LLC) protocol, already mentioned in Section 4.2.2. As with PPP and SIP level 3, LLC is used for encapsulating IP messages prior to exposing them to the reality of a physical channel.

Ethernet information issues

IEEE 802.2 LLC, like PPP, is in the HDLC family of protocols, so its messages have the standard HDLC-style format. However, they have a slightly more variant form than that of PPP. Here, the use of the connectionless LLC1 protocol will be assumed, since this is the case in the majority of LAN installations. Also, an extension of the LLC header, called the Sub-Network Access Protocol (SNAP) is assumed. Again, this is the norm when transmitting IP traffic over an IEEE 802 LAN. The LLC1 plus SNAP message format is shown in Figure 9.13. Note that the eight-byte LLC/SNAP header contains constant values. In general, the LLC identifier fields contain seven-bit identifiers, identifying processes within the communicating computers. The least significant bit of the destination identifier indicates whether it is an individual identifier or a group identifier, and the least significant bit of the source identifier indicates whether the message is a command or a response. In this case, the constant values of 170 indicate that a SNAP communication is taking place.

All of the messages are of the HDLC unnumbered information type (as used by PPP), and so have the control field set to the constant value **00000011**. Note that the LLC1 message does not end with the usual 16-bit cyclic redundancy code; this is because the underlying IEEE 802 network protocol has a 32-bit CRC to allow error detection. The SNAP identifier of zero indicates that the following two bytes

LLC destination identifier = 170	1 byte
LLC source identifier = 170	1 byte
LLC control = **00000011**	1 byte
SNAP identifier = 0	3 bytes
SNAP EtherType = 2048	2 bytes
Information content	

Figure 9.13 Format of LLC1 plus SNAP message

contain an ethernet type code, and the value of 2048 indicates that the message contains an IP message. These type codes originate from the proprietary Ethernet, which carries the 16-bit codes in the field that is used for the message length in IEEE 802.3 messages. The values used are allocated by Xerox, the originators of Ethernet.

For transmission over an IEEE 802.3 Ethernet style network, the LLC message is encapsulated in a message of the form shown in Figure 6.6. That is, assuming 48-bit computer identifiers are used, a seven-byte preamble, a start byte and a 14-byte header are prepended and a four-byte CRC trailer is appended. For TCP/IP messages, which have a minimum length of 40 bytes, padding of the 802.3 message will never be needed, since it is certain to exceed the minimum length of 64 bytes. The maximum length of an 802.3 message (excluding preamble and start byte) is 1518 bytes so, excluding the 18-byte 802.3 header and trailer and the 8-byte LLC/SNAP header, the maximum IP message size is 1492 bytes. This cannot be exceeded, since there is no segmentation mechanism.

Ethernet time issues

The parameters for an IEEE 802.3 standard Ethernet are a 10 Mbits per second transmission rate, with a latency of only a few microseconds. For a TCP/IP message with the default maximum length of 576 bytes, together with a 26-byte encapsulation overhead, plus preamble and start byte, the length of the communication time period is around 500 microseconds. The unknown factor is the delay before message transmission can start. The CSMA/CD mechanism means that, first, any existing transmission must be allowed to finish and, second, contention may occur with other computers also wishing to transmit messages. Therefore, the delay depends on the overall loading of the Ethernet.

This loss of quality due to multiplexing of a physical channel is the same sort of effect as might be a problem on long distance channels. The worsening factor is that the CSMA/CD mechanism may collapse under high load, with most of the capacity being wasted on collisions and their detection. In practice, a well-designed network layout should ensure that important servers, such as a WWW server or an Internet switch, are not placed on LANs with a lot of competing computers.

Ethernet space issues

The IEEE 802.3 messages carry source and destination identifiers, which are assigned to the communicating computers by the manufacturers of their Ethernet interfaces. Thus, they are globally unique and certainly serve the purpose of identifying the computers on the particular Ethernet. The LLC identifiers, here used just to indicate that SNAP information follows, are somewhat akin to the notion of well-known port numbers in TCP and UDP. They allow different identifiers within one computer, and so could be a vehicle for multiplexing, but are actually used more as a type field describing the type of information in the message.

Summary of the Ethernet issues

IP messages sent over an Ethernet are directly encapsulated into Ethernet messages, with no segmentation or concatenation involved. For an IEEE 802.3 Ethernet, the encapsulation has two stages. First, the IP messages are placed in an LLC1 message. Second, the LLC1 message is placed in an IEEE 802.3 message for transmission. On an Ethernet with low loading, transmission should be able to begin immediately in many cases, and then complete within one millisecond. However, on a busy Ethernet, transmission may be delayed significantly by other channels multiplexed on to the same multipeer channel.

9.7 OVERALL COMMUNICATIONS

The overall communication involved in fetching a WWW page has now been decomposed, employing realistic practical networking assumptions, to the point where it is implemented by the physical transmission of bit sequences over physical channels: private or leased cables, telephone channels or telecommunications services. As well as indicating exactly how this implementation can be achieved, this decomposition aids in understanding the quality that is achievable for the required communication. There are many technical factors that may influence the quality as observed by the WWW user at his or her home computer. To summarize these, a brief reprise of the overall information, time and space issues is given below.

Overall information issues

In fetching a WWW page, there are two essential pieces of information being shared. First, a URL is given to the WWW server by the client. Second, a page representation is given to the WWW client by the server. This representation consists of an HTML document, which may include one or more extra pictures or Java applets. The speed of fetching the page necessarily depends on the size of the page's representation. In particular, if it includes one or more large pictures, then many bits of information have to be transmitted, even if the picture is in a compressed form. This is a lesson that many WWW page providers have been slow to learn. The attraction of decorating textual pages with pictures exceeds the inclination to minimize the amount of information communicated. More knowledgeable providers decorate their pages with small pictures of, for example, postage stamp size.

Aside from the essential information, overheads are imposed by the various protocols used to move it. The first of these comes from HTTP. When the URL is sent from client to server, it is embedded in the lines of text that form the HTTP request. When the WWW page representation is sent from server to client, it is preceded by the lines of text that form the header of the HTTP response. In the case of the request, the overhead is many times the size of the URL information sent; however, the total size of the request is still relatively small. In the case of the response, the overhead is very small compared with the typical amount of page information sent, even if the page is purely textual.

The HTTP/1.0 handshake is implemented using the TCP/IP service, which involves various overheads. First, there is the TCP information exchanged to establish a connection and tear down a connection. These overheads are less significant if the size of the HTTP response is relatively large, since their cost can be amortized over the cost of transferring HTTP information. For this, there is the TCP/IP header on each information-carrying message — 40 bytes per 536 bytes of HTTP information, if the default maximum TCP/IP message size is used. There is also TCP acknowledgement information sent back when there is no scope for piggy-backing. In the case of a PPP-implemented IP channel, the 40 bytes of header information may be compressed to as few as three bytes, but this is not the case on most other types of channel. When acceptable to both client and server, the best way to minimize the TCP/IP header overhead is to minimize the number of messages needed, by agreeing on a larger than default maximum message size.

The final protocol overheads come from transporting TCP/IP messages over physical channels between computers. PPP, SIP level 3 and LLC1 were three examples of encapsulation looked at here. In the case of SIP level 3, messages are fragmented and further encapsulated in SIP level 2 messages. These might be encapsulated further within the SMDS network. In the case of LLC1, messages were further encapsulated in IEEE 802.3 Ethernet messages. The size of the overhead varies, depending on protocol, but is a penalty that applies to each TCP/IP message transmitted.

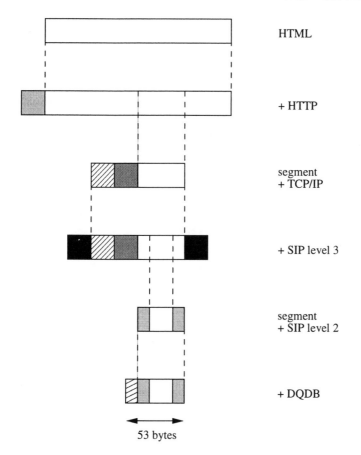

Figure 9.14 Fragment of WWW page in a 53-byte message

To illustrate all of the various information overheads, Figure 9.14 shows how 44 bytes of the HTML representation of a WWW page might be seen by someone observing it, after various encapsulations, when it is contained in a 53-byte DQDB message passing through an interface to an SMDS network. Note that the size of the headers and trailers depicted is not to scale relative to the information content. The situation is reminiscent of an iceberg: beneath the surface object seen by an observer, there is much hidden. An alternative analogy, perhaps more appropriate, is a swan: a serene presence on the water, but with frantically paddling webbed feet under the water. The point is, of course, that although all of the extra information that is exchanged by various computers is an overhead to the WWW user, it is all necessary to present the illusion of global web-like information connectivity.

Overall time issues

The overall time period for fetching a WWW page is affected by numerous factors, many of them related to the different protocols used to implement the communication. However, one other main factor is the behaviour of the WWW client and server computers themselves. The home computer is responsible for managing the communication over the telephone line, and for interpreting the WWW page information received. Normally, it is reasonable to assume that it can devote most of its processing capacity to these tasks, to carry them out as speedily as possible. However, if the user has the computer doing other things at the same time, then the WWW operation will suffer.

The server computer is responsible for managing the communication over its Ethernet, and for retrieving and transmitting the WWW page information. Similar issues apply as for the home computer but, if the server is a specialized computer, then it will not be undertaking any different work at the same time. However, the problem is that it may be serving several, possibly very many, clients simultaneously. This will result in competition for its processing power, its storage accessing and its IP channel through the Ethernet. Therefore, significant time delays can occur when a popular WWW site is being accessed. The delays will vary depending on the time of day when the WWW access is made. They are likely to be worst during normal working hours in the time zone of the server.

There are certain absolute limits on the rate at which information can be transmitted. If there is a physical separation of 10 000 km between the WWW client and the WWW server, then the speed of light imposes a minimum latency of over 30 milliseconds on any transmission. This limit assumes a direct link, whereas in fact the link is formed from a number of hops between switching computers. Switching delays, and indirect routes that are not as the crow flies, mean that the true latency will be rather more than this. The rate of transmission depends on the quality of the physical channels used for the hops. The overall sustained rate between client and server can be no better than the rate of the slowest channel. In this case, the channel with the slowest physical rate is most likely the telephone channel to the home computer — as already mentioned, this could be about 3000 bytes per second at best.

Of course, physical latencies and transmission rates give the complete picture only when physical channels are being dedicated to the communication being studied. Where there is sharing of channels through multiplexing, latencies increase and rates decrease. This effect depends on the time of day that communication takes place. In the case of a trans-Atlantic channel (or even a trans-USA or trans-European channel) that passes through different time zones, peak loadings do not have a simple relationship to the time of day at one or other end of the channel.

There are two standard ways of ameliorating the effect of physical transmission and processing delays, both familiar from the general world of computer architecture: cacheing and pipelining. For a WWW page access, cacheing comes from having a copy of the required page stored in a location that is more local to the

WWW client than the WWW server is. *In extremis,* the client might have a copy itself, saved from an earlier access. It then just needs a small HTTP handshake with the server to ascertain that this copy is still up to date and, if it is, the page information need not be fetched again. More generally, a WWW proxy between the client and the server may be able to provide a copy of the required page more locally. In this case, the proxy can check with the server that the copy of the page is still fresh.

Pipelining is possible by using the fact that the channel between client and server is implemented using a series of inter-connected channels between IP switches. The ideal would be to ensure that all of these channels are kept active simultaneously, and at their maximum rates. In theory, this is feasible if all of the channels have roughly the same transmission rates. Obviously, if any one is significantly slower, then it presents a bottleneck in the pipeline, and the others are slowed to the same rate. In practice, matters are more complex, due to the use of TCP over IP and to the fact that HTTP/1.0 makes inefficient use of TCP, as described earlier in Section 9.3. The scope for pipelining is constrained by the TCP transmission windows used between the WWW client and server. These constrain the total amount of information that can be in the pipeline at any time.

The standard TCP window size is expressed using 16 bits, which means that it has a maximum value of 65 535 bytes. The theoretical bound on the amount of information buffered by a channel is the product of its rate and latency. For most conventional channels, this is less then 65 535. For example, using the parameters for a 10 BASE 2 Ethernet given in Section 6.4.1, the bound is just over one byte only. However, for very high speed channels or very long distance channels, the TCP window size may not be adequate. For example, a 45 Mbits per second long distance channel with a latency of 30 milliseconds has a capacity of 125 000 bytes. This could only be about half used by the standard TCP window. To cope with this problem, there is an optional extension to TCP. If both parties agree at the time of connection establishment, scaling of the window size field can be used: the 16-bit values can be scaled up by an agreed constant power of two. For example, a scaling factor of 16 means that an effective 20-bit window size can be used. Ideally, a scaling factor could be used to ensure that the maximum window size is always greater than the sum of the capacities of all of the hops between IP switches.

Even if the maximum window size is big enough to allow full pipelining, there is still the problem that HTTP/1.0 causes the slow start algorithm to be used for each HTTP handshake, and therefore communications begin with a small window size: one maximum TCP/IP message size. Thus, at the initial stage, pipelining is not possible at all, since only one message is in transit. The HTTP handshake might well be completed before the TCP window reaches its maximum allowed size.

The situation worsens when IP messages are lost in transit. This might be due to transmission errors, the telephone line being the main likely culprit, since other physical channels are likely to be very reliable. Other losses will be due to congestion at some point on the route, due to excessive attempted multiplexing. Whenever losses occur, further delays are introduced, first by the need for timing out and retransmission, and second by the fact that the working TCP window size is reduced after losses are experienced.

In summary, for the overall communication time, there are many relevant factors. In most practical cases, delays that are noticeable to a user are likely to be caused by three main things:

- slow telephone line transmission;
- trans-Atlantic channel congestion; and
- server slowness.

There is no straightforward solution to the first of these, apart from seeking a faster ISDN channel or awaiting high speed home networking via cable. The second and third problems are best addressed by wise choice of the time of day chosen for the communication — ideally when both the European country and the USA are not at work or, even better, when both are asleep. In the future, HTTP/1.1 will replace HTTP/1.0, and is tuned to TCP efficiency, which gives improved performance by eliminating unnecessary protocol-related delays.

Overall space issues

There are two basic space issues in this example. First, the WWW client computer has to find the Internet identifier of a WWW server computer for the page required. Second, a channel between the client and the server has to be implemented. The first issue is very straightforward, given the way in which URLs are constructed. The URL contains a DNS name for the server, and this can be mapped to an IP identifier using the standard Internet directory service. Although simple, this scheme has a particular pitfall: the unique identifier for the page incorporates a physical address for the page. Thus, identical pages might have different identifiers just because they are stored in different places. Further, when a page's location changes, its identifier must be changed as well. Future Uniform Resource Identifiers for WWW-style objects are likely to be names, rather than locations. This answers the deficiencies of URLs, but means that a more sophisticated directory system is needed, to allow locations to be determined.

The channel between client and server can be viewed at various levels. At the top level, it is a channel for sharing a URL and the WWW page contents identified by that URL. This channel is implemented by an HTTP channel that allows the sending of an HTTP request in one direction, then the sending of an HTTP response in the other direction. In turn, the HTTP channel is implemented by a TCP channel that allows the transmission of byte sequences in either direction. The TCP channel, possibly multiplexed with other TCP channels, is implemented by an IP channel that allows the transmission of TCP/IP messages in either direction.

The IP channel between the WWW client and server is implemented by a switching network that allows IP messages to be routed through the Internet. This relies on the use of routing protocols to guide switches on how to route the message as it passes through towards its destination. For each hop between computers, the IP channel, possibly multiplexed with other IP channels, is implemented by a hop channel that allows the transmission of IP messages from the current switch to an adjacent switch. Each such hop channel may be directly implemented using a physical transmission medium. Alternatively, some type of network might be used to provide the channel; eventually, this is underpinned by some physical transmission media.

9.8 CHAPTER SUMMARY

This chapter has given a detailed account of the implementation of a popular type of communication: accessing a World Wide Web page, using the technology and protocols available in early 1997. The Hypertext Mark-up Language (HTML) and the Hypertext Transfer Protocol (HTTP) are the two agreed protocols that are specific to the application. The standard Internet TCP/IP protocols are then used to support a reliable conversation between WWW client and WWW server using HTML and HTTP. The Internet identifier of the WWW server is inferred by the WWW client using the Uniform Resource Locator (URL) for the required page. The speed of access depends, among other things, on the number of simultaneous clients being handled by the server.

The TCP/IP communication is supported by the standard message routing mechanisms over the Internet. This makes use of relevant physical channels that exist between computers. In this example, a slow telephone channel with modems was included to connect the WWW client to an Internet provider; the Point-to-Point Protocol (PPP) was used to transport IP messages over this channel. An Ethernet LAN was included to connect the WWW server to an Internet provider; the IEEE 802.2 LLC1 pprotocol with SNAP was used to transport IP messages over this LAN. In between the two Internet providers, Internet Exchanges and other switches are needed, together with channels between them. These channels may include LANs, MANs, leased lines and telecommunication services. In the example, one of the channels must traverse the Atlantic Ocean, which introduces delays due to distance and congestion.

9.9 EXERCISES AND FURTHER READING

9.1 Obtain access to the World Wide Web, either using Internet access already available to you, or by visiting a 'cybercafe' that provides access. Do some 'Web surfing' — that is, dip into various WWW pages and hop along new links frequently — and also make use of at least one search engine. While doing so, note the URL of each page you use and, in particular, deduce where the page is actually stored. Investigate the access speed differences for pages stored in different places.

9.2 Use the WWW to find out about the Virtual Reality Modelling Language (VRML), which is the three-dimensional generalization of the two-dimensional HTML. It allows the creation of virtual worlds containing three-dimensional scenes and objects, and the second version of VRML — VRML 2.0 — was finalized in August 1996. If you have access to a VRML browser, try out its capabilities.

9.3 Read RFC 1737, which contains a discussion of the minimum set of requirements for Uniform Resource Names in the Internet.

9.4 Find out, from RFC 1738, the appropriate syntax for URLs that begin with protocol names other than 'http', and relate this syntax to the requirements of each protocol allowed. Look out for examples of these other types of URLs when surfing the Web.

9.5 Try producing your own WWW page(s) by constructing HTML representations. A WWW authoring tool that includes an HTML editor would be useful when doing this.

9.6 Collect examples of handlers that make use of the CGI standard, to see what they actually to.

9.7 If you are familiar with computer programming, try writing some Java programs, and then exercise them through a WWW browser.

9.8 Investigate the Platform for Internet Content Selection (PICS) which supports a content rating system (on violence, language and nudity/sex scales) for WWW pages, in order to protect the innocence of children and/or parents.

9.9 Read RFC 2068, which describes HTTP version 1.1. Compare the header fields used in HTTP requests and responses with the header fields used in Internet electronic mail (RFC 822) and MIME (RFC 2045), and explain the differences.

9.10 Check whether your WWW browser is configured so that it makes use of proxies when making page requests.

9.11 Consult Figure 9.5. Using a WWW browser, make a request for a WWW page that contains at least two pictures, and look for evidence of the implementation steps shown in the figure.

9.12 Read RFC 793, which describes TCP. In particular, study Figure 6, which shows a state diagram for connection establishment and closing.

9.13 Read RFC 1323, which describes TCP extensions to support transmission over paths with high bandwidth*delay products.

9.14 Search the WWW for information on the status of TCP next generation (TCPng), that is, an updated version of TCP, just as IPv6 is an updated version of IPv4.

9.15 Find information on Next Generation HTTP (NG-HTTP), and explain the ways in which it uses the TCP service more efficiently.

9.16 Consult the most up-to-date version of the 'Assigned Numbers' RFC, to find the list of well-known port numbers for TCP and UDP. Also, look at the list of 'registered ports', that is, port numbers accessible to ordinary users, but which have a commonly understood significance.

9.17 If you have access to a telephone link to the Internet, investigate whether you can discover the maximum number of IP messages that can be transmitted per second. Also, how is the IP identifier of your computer allocated?

9.18 Compare RFC 1548, which describes PPP, with RFC 1055, which describes the SLIP (Serial Link IP) protocol, a predecessor of PPP still used in practice. What are the main improvements made by PPP?

9.19 As can be dicovered from RFC 1994, CHAP is one authentication protocol that can be used with PPP. An alternative is a lighter-weight protocol called PAP (Password Authentication Protocol). Explain why CHAP is more secure than PAP.

9.20 Investigate the structure of the global Internet. Find out where main Internet Exchanges are located, and how they are inter-connected. In particular, determine what inter-continental channels are available, and the transmission rates that they support.

9.21 Why do you think that SIP level 3 messages have their sequence number and length at both the beginning and the end?

9.22 Comment on how the timing conclusions of the case study might be altered if a 100 Mbits per second ethernet had been used at the server computer end, rather than a 10 Mbits per second one.

9.23 Describe the effect of adding an FDDI backbone at the server computer end, between the Internet switch and the ethernet of the server.

9.24 Draw pictures similar to Figure 9.14 for the cases where the IP message is being carried (a) by PPP over a telephone line, and (b) by LLC1/SNAP over an ethernet.

9.25 Repeat Exercise 9.1, but attempt to diagnose what the major cause(s) of any delays are.

9.26 Carry out a case study similar to that in this chapter, except for the Internet file transfer protocol (FTP), rather than for HTTP. What are your main conclusions about the effect of communications on the ease of transferring files from one computer to another?

Further reading

Numerous books about the World Wide Web (and 'The Internet') have appeared since the mid-1990s. These are more geared towards the potential user, rather than the computer communications specialist. One book that takes a more insider view is *The World Wide Web: Beneath the Surf* by Handley and Crowcroft (UCL Press 1995). A paper written by the creators of the WWW is "The World Wide Web" by Berners-Lee, Caillau, Loutonen, Nielsen and Secret, published in Communications of the ACM, **37**, August 1994. The TCP protocol and, of course, other aspects of the Internet's operation, are covered extensively in *Internetworking with TCP/IP*

Volume 1 by Comer. Mogul's work on the inefficiencies of HTTP over TCP can be found in his paper "The Case for Persistent-Connection HTTP", in the proceedings of SIGCOMM-95.

CHAPTER TEN

CASE STUDY 2: CONTROLLING A MANUFACTURING DEVICE

The main topics in this case study about controlling a manufacturing device are:

- The Manufacturing Automation Protocol project
- The Manufacturing Message Standard (MMS) protocol for communication between factory floor devices
- Implementing MMS using ISO standard protocols: FullMAP and MiniMAP
- Use of ISO standard application, presentation and session protocols
- Use of ISO standard transport and connectionless network protocols
- IEEE token bus LANs as the physical technology of MAP

10.1 INTRODUCTION

The first case study, in Chapter 9, concerned a communications application that is likely to be familiar to, and probably experienced by, most readers of this book. In addition, it involved the use of TCP/IP, the pivotal Internet protocols, as well as a range of typical physical communication systems. The other two case studies, in this chapter and in Chapter 11, look at two applications that are less familiar to the general computer user.

Both of these applications make use of communication facilities that are different from the standard Internet provision, and so they give a contrast. One particular feature of both is the need for guaranteed real-time communication, in contrast to the best effort timing of World Wide Web page access. The level of descriptive detail in these two chapters is much less than in the first case study, reflecting the fact that readers are less likely to encounter these applications in everyday life.

In this chapter, an application from the world of Computer Integrated Manufacturing (CIM) is examined. CIM is concerned with the use of computers in designing, planning, dispatching and controlling manufacturing operations. The application is not one of directly implementing CIM, rather it is one of implementing appropriate factory communications that support implementation of CIM. It centres around an implementation strategy based on the Manufacturing Automation Protocol (MAP) standard.

MAP provides a way of connecting together both general-purpose computers and special-purpose factory automation devices, such as robots, programmable controllers and automatic guided vehicles. It also provides a way of integrating the implementation of communications within a factory with the implementation of other commmunications within the owning organization, and beyond.

The MAP project originated at the end of the 1970s, when General Motors realized that half of its automation budget was being spent on developing special-case interfaces between automation devices that used different proprietary communication protocols. The first version of MAP was just a procurement specification, in which General Motors defined the protocols that had to be implemented in any piece of equipment that it purchased. The result was that the cost of equipment increased, as suppliers produced specialized versions that matched the MAP specification rather than used their normal proprietary protocols. Therefore, General Motors founded a MAP User Group in 1984, in order to turn the MAP project into a standardization exercise that had wide industrial support. The aim was for all factory equipment to have a common standardized interface, to allow companies to integrate easily equipment from many different suppliers.

In parallel with General Motors and its MAP exercise, Boeing had identified a similar incompatibilty problem with its office computing systems. This led to the Technical and Office Protocol (TOP (Technical and Office Protocol)) project, and a TOP User Group, which developed a standard solution along very similar lines to MAP. Indeed, the MAP and TOP groups worked closely together, to ensure compatability between the two exercises. This was wise, given that both factory automation and office automation are of importance to many companies. The central differences between MAP and TOP concern information and time issues in the user application. Different types of information are exchanged in the different environments. Real-time operation is important in many factory floor situation, but is less of an issue in office situations.

The central policy of MAP and TOP was to make use of communication protocols that were international standards, or were in the process of being internationally standardized. If appropriate protocols did not exist, then new protocols would be defined, but with a view to submitting them as candidates for international standardization. Thus, the intention was to develop solutions that used standard technology, and so would not be expensive for manufacturers to implement. All of this is completely commendable.

The only problem is that MAP and TOP were being developed at a time when, although the early Internet and its protocols existed, few people thought or realized that it would become the *de facto* international standard for communication communications. Rather, the efforts of the International Organization for Standardization (ISO), under its Open Systems Interconnection (OSI) strategy, seemed set to dominate the world. ISO and its OSI standards are discussed in Chapter 12. However, a historical summary is that most ISO standards were slow to mature and, further, manufacturers were slow to adopt them.

In contrast, Internet standards existed, were freely available, and free implementations were also readily available. This fact was not foreseen when MAP and TOP were being standardized. The MAP and TOP User Groups were not the only bodies wrong-footed by the eclipse of OSI by the Internet. Another notable victim was the US government which, although the major funder of the Internet work, also announced a commitment to ISO standards through its GOSIP (Government OSI Profile) specifications for equipment suppliers.

The fruits of the MAP project have not been wasted, however. One of these was a new protocol for the exchange of messages between factory automation devices. Another was a new type of LAN appropriate for the factory environment. The OSI protocols used as the implementation glue between the LAN and the message exchange protocol are the OSI equivalents of TCP and IP; indeed, both were very strongly influenced by TCP and IP. Thus, by substituting TCP/IP for these protocols, the MAP standard implementation strategy for factory messaging is very Internet compatible. MAP also includes the use of OSI protocols to support other applications, such as file transfers and directory services. Given that these have Internet equivalents, which are in very common use throughout the world, there is more of a compatibility problem. In this case study, however, the focus is on the factory messaging side only.

10.2 THE PROBLEM: INFORMATION, TIME AND SPACE ISSUES

The central problem is allowing the exchange of messages between two factory floor computers, in order to support the needs of Computer Integrated Manufacturing. The first point to note is that the computers might have widely varying complexities, ranging from specialized manufacturing equipment, through workstations for controlling the equipment, to minicomputers and mainframe computers. The relationships between these various types of computer will vary from factory to factory, but certain generic communication needs are typical of many manufacturing environments:

- workstation communicating with the devices it controls;
- process control computer communicating with workstations; and
- main computers communicating with process control computers.

This is a hierarchical view, assuming devices controlled by workstations, which are controlled by process control computers, which are controlled by main computers. It corresponds to a certain way of organizing manufacturing in a factory:

- groups of inter-related devices, each group with a single controlling workstation;
- manufacturing cells, each capable of performing one or more processes using its devices and their controllers; and
- manufacturing facility, possibly divided into different shops, each containing different manufacturing cells.

To allow cooperation between controllers, cells, shops and facilities, communication is also necessary among peers: workstation to workstation, control computer to control computer, etc.

The information, time and space characteristics can be summarized in general terms, according to level in the hierarchy. The higher the level, the more complex the information, the more asynchronous the timing, and the larger the physical distance. The lower the level, the simpler the information, the more synchronous the timing, and the smaller the physical distance. MAP provides a single solution, which can be tuned to the particular levels of communication that are required for a particular factory.

MAP information issues

The problem is to allow the exchange of information between two computers A main issue is to define what types of information exchange are desirable to support CIM. The Manufacturing Message Specification (MMS), is a new protocol developed as part of the MAP project. It includes many different message types, not just for the exchange of information in itself, but also for the coordination and synchronization of activities. That is, there are control messages as well as data messages.

There is most scope for using MMS at the factory cell level, and below. Some higher level communications are better dealt with by other standard protocols, for example, a file transfer protocol or a database transaction protocol, which are not being considered in this case study. The support of all different types of MMS message is rather ambitious for factory floor devices containing simple computers. At a result, eight different groupings of message types have been defined, each targeted at the particular needs of real manufacturing devices. The simplest of these includes only five message types.

A client-server model underpins MMS, with the client and server roles matching master and slave roles. Thus, for example, a factory floor device would be a server and its controlling workstation would be a client. The information exchange in a communication involves interactions taking place between a client and a server. A client sends request messages to a server, and the server sends back response messages to the client. This style of operation is familiar from Chapter 9, where accessing a WWW page involved a request-response mechanism between WWW clients and servers.

The situation is more complex here, because servers are not just responsible for sending back information in responses, but are also responsible for carrying out actions. It is therefore necessary to have an abstract notion of the capabilities of a manufacturing device, to allow MMS messages to be defined for controlling it. The associated Virtual Manufacturing Device (VMD) standard is mentioned very briefly in Chapter 2 as an example of an abstract computing facility. It is similar in concept to things such as virtual filestores and virtual terminals. An MMS server receives MMS requests for operations on a VMD, and maps these on to real operations performed by a real manufacturing device. It sends back MMS responses based on responses from the real device.

MMS defines a range of different classes of object that can be included in a VMD and, for each class, there is a range of services that can be requested. Support for the different MMS message types can be restricted, first by limiting the number of object classes supported, and second by limiting the number of services supported for each class of object. A full description of the different object classes is considerably beyond the scope of this book. The main classes are:

- domains;
- program invocations;
- memories: programs and data sets;
- variables: unnamed, named, scattered, lists, types;
- semaphores;
- events;
- journals; and
- operator terminals.

These have various different characters. Domains and program invocations are top-level components of VMDs. A domain is an object that contains all of the resources required to carry out a particular application of the VMD. A domain may contain other sub-objects that contribute to the overall function. A domain may be permanently present in a VMD, or may be dynamically created as required. A program invocation represents a collection of cooperating tasks being executed within one or more domains. Each program invocation is created, then evolves through a series of states, and then is destroyed. As an example, a domain might correspond to a robot arm, and a program invocation within that domain might correspond to the tasks needed to cause a particular movement of the robot arm.

The other classes of object have rather more supporting roles. Memories allow the contents of domains to be loaded to and from files. Variables allow information to be read and written to and from VMDs. This includes reading status information and writing control commands. Semaphore objects give support for synchronization of different tasks in a VMD, and event objects give support for monitoring and handling events in the VMD. Finally, journal and operator objects support human users, through the writing of journal files and the handling of interactive terminals respectively.

The above VMD mechanism is very general, by design. Further companion standards for MMS have been defined by specialist bodies for different application areas, including robotics, process control, numerically controlled tools and programmable controllers. These standards define relevant extensions or tailorings of the basic VMD objects and services. For example, the robot companion standard defines three domains: a robot arm domain, a calibration procedure domain and a safety equipment domain. There are program invocations for robot arm manipulation and for performing calibration procedures. As well as specialist semaphores and events, there is a collection of named variables. These can be read to give information on things such as the type of robot arm and its current position and status.

Despite the complexity of the VMD model, most MMS messages are short and simple, both for requests and responses. The complexity comes from the variety of different types of message allowed. In fact, there are 86 different types of request message, 82 of which require a response. A typical request contains a service type and a few, if any, parameters for that service. A typical response contains status information related to the request made. The parameters and status information are values expressed using normal high-level programming language data types. Thus, a request-response interaction is similar in information-sharing character to a function call in a programming language. Some parameters are passed to the function, the function performs one or more actions, and some results are passed back from the function.

A more detailed description of MMS messages is omitted here, since a better understanding would require an unnecessary deeper exploration of the VMD model. This is more relevant to a study of CIM, rather than a study of computer communications. Here, it is enough to know that MMS messages can be represented by byte sequences, as explained in Section 10.3, and that most messages are short. Readers interested in finding out more about MMS are warned that the two ISO standards documenting MMS are well over 400 pages long in total.

MAP time issues

Absolute bounds on the time periods of some MMS interactions are essential in a real-time factory environment. For example, devices such as actuators and sensors require regular cycles of commands and readings respectively. This involves writing and reading VMD variables at constant intervals. The duration of such communications is short, since reading a sensor typically only requires sharing a few bytes of information. What matters is that the starting time of the communication is predictable. There will be other, less time critical, communications involving such devices, for example, loading and unloading data and programs. These communications have significantly longer durations, but can be more flexible in terms of starting time.

Other communications involving more sophisticated computers may also be time critical. MMS interactions connected with VMD events and semaphores require prompt communication. For example, a particular event might correspond to an alarm being signalled by a sensor, and it is important that an appropriate control computer handles the event as quickly as possible. When a semaphore is being used to synchronize two tasks, then prompt communication is desirable so that execution of the tasks is not delayed unduly.

In order to ensure prompt communication when required, it is necessary to minimize the communication overheads. There are several aspects to this. First, the physical communication time must be kept short. Second, the protocol processing must be kept as simple as possible. Third, if a communication channel is being multiplexed with others, then any delays introduced must be modest and bounded. As long as these overheads are kept under control, MMS interactions can be made to happen at the times required, and also to proceed quickly.

The basic time package used by MAP for communication between two factory computers is a connection. The information sharing during the lifetime of the connection consists of a series of handshakes, each corresponding to a client request and a server response. A few types of MMS request do not require responses, and so these are straightforward unsegmented communications rather than handshakes. The connection supplies an overall context for the dialogue. Note the contrast with the World Wide Web, where each HTTP handshake is treated as a separate connection, something that leads to inefficiency.

Typically, a client will establish a connection to a server. However, in general, any computer can establish a connection with any other. For example, some computers may be both clients and servers. When a connection is established, some contextual matters are agreed. These include a limit on the number of requests that can be sent ahead before responses are received, that is, on the extent to which overlapping of handshakes over time is allowed. This is a simple MMS flow control mechanism, and the limit can be different for the two computers involved. Another matter that is agreed at connection establishment is the range of objects and services that are supported.

MAP space issues

The space for each communication is straightforward: two computers and a channel between them. The channel is used for the exchange of MMS messages. Each such communication takes place against the overall background of all communications within a factory, and possibly beyond. In particular, the channel used for a particular communication may be multiplexed with other channels when it is implemented. This will cause interference, principally with temporal behaviour. There may be other intentional interference, for example, when several cooperating computers are communicating in order to synchronize their behaviour.

The physical size of the communication space is small. Within a manufacturing cell in a factory, devices and controllers will be close to one another, with distances measured in tens of metres. Overall, any communication spaces within a factory are small enough to be classified as being local area. In some cases, MMS messaging might be used over an entire manufacturing site, which might stretch the space into the metropolitan area category. The effect of the physical space being limited in size, together with the fact that the physical communication channels are under the control of the factory owner, mean that the quality of communications is very predictable. This is in distinct contrast to the WWW case study, where the communication space might span the world, and be shared with numerous others.

In MAP, computers are not identified by their network identifiers. Instead, independent names are used, in order to introduce location independence. To make this work, MAP requires a directory service which is used to map names to network identifiers. The directory service is based on a directory information base that stores details of objects and names for these objects. In particular, computers and their network identifiers are one type of object. Computers interact with the directory service in another client-server relationship. A query about a name is sent to the server, and it responds with information associated with that name. Note that an advantage of the connection-oriented style for MMS communications is that computer names need only be mapped to network identifiers before connection establishment, not every time that an MMS message is sent. Thus, name-network identifier mappings are another part of the context of a connection.

Summary of the problem

Communication between factory floor computers is needed to support the needs of Computer Integrated Manufacturing. With MAP, computers act as clients and/or servers, using and supporting operations on tailored versions of an abstract Virtual Manufacturing Device (VMD). These operations are mapped to real operations on real devices. Requests from clients to servers, and responses from servers to clients, are sent using Manufacturing Message Service (MMS) messages. These convey parameters and results expressed as values from high-level language data types. A connection is established between any pair of computers that needs to communicate, and then a sequence of MMS handshakes takes place within that connection. Handshakes may overlap in time, up to a limit agreed by the computers. For some factory floor devices, the communications will be time critical. The physical separation of communicating computers is relatively small, which aids the implementation of fast and reliable communications.

10.3 FULLMAP IMPLEMENTATION

In this section, the **FullMAP** implementation of an MMS communication is described. This involves using a collection of different ISO standard protocols, each implemented in turn using another. The result is a rather complex implementation, with inevitable overheads for protocol handling. This is good as a general solution to enable communication between different computers, but it is poor for efficient operation. In Section 10.4, the stripped-down **MiniMAP** implementation is described. This is used for time critical communications, and for less sophisticated computers. FullMAP is more appropriate for use between controller computers and other minicomputers or mainframe computers. Some computers may support both MiniMAP, for communicating with simple devices, and FullMAP, for communicating with peers or masters. This dual implementation is termed the **Enhanced Performance Architecture** (EPA) of MAP.

In summary, the implementation task is to enable the exchange of MMS messages between two computers, within the context of a connection, and satisfying any required timing constraints. The implementation makes use of one or more physical channels that are capable of transmitting bits within a factory, As will be seen, the FullMAP implementation requires a stack of eight separate protocols, each helping with a stage of implementation (and some helping distinctly more than others). This approach was not the invention of MAP's designers, but rather was a consequence of following the ISO OSI standard. The nature of this standard is explained in Section 12.3. The description of how MAP uses various OSI protocols is kept brief here — a full account would require much space, indeed the relevant standards documents extend to many thousands of pages.

At the centre of the implementation is a connection-oriented service that can provide a reliable channel between two computers. This channel is used for exchanging messages, which are just sequences of bytes. Following ISO terminology, the service will be described as the **transport service**. It is akin to the service supplied by the TCP/IP protocols. The transport service will be used as the basis for implementing the required MMS communications, with three protocols being involved. The transport service itself will be implemented using appropriate networking and physical channels. The implementation of the MMS service will be described first, followed by the implementation of the transport service.

MMS implementation using transport service

The problems solved by the implementation of MMS using the transport service are concerned with information and time issues. There are no space issues, since the transport service supports a space consisting of two computers and a channel between them, which is exactly what is required for the MMS communication.

The major information issue is that MMS messages carry parameters and results expressed using high-level language data types, whereas the transport service allows the transmission of byte sequences. The formats of MMS messages are defined using the constructs of ASN.1, and this ensures agreement on the syntax of messages between communicating computers. ASN.1 is described on page 40 in Chapter 2. Then, for actual communication, the standard Basic Encoding Rules (BER) associated with ASN.1 are used for representing MMS messages in terms of byte sequences.

This allows messages to be sent using the transport service. The general approach here is common to other ISO standard protocols for applications, for example, its file transfer and virtual terminal protocols. The approach is also followed by some of the more modern Internet application protocols, notably its network management facilities.

The time issues centre on the overall time package of MMS communications. Each MMS connection between two computers is implemented using a transport service connection. It is allowable for a sequence of non-overlapping MMS connections between the same two computers to take place within the time period of one transport connection. However, in the simplest case, the time period of an MMS connection is the same as that for its supporting transport connection. Within an MMS connection, messages containing requests and messages containing responses are sent. In most cases, where messages are short, the communication of one MMS message is implemented by the communication of one bit sequence. However, if required or convenient, a message communication may be segmented, so that successive components of the message are sent as successive bit sequences.

The three protocols used for the implementation play three different roles. The first is concerned with establishing the MMS connection (which includes the negotiation of an agreement on the way the MMS connection will be used) and later with closing the connection. The second protocol is concerned with the information problem: agreeing the syntax of the messages exchanged, and how they are represented as byte sequences. The third protocol is concerned with the time problems: the relationship of the MMS connection time period to the transport connection time period, and the timing of messages sent within the MMS connection.

Given that the information and time implementation arrangements of MAP are more or less fixed, there is no particular need for the second and third protocols. However, note that these protocols can be used to implement more interesting dynamic behaviour, which could be exploited by more sophisticated MAP implementations on a collection of computers that all agree to use the extra capabilities. Following ISO terminology, the three protocols will be termed an **application protocol**, a **presentation protocol** and a **session protocol** respectively.

The explanations of the three protocols that follow might seem somewhat complex. However, as will be seen after the three separate explanations, the combined effect in terms of messages actually sent using the transport service is very standard for a single connection-oriented protocol.

Application protocol

The protocol used is the ISO standard **Association Control Service Element** (ACSE) protocol. The word **association** is used by ISO to describe a connection that is a communication between two application processes. Here, this means a communication between two processes that are exchanging MMS messages.

The ACSE protocol involves the exchange of 'application protocol data units' in ISO-speak. These are just messages with a format designed for the protocol and, like MMS messages, this format is expressed using the standard ASN.1 data types. They are represented as byte sequences using the ASN.1 Basic Encoding Rules. The protocol is very simple. It has one handshake for establishing a connection, one handshake for normally terminating a connection, and one unsegmented message communication for abnormally terminating a connection.

To establish a connection, one computer sends an **AARQ** (application association request) message to the other. It responds with either an **AARE+** (application association response positive) or an **AARE-** (application association response negative) message, depending on whether or not it wishes to accept the connection. The AARQ message contains various information, including:

- ACSE protocol version number;
- identifiers of the communicating ACSE processes on each computer;
- identifier of the application (here, ISO 9506 MMS); and
- user information: here, limits on the number of outstanding MMS requests, and the range of MMS services supported.

The AARE+ message also contains information of this type, and this gives the final negotiated parameters chosen by the respondent on the basis of the information supplied by the requester.

The ACSE protocol is not involved in the sending of messages within the connection. It is next used to close the connection. One computer sends an **RLRQ** (release request) message to the other, which responds with either an **RLRE+** (release response positive) or an **RLRE-** (release response negative) message. In the event of an abnormal termination due to a failure in the communication service underpinning ACSE, an **ABRT** (abort) message is sent to each computer, indicating that the connection has been abruptly terminated. The communicating computers can also cause abnormal terminations, but using the presentation protocol (described next), rather than the ACSE protocol.

Presentation protocol

The protocol used is the ISO standard Presentation protocol. This is concerned with managing the **presentation context** used within a connection. A presentation context is an association between the abstract syntax used to describe messages (for example, the ASN.1 syntax of MMS messages) and the transfer syntax used to communicate messages (for example, the byte sequences used to represent ASN.1 data types). When a connection is established, a set of one or more presentation contexts to be used is agreed. During the connection, presentation contexts can be added to, or removed from, the agreed set. In the case of MAP, there is only one agreed presentation context, so the full capabilities of the presentation protocol are not required.

The protocol involves the exchange of 'presentation protocol data units' (using ISO-speak). These are just messages with a format designed for the protocol and which are expressed using the standard ASN.1 data types. They are transmitted as byte sequences using the ASN.1 Basic Encoding Rules. An unusual feature of the message format is that it does not contain a field indicating the type of the message. This is because all presentation protocol messages are encapsulated inside session protocol messages, and the type of the presentation protocol message is inferred from the type of the session protocol message.

The protocol has one handshake for establishing a connection, one message for sending information within a connection, and two types of message for abnormally terminating a connection. There is also a handshake for altering the context set during a connection, but this is not needed here.

To establish a connection, one computer sends a **CP** (connect presentation) message to the other. It responds with either a **CPA** (connect presentation accept) or a **CPR** (connect presentation reject) message, depending on whether or not it wishes to accept the connection. The CP message contains various information, including:

- presentation protocol version number;
- presentation contexts to be used;
- range of presentation services supported; and
- user information: here, the ACSE protocol AARQ message.

The CPA message also contains information of this type, which gives the final negotiated parameters chosen by the respondent on the basis of the information supplied by the requester. As user information, it contains the ACSE protocol AARE+ response. Thus, the ACSE messages are encapsulated inside the presentation protocol messages.

During the lifetime of the connection, information is sent between the computers inside presentation protocol **TD** (transfer data) messages. These messages contain one piece of additional control information: a presentation context identifier. This indicates which presentation context has been used for the information in the message, so that the recipient can interpret it correctly. In the case of MAP, this identifier is redundant, since the same context is always used. Note that there is no handshaking for information transfer. This is not necessary, since the protocol is implemented using a reliable underlying transport service.

The presentation protocol is not involved when a connection is terminated normally. However, it is used for abnormal terminations. An **ARU** (abnormal release user) message can be sent by either computer, and this causes the connection to be terminated immediately, without a handshake. When there is an abnormal termination due to a failure in the communication service supporting the presentation protocol, an **ARP** (abnormal release provider) message is sent to each computer. As user information, the ARP message carries an encapsulated ACSE protocol ABRT message.

Thus, as far as termination of a connection is concerned, there is a division of responsibility between the ACSE protocol and the presentation protocol. For normal termination by one of the computers, an ACSE handshake is used. For abnormal termination by one of the computers, a presentation message is used. For abnormal termination due to the failure of the communication service, both protocols are involved.

Session protocol

The protocol used is the ISO standard connection-oriented Session protocol. As the name suggests, there is also a standard connectionless protocol. The connection-oriented session protocol has a range of different functions concerned with managing a dialogue between two computers. However, most of these are not used by MAP, which simplifies matters considerably here. The only features needed are those for connection establishment, information transfer within a connection and termination of a connection. These are already somewhat familiar from the previous two protocols. For interest, note that the session protocol supports mechanisms for policing simplex or half duplex behaviour over the underlying full duplex channel. It also contains the apparatus for dividing connection time periods into activities and smaller units, as described on page 97 inChapter 3. In total, there are 36 different types of session protocol message, only 10 of which are mentioned here.

The session protocol is directly supported by the transport service, and its messages have a standard format decribed in terms of a byte sequence representation. These messages are called 'session protocol data units' in ISO-speak. The general format is:

- message type (one byte);

- message length (one byte);
- message parameters (variable length);
- user information (if appropriate).

If a message is longer than 254 bytes, then the length field contains the value 255 and is immediately followed by an extra two-byte length field. This allows messages to be up to 65 535 bytes long. The session protocol message format is the first one here that has a direct encoding in terms of byte sequences. The others — MMS, ACSE and presentation protocol — are defined in terms of ASN.1 and then encoded using its Basic Encoding Rules.

To establish a connection, one computer sends a **CN** (connect) message to the other. It responds with either an **AC** (accept) or an **RF** (refuse) message, depending on whether or not it wishes to accept the connection. The CN message contains various information, including:

- session protocol version number;
- unique connection identifier;
- maximum session protocol message size;
- initial settings for duplexness, activities, etc.;
- range of session services supported; and
- user information: here, the presentation protocol CP message.

The AC message also contains information of this type, which gives the final negotiated parameters chosen by the respondent on the basis of the information supplied by the requester. As user information, it contains the presentation protocol CPA response. Thus, the ACSE messages are encapsulated inside the presentation protocol messages, which are encapsulated inside the session protocol messages. In fact, *all* ACSE or presentation protocol messages are encapsulated within session protocol messages before transmission using the transport service. The maximum length of the user information field of a CN message was 512 bytes in the first version of the session protocol, and 10 240 bytes in the second version. If the encapsulated CP message is longer than the limit, then its communication can be segmented, with subsequent portions being sent immediately afterwards in **CDO** (connect data overflow) messages. Permission for each CDO message to be sent is given by the respondent sending an **OA** (overflow accept) message.

During the lifetime of the connection, information is sent between the computers inside session protocol **DT** (data) messages. As for the presentation protocol, there is no handshaking for information transfer because the protocol is implemented using a reliable transport service. If there is a restriction on the maximum allowed size of session messages, and a DT message is too long, its communication can be segmented into a sequence of sub-communications of small enough DT messages. Here, a particular circumstance is that all presentation protocol messages encapsulated in DT messages are TD (transfer data) messages. However, note that other presentation protocol message types, not used for MMS, may also be encapsulated in DT messages. The session protocol also has three other types of specialist information-carrying messages apart from DT, but these are not necessary for MMS either.

The session protocol has a handshake for normal connection termination and for abnormal session termination. The first involves one computer sending an **FN** (finish) message to the other, which responds with a **DN** (disconnect) message. The second involves one computer sending an **AB** message to the other, which responds with an **AA** (abort accept) message. This handshake is unlike the cases seen for the other two protocols, which just involve a single message communication. The reason is to ensure that the aborter knows when no further session protocol messages are going to arrive. This is so that it can begin a new session connection within the same transport connection time period if it wants to.

Summary of application, presentation and session protocols

The above three protocol descriptions will have conveyed an impression of similar things being done in each of the protocols. This was deliberate. A single protocol to manage an MMS connection would have been a rather simpler alternative. However, to stress the point again, the complexity comes from the fact that three rather general-purpose ISO standard protocols are being used to solve one specific problem. This inevitably leads to some redundancy. In the overall context of FullMAP, the general-purpose protocols are also used to support, in addition to MMS, the ISO standard protocols for the directory service, file transfers and network management.

As a visual summary of the preceding description, Figure 10.1 shows the messages exchanges that take place for (a) connection establishment, (b) information transfer within an established connection and (c) normal connection termination. The figure summarizes how an MMS connection is implemented using messages sent over the channel supplied by the underlying transport service. In this combined form, the true underlying protocol shines through:

- an establishment handshake;
- message transmission; and
- a termination handshake.

There are no excess messages transmitted. Complication arises because the format of the messages is more elaborate than necessary. Thus, the number of messages is appropriate, but the size of the messages, and the time needed to process them, is larger than necessary.

Transport service implementation

The problems solved by the transport service implementation are concerned with time and space issues. There are no major information issues, since physical communication channels transmit bit sequences, and the messages used by the

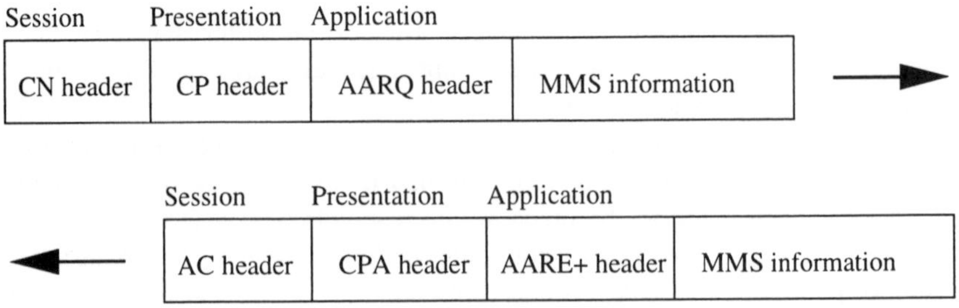

Session Presentation Application

CN header	CP header	AARQ header	MMS information

→

Session Presentation Application

AC header	CPA header	AARE+ header	MMS information

←

(a) Messages sent for connection request and positive connection response

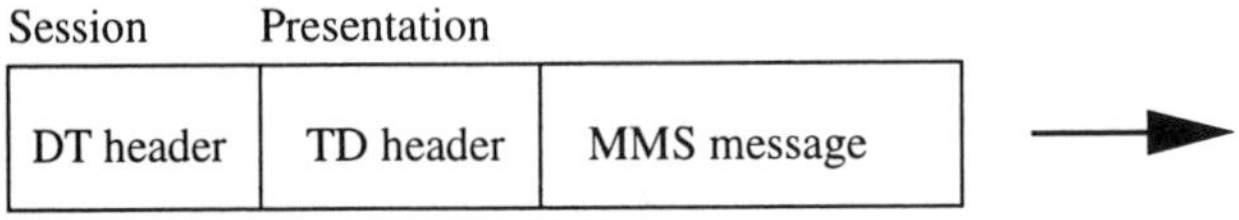

Session Presentation

DT header	TD header	MMS message

→

(b) Message sent for transfer of MMS message within connection

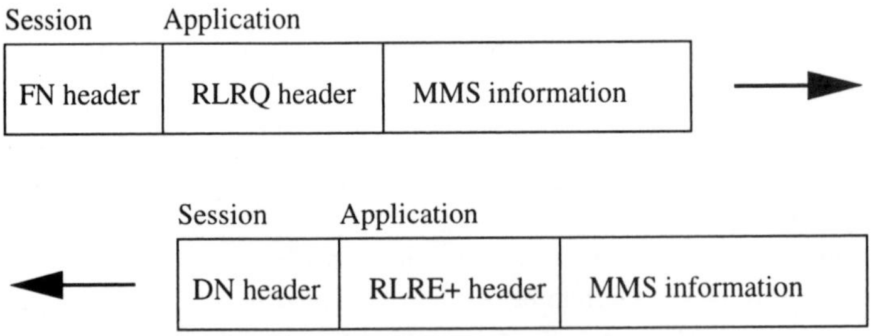

Session Application

FN header	RLRQ header	MMS information

→

Session Application

DN header	RLRE+ header	MMS information

←

(c) Messages sent for disconnection request and positive disconnection response

Figure 10.1 Message exchanges for MMS connection management

transport protocol, and those protocols that support it, are described in terms of byte sequences.

The requirement of the transport service is that it should provide a reliable full duplex channel capable of transmitting byte sequences between two computers. For FullMAP, a further feature of the service is that it should be connection-oriented, in order to supply a context within which communication of information takes place. This context includes things such as the full identifiers of the communicating computers, and the quality of service offered within the connection. The time issue is to supply a service of the appropriate quality: connection-oriented, reliable, and with adequate latency and delays. The main space issue is to supply a channel between any two computers, using some collection of physical channels between them. A second space issue is to deal with multiplexing, and perhaps splitting, of channels.

To the most extent, these problems can be solved by an implementation using the Internet TCP/IP protocols, as seen in Chapter 9. TCP supports stream-oriented connections, rather than unit-oriented connections, as its basic service. However, use of the TCP 'push' or 'urgent' facilities can create a unit-oriented service, if required. The only significant difference here from the situation in Chapter 9 is that absolute time bounds are needed for some communications. However, this is counter-balanced by the fact that the communication space is very much more local — within a factory, rather than spanning the world.

FullMAP is based on ISO standard protocols, so TCP/IP is not used. However, the protocols that are used are both very strongly based on TCP and IP. The ISO standard transport protocol, which is analogous to TCP, is used to implement the reliable connection-oriented service. The most important features of this protocol were described in Section 4.5.2. In turn, this protocol uses a service implemented by the ISO standard connectionless inter-network protocol, which is analogous to IP and supplies a best effort connectionless service between arbitrary pairs of computers. This protocol was described in Section 8.3.3.

In principle, communication between computers located anywhere is possible using the ISO inter-network protocol. Indeed, this is a feature of the design of the FullMAP implementation strategy. An organization may wish to have coordinated communications among all of its sites, throughout one country or perhaps throughout the world. It is technically feasible to use the MMS protocol between arbitrary sites, but it is not practicable to use it as a means for controlling factory floor devices from a long distance away. Thus, use of MMS is restricted to particular factories or, at most, to one particular manufacturing site. However, other application protocols included as part of MAP are not time-critical, so they could make use of the general-purpose inter-networking facility. For example, files might be transferred from one part of an organization to another. So, as with the presentation and session protocols, the inter-network protocol is over-rich for MMS purposes, but has a part to play in the overall FullMAP design.

For this case study, the physical channel arrangements are restricted to those used within one factory. These fall within the local area category of networking.

A normal arrangement is for there to be one backbone network, connecting the larger computers and also other, more local, networks within manufacturing cells. The token bus message broadcasting network, described in Section 6.4.4, was developed as part of the MAP project. This was because neither of the two main established types of network — ethernet and token ring — had exactly the characteristics required for the factory environment. Token bus is essentially a mating of these two. The use of a bus was deemed desirable, to reflect existing factory cabling for voice and video transfer. The use of a token was deemed desirable, to obtain guarantees on communication latency.

The implementation of the transport service over factory floor token bus networks involves several protocols. The transport protocol is concerned with implementing the desired time package, given a best effort connectionless service between two computers, and is briefly considered first. The inter-network protocol and two token bus-related protocols are concerned with implementing the connectionless service, and are considered after this.

Transport protocol

The ISO standard connection-oriented transport protocol has five variants, used depending on the underlying communication service, as explained on page 133 in Chapter 4. Here, the TP4 protocol is the relevant variant. Its basic mechanisms for implementing a reliable connection-oriented service are described in Section 4.5.2.

A range of appropriate formats for transport protocol messages, 'transport protocol data units' in ISO-speak, is defined. The general format is:

- message length (one byte);
- message type (four bits);
- compulsory message parameters (four bits, then fixed length);
- optional message parameters (variable length);
- user information (if appropriate).

There are 10 different types of message. This is interestingly few, compared with the 36 different message types of the session protocol, which has rather less implementation work to do.

In TP4, a three-way handshake is used to establish a transport connection, as in TCP. One computer sends a **CR** (connect request) message to the other. It replies with either a **CC** (connect confirm) message as a positive response or a **DR** (disconnect request) message as a negative response. If a CC message is confirmed, the first computer sends an **AK** (acknowledge) message to complete the handshake. The CR message contains various information, including:

- transport protocol version number;
- transport protocol class (here, class 4);
- requester's connection identifier;

- choice of seven-bit or 31-bit serial numbers (31-bit for MAP);
- maximum information-carrying message size;
- initial sliding window size;
- estimated acknowledgement delay time;
- checksum (present in all TP4 messages).

The CC message also contains information of this type, which gives the final negotiated parameters chosen by the respondent on the basis of the information supplied by the requester. It also contains the responder's connection identifier. Together, the requester's and responder's identifiers, each 16 bits long, uniquely identify the connection. For each connection, serial numbers begin at zero, and the connection identifiers can be seen as being a more significant component prepended to the serial numbers. CR and CC messages can also carry information to negotiate delay, throughput, error rate, priority and security requirements, but this capability is not used in MAP.

Note that the transport protocol CR message does not carry the session protocol CN message as user information. The reliable transport connection is first established with a CR-CC exchange, and then the session connection establishment is carried out using the normal information-carrying capability of the transport connection. Indeed, all session protocol messages are carried within the reliable information-carrying messages of the transport protocol. This also makes it straightforward to have several consecutive session connections within one transport connection.

Information is transmitted within the transport connection using **DT** (data) messages. These also carry serial numbers and checksums. DT messages carry session protocol messages in MAP. If a session protocol message is too large to fit a maximum-size DT message, its communication is segmented over a series of consecutive sub-communications of smaller-sized DT messages. DT messages carry a one-bit field to indicate whether or not the message contains the final component of a unit of user information. The maximum size of DT messages is negotiated when the connection is established. When the transport protocol is supported by the connectionless network protocol, as here, the maximum allowed size is 64 512 bytes.

DT messages are acknowledged by sending back **AK** (acknowledge) messages. Acknowledgements are not piggy-backed on information-carrying messages, as in TCP and other protocols. However, a similar sort of effect can be achieved, since a DT message communication can be concatenated with an AK message communication, resulting in the transmission of only one message. DT and AK messages carry their destination's connection identifier. This ensures that the serial numbers carried refer to the correct connection, and is why it is safe for serial numbers to begin from zero for each connection. Timeouts, based on the estimated acknowledgement delay times specified at connection establishment, are used to deal with lost DT or AK messages.

As well as the flow-controlled DT-AK information transfer mechanism, there is also an **expedited data** mechanism. This allows urgent information to be sent, regardless of the state of the sliding window mechanism. In essence, this supplies a second channel, in addition to the normal information-carrying channel. Information is sent in an **ED** (expedited data) message, and an acknowledgement is sent back in an **EA** (expedited acknowledge) message. These messages carry serial numbers, not related to the normal serial numbers, so that ED messages are sequenced, and so that ED and EA messages can be matched up. The expedited service can be used for some session protocol messages, for example, the AB (abort) message, which should be delivered promptly.

To terminate a connection, one computer sends a **DR** (disconnect request) message to the other. It responds with a **DC** (disconnect confirm) message. Recall that a DR message is also used to refuse a request for connection establishment; this use of DR does not require a DC response. After a connection has been terminated, its connection reference identifiers are frozen by both computers for a time long enough to ensure that no historic duplicate messages for the connection are able to arrive to cause confusion. After this period, the identifiers are defrosted, and may be used again safely.

Network protocols

Inter-network protocol

The ISO standard connectionless network protocol (CLNP) is used to directly support the transport protocol. In the environment chosen, there is no need for any of its associated routing protocols, that is, the ES-IS and IS-IS protocols described in Section 8.3.3. This is because the collection of token buses, inter-connected by bridges, supplies a network with connectivity between any pair of attached computers. Therefore, all computers can be regarded as end systems, with no intermediate systems being necessary.

The general format of a CLNP message was described in Section 8.3.3, as was the way in which its header fields are used to implement the required information, time and space features. Data (DT) type messages are used to transfer transport protocol messages. If a transport protocol message is larger than the maximum size allowed in a DT message, then the CLNP fragmentation mechanism can be used. The restriction on CLNP message size comes from the fact that each CLNP message must fit into one token bus message. This limit is discussed later in this section. There is no mechanism for segmentation of communications below the CLNP.

Note, in passing, that the communication of a large MMS message might encounter segmentation in three places: across session protocol DT messages, across transport protocol DT messages and across CLNP messages. There is some scope for concatenation, since transport protocol messages can be concatenated within CLNP messages. It is also possible to concatenate the communication of some types of session protocol messages within transport protocol messages, but this does not apply to the types used in MAP.

Each CLNP message carries a destination and source identifier. MAP defines a format for these identifiers, based on the ISO guidelines for inter-network identifiers. The general format of these ISO NSAP identifiers is described on page 145 in Chapter 5. These identifiers can be up to 20 bytes long, and have three components:

- authority and format identifier (AFI);
- initial domain identifier (IDI); and
- domain specific part (DSP).

For MAP, the AFI and IDI are used to identify a particular organization that is using MAP. For example, if the public PSPDN packet switching service is used to connect organizations, then an X.121 identifier would be used to identify an organization. In this case, the AFI would be equal to 39, denoting an X.121 IDI and a binary-encoded DSP. Note that, when using the reliable connection-oriented PSPDN, its quality of service is de-enriched to implement the connectionless network protocol.

The DSP is used for computer identification. For MAP, it is nine bytes long and has three components, to give a three-level hierarchical identifier scheme within an organization. The first component is an IS-IS identifier, used to specify a particular network within an organization inter-network. The second component is an ES-IS identifier, used to specify a particular computer within a network. The total length of these two components is eight bytes, but the relative sizes used can be selected by the organization to suit the structure of its networks. The third component, which is one byte long, is used to identify a process within the computer.

When CLNP messages are only being sent within one local area network, as here, these full-blown identifiers are not really necessary. Apart from the last byte, each standard identifier will map directly to the physical identifier of a computer on the network. The final byte allows discrimination between different processes on that computer. The full generality of the MAP identifier scheme is of use when communication is required beyond a single factory or even, perhaps, between different manufacturing organizations.

Given that the CLNP identifiers are not needed and that, if transport protocol messages are kept sufficiently short, the CLNP fragmentation facility is not needed, matters can be greatly simplified. MAP allows the use of the **inactive CLNP** rather than the full CLNP. Inactive CLNP messages are assumed to be information-carrying messages (i.e., DT type messages) and the header is reduced to just the first byte. This is the protocol identifier field and, instead of carrying the value 129 used for full CLNP, this carries the value 0 to denote inactive CLNP. The use of inactive CLNP is entirely suitable for the case study under examination here.

Link control protocol

The CLNP connectionless service is not directly implemented in terms of the physical token bus service. The IEEE 802.2 Link Level Control (LLC) protocol is used as an intermediary. Although devised by the IEEE, this protocol was later adopted by ISO as one of its standards. The use of the LLC protocol is exactly the same as seen in the World Wide Web case study in Chapter 9, where LLC was used as an intermediary between IP and transmission using an ethernet service. As there, FullMAP uses the connectionless LLC1 protocol. In contrast though, the Sub-Network Access Protocol (SNAP) extension of LLC is not used. The effect of the LLC involvement is that each CLNP message is encapsulated inside an LLC1 message.

Each LLC1 message has a three-byte header. The first byte identifies the LLC protocol process at the destination and the second byte identifies the LCC protocol process at the source. The value 254 is used in both cases to indicate that a CLNP communication is taking place. The third byte always has the value 3 to signify that the message is an LLC **UI** (unnumbered information) message.

Therefore, the role of LLC is to assist in the handling of different information types. Its header indicates how the user information following is to be interpreted, in this case that it is a byte sequence representing a CLNP message. This discrimination is necessary if the physical token bus communication channel is also being used to carry traffic from protocols other than CLNP.

Token bus protocol

The characteristics of the IEEE 802.4 standard token bus network were described in Section 6.4.4. As already mentioned, this standard was developed as part of the MAP project and, after adoption by the IEEE as a standard, was also adopted by ISO. Thus, from the MMS protocol downwards to the token bus protocol, the FullMAP implementation employs ISO standard protocols.

The information feature of interest here is that a token bus message can carry up to 8191 bytes between its start byte and stop byte. Assuming the IEEE standard 48-bit identifier scheme is used on the token bus network, then 17 of these bytes are used for token bus control information (frame control byte, destination identifier, source identifier and CRC-32 cyclic redundancy code). Another three bytes are needed for the LLC1 header. Thus, the maximum size of CLNP message carried is 8171 bytes.

The required absolute timing for MMS communications derives from the timing of the token bus network. Had an ethernet style of network been used, there would have been the possibility of arbitrary delays, due to contention with other communications sharing the physical channel. The token bus network has target token rotation times, which allow guarantees to be made on the maximum delay before a communication can start. The setting of the target token rotation times for a network depends on the number of connected computers, and how time critical the service requirement is.

After the bounded initial delay before a communication starts, the communication proceeds at the physical rate of the token bus. This rate is 10 Mbits per second in the standard FullMAP specification. Of course, this is not the rate at which raw MMS information is sent. Each MMS message is preceded by six different protocol headers, and followed by the token bus CRC trailer. Sending messages also introduces the need for transport protocol acknowledgement messages, as a further overhead.

The physical communication channel for a token bus network is a length of coaxial cable or fibre optic cable, with broadband transmission being used. This is beneficial in typical factory environments for various reasons.

First, a bus style of cabling is natural for linear assembly lines. Second, such cabling may already exist for other purposes, such as voice and video, and so can be shared with the token bus transmissions using frequency division multiplexing. Third, the broadband signalling allows longer distances to be spanned, and more computers to be attached.

For an arrangement where there is one token bus backbone, with other smaller networks connected, bridges can be used. Each bridge is a computer attached to two or more token bus networks. A bridge observes all messages passing on the buses to which it is attached, and forwards messages to other buses as appropriate. This is done at the level of the token bus protocol, using the destination identifiers carried by its messages. Thus, there is no need for any higher level intervention, for example, by the CLNP protocol.

Summary of transport and network protocols

The transport protocol and the three network protocols make a valuable contribution, by implementing a reliable transport service, between two computers. In the case study used here, the connectionless inter-network protocol and the logical link control protocol do not contribute a great deal since, essentially, transport protocol messages are being directly transmitted encapulsated in token bus messages. However, they do open up the possibility of extensions to include inter-networking, and also sharing of the token bus with other types of traffic. The transport protocol is essential to supply the required reliable time package, and the token bus protocol is essential to supply the required absolute time and space characteristics.

As a visual summary of the preceding description, Figure 10.2 shows the message exchanges that take place for (a) connection establishment, (b) information transfer within an established connection and (c) normal connection termination. The figure summarizes how a transport connection is implemented using messages sent over the underlying physical channel. There is an establishment handshake, information transfer handshakes and a termination handshake. Note that the information handshake is not a strict DT-AK (data-acknowledge) exchange — the AK messages are sent as necessary to manage the sliding window mechanism.

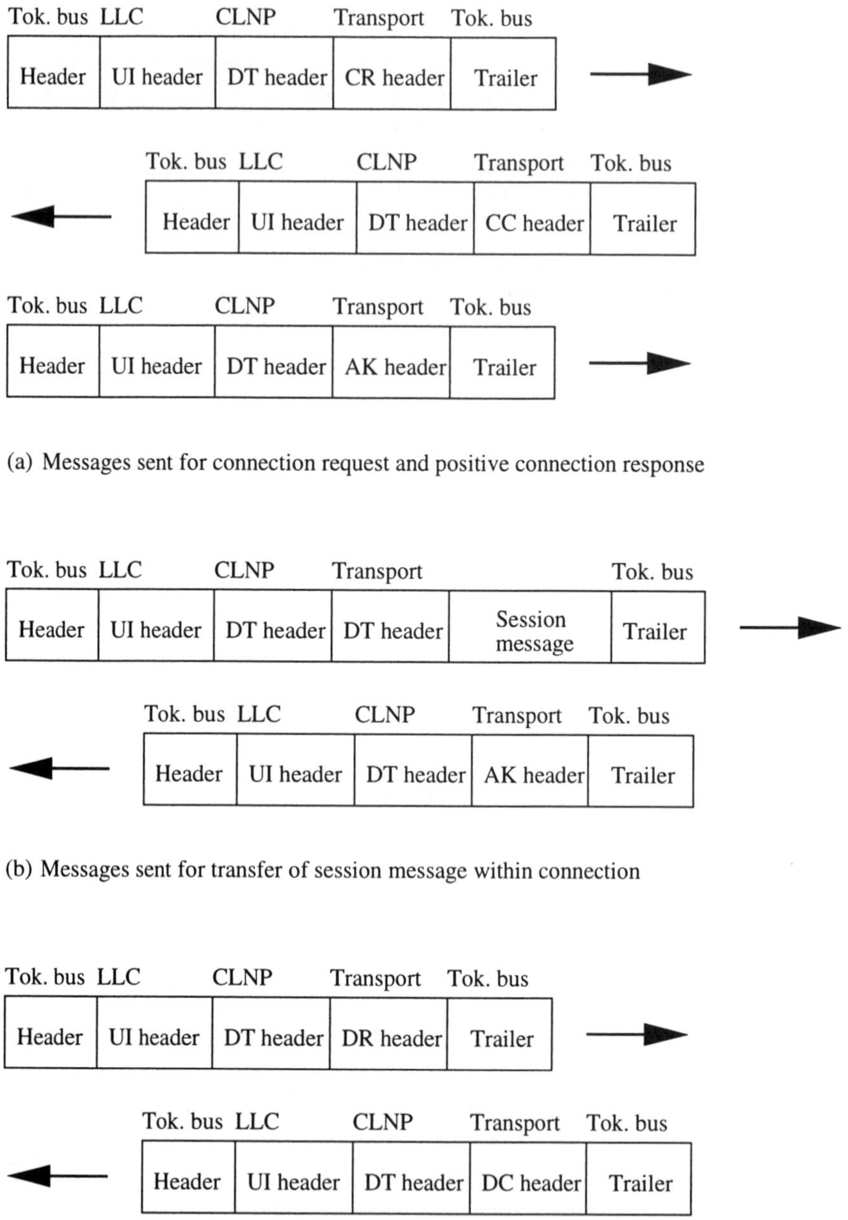

Tok. bus LLC   CLNP   Transport   Tok. bus

Header	UI header	DT header	CR header	Trailer

➡️

Tok. bus LLC   CLNP   Transport   Tok. bus

⬅️

Header	UI header	DT header	CC header	Trailer

Tok. bus LLC   CLNP   Transport   Tok. bus

Header	UI header	DT header	AK header	Trailer

➡️

(a) Messages sent for connection request and positive connection response

Tok. bus LLC   CLNP   Transport   Tok. bus

Header	UI header	DT header	DT header	Session message	Trailer

➡️

Tok. bus LLC   CLNP   Transport   Tok. bus

⬅️

Header	UI header	DT header	AK header	Trailer

(b) Messages sent for transfer of session message within connection

Tok. bus LLC   CLNP   Transport   Tok. bus

Header	UI header	DT header	DR header	Trailer

➡️

Tok. bus LLC   CLNP   Transport   Tok. bus

⬅️

Header	UI header	DT header	DC header	Trailer

(c) Messages sent for disconnection request and positive disconnection response

Figure 10.2 Message exchanges for transport connection management

Overall, the behaviour is very similar to that shown in Figure 10.1 and, indeed, to any implementation of a connection-oriented time package. Each message transmitted in Figure 10.1 involves sending one or more of the information-carrying messages of Figure 10.2.

Summary of FullMAP MMS communication implementation

The FullMAP implementation involves the use of seven protocols in order to achieve an MMS communication. All of these are ISO standard protocols:
- the ACSE application protocol;
- the presentation protocol;
- the connection-oriented session protocol;
- the TP4 connection-oriented transport protocol;
- the connectionless network protocol;
- the LLC1 logical link control protocol;
- the token bus protocol.

Four of these protocols do not perform a very significant function: the presentation, session, network and logical link control protocols. This is because extra features that they possess are not actually used or needed when implementing an MMS communication using physical communications over a token bus network, or a collection of bridged token bus networks. However, the fact that these ISO standard protocols are included means that inter-working with less specialized applications or communication spaces is possible.

A common factor to these protocols, not discussed so far, is that of computer identification. The identifiers used in token bus protocol messages are the physical network identifiers for computers. However, within computers, there is usually more than one process involved in communication, and the communication channels of the different processes have to be multiplexed on to the same physical communication channel. Thus, there is a need for extra identification to specify particular processes within the computers. This need is accommodated by all of the other protocols. Note that this idea has already been seen in Chapter 9 in the discussion of TCP and UDP ports. Ports are used to identify different multiplexed channels involving a single computer.

In the LLC protocol, the first two bytes of each message are used for sub-identifiers, to allow multiplexing alongside LLC. In the full CLNP protocol, the last byte of the standard NSAP identifier is used for a sub-identifier, to allow multiplexing. In the transport, session and presentation protocols, the messages for connection establishment contain sub-identifiers to specify the process involved in the protocol conversation. The ACSE connection establishment message contains a name that identifies the MMS process that is communicating. The identifier that corresponds to this name is derived by combining:

- the presentation protocol sub-identifier;
- the session protocol sub-identifier;

- the transport protocol sub-identifier; and
- the CLNP identifier.

The resulting identifier is a unique global indentifer for the application process within the computer. For local implementation purposes in the environment of the case study, the CLNP identifier is, in turn, mapped on to a combination of:

- the CLNP sub-identifier;
- the LLC sub-identifier; and
- the token bus physical network identifier.

In other environments, different mappings would be used.

Sharing of channels, caused by several processes executing at the same time on one computer, is the only potential cause of unexpected delays to MMS communications within a factory. The token bus multiplexing scheme ensures fairness and a bound on latency when several channels are multiplexed on the same physical channel. Ensuring that the computers in a controller or a manufacturing device are not overloaded with communication activity is a computer system management activity, rather than a computer communications activity.

10.4 MINIMAP SIMPLIFIED IMPLEMENTATION

Given that there is a significant amount of redundancy in the FullMAP implemention, a slimmed-down version is available. The MiniMAP implementation is intended for simpler computers within manufacturing cells, and supplies just the essentials needed to support MMS messaging. The protocols are simpler to handle, which means less processing and faster handling of MMS messages. MiniMAP also supports two other necessary applications: the directory service and network management. MiniMAP implementations are used over single token buses connecting controlling computers and manufacturing devices. Some controlling computers have both a MiniMAP implementation used to communicate with their slaves, and also a FullMAP implementation for communicating with their peers and masters — this mixed implementation is the Enhanced Performance Architecture.

MiniMAP does not just strip away the FullMAP protocols that serve little or no purpose, but goes further by stripping away the connection-orientation of the implementation. Essentially, the sending of each MMS message is implemented using a reliable connectionless service supplied by the token bus network. Therefore, all five protocols used above the logical link control protocol are removed. The two main losses are the implementation of MMS connections that was given by the ACSE protocol, and the implementation of a reliable service given by the transport protocol. The first of these losses is dealt with by introducing a minimalist special-purpose protocol for MMS connection establishment and termination.

The second of these losses is dealt with by the use of different versions of the logical link control and token ring protocols.

MiniMAP makes use of the LLC3 version of the IEEE 802.2 logical link control protocol, which is the acknowledged connectionless variant. In turn, this makes use of an **immediate response** feature allowed by the IEEE 802.4 token bus protocol. The LLC3 protocol involves the use of an AC (acknowledged connectionless) message type, rather than the UI (unnumbered information) message type used by LLC1. One computer sends an information-carrying AC message to another, which then acknowledges it by sending back an AC message. Each message carries a one-bit sequence number, so that an acknowledgement message can be matched up with the information-carrying message that it refers to.

The standard way of transmitting messages on a token bus network is called **request with no response** mode. This allows an unreliable connectionless service to be offered. There is an alternative way, called **request with response** mode. If a computer transmits a message with a header bit set to indicate this mode, the recipient must send back an acknowledgement message immediately. The token does not change hands to allow this. The token-holder, who transmitted the original message, just waits for the response message to arrive. If no response is received promptly, a timeout occurs and the token-holder retransmits its message. There is a limit on the maximum number of retransmissions before the communication is deemed to have failed. This extra mode means that the token bus network can supply a service that meets the needs of the LLC3 protocol.

The simplification involved in MiniMAP does not stop with the token bus protocol. It extends to the physical bus as well. Instead of the 10 Mbits per second broadband transmission used for FullMAP, 5 Mbits per second carrierband transmission is used instead. Because the transmission frequency does not need to be restricted to a narrow bandwidth, the electronics are simpler and relatively inexpensive. Also, bidirectional transmission along the bus can be used, avoiding the need for a headend.

In summary, MiniMAP can be seen as the special-purpose efficient solution to factory messaging, albeit one that is consistent with ISO standards. The MMS protocol was developed for the MAP project, as was the token bus network, together with its support for an acknowledged connectionless service.

10.5 CHAPTER SUMMARY

Computer integrated manufacturing requires communication between factory floor computers and devices supplied by many different manufacturers. The Manufacturing Automation Protocols (MAP) project, initiated by General Motors, devised an implementation strategy for the necessary communications based on internationally standard ISO protocols. At the time the work was done, this seemed the obvious way to proceed, but unfortunately did not foresee the dramatic rise of the

Internet and its protocols. As a result, MAP includes some protocols that have not been widely adopted in the computer communications world.

The Manufacturing Message Standard (MMS) defines a wide range of message types to support client-server interactions between computers, particularly between controllers and manufacturing devices. This is based on an abstract Virtual Manufacturing Device model. With the FullMAP implementation, the connection-oriented MMS communications are implemented using, in turn, ISO standard application, presentation, session, transport, network, logical link and token bus, protocols. Some of these protocols are largely redundant, but they do allow the use of other applications or networking that extends beyond one factory. The MiniMAP implementation is a much stripped down version that eliminates five of the seven protocols. It is suitable for use in local manufacturing cells within a factory.

The time critical requirements of MMS messaging are met by the use of token bus local area networks to provide physical conectivity. Maximum token holding times ensure that there is a guaranteed bound on the delay before a communication can take place. Most MMS messages are short and so, despite the overheads introduced by various protocol headers, transmission is fast using the 5 or 10 Mbits per second transmission rate of the token bus network.

10.6 EXERCISES AND FURTHER READING

10.1 If you are lucky enough to have access to an environment in which MAP is used, find out what kinds of manufacturing devices are involved, and what MMS communication patterns exist.

10.2 Find out about any recent activities of the MAP User Group

10.3 The Technical and Office Protocol (TOP) project uses a standard IEEE 802.3 ethernet as its physical basis, rather than the token bus used for MAP. Why is this?

10.4 Investigate the details of the Virtual Manufacturing Device (VMD) specification, and comment on how the various object classes have relevance to different types of real manufacturing devices.

10.5 What other applications are included in MAP, apart from MMS?

10.6 The ISO presentation protocol includes the notion of a presentation context. Explain how this notion also occurs in the Internet MIME standard for electronic mail.

10.7 Trace the ancestry of the ISO session protocol, in particular the influence of ITU-T information service standards.

10.8 The ISO session protocol involves the exchange of 'tokens' to control various features. Find out about the general token mechanism, and list all of its uses in the protocol.

10.9 Carry out a detailed comparison of the ISO TP4 transport protocol and the Internet TCP protocol. What are the significant differences?

10.10 Check that it is possible to convey TP4 messages inside Internet IP messages by looking up the IP protocol number that is assigned to TP4.

10.11 Give reasons why a token bus was specified for MAP, rather than a token ring.

10.12 The presentation and session protocols both have a sub-identifier, to specify a process responsible for a given connection. Explain why it is not really necessary to have separate sub-identifiers for each protocol.

10.13 MiniMAP makes use of the request with response mode of the token bus protocol. Verify that the original Ethernet protocol had a similar mechanism.

10.14 Discuss the message transmission times, and delays before message transmission, that are likely on a 10-computer MiniMAP network. How would these times be affected if FullMAP was used?

10.15 Suggest alternative ways in which the FullMAP collection of protocols might be slimmed down, while retaining the feature of guaranteed reliable message transmission.

Further reading

A specialist textbook that covers MAP (and its close relative TOP) is *MAP and TOP Communications* by Valenzano, Demartini and Ciminiera (Addison-Wesley 1992). Most general textbooks have a description of the ISO standard protocols, although more modern editions are beginning to omit much of the detail, given the relative disuse of these protocols. The full details of MMS are available in ISO standards documents, but these are heavy work for the casual reader. There is not much, if any, research literature on MAP, largely because the project was deliberately based on safe technology, rather than attempting to advance the state of the art.

ELEVEN

CASE STUDY 3: MAKING A VIDEO TELEPHONE CALL

The main topics in this case study about video telephony are:

- video telephony in the context of teleconferencing
- compression of a video signal
- management of an ATM connection
- use of the AAL5 ATM adaptation layer
- implementation of ATM using STM

11.1 INTRODUCTION

This third case study is concerned with a communications application that can be seen as a direct successor to the traditional telephone call. It involves two computers, each equipped with video input and output. These are used as the end points of a video telephone call, and are connected together via a public B-ISDN service based on ATM technology. This is directly analogous to two telephone handsets that are connected together via the plain old telephone system. The situation in this case study differs from that in the other two case studies, in that a black box network is inserted between the two video telephone callers. The two computers interact with this black box, rather than with each other.

In looking at this particular application, there will be opportunities to briefly examine related side-issues. These include the general issue of multimedia multipeer teleconferencing, of which this is merely a special case with one medium — video information — and two peers. There is also the issue of how this direct ATM networking might interact with other types of networking, in particular with the Internet and with conventional local area networks. At the time of writing, in 1997, both of these issues were still in a state of flux, the province of designers and working groups, rather than providers of public services. Indeed, direct access to ATM networking itself was still available to only a few experimentalists as a public service over long distances. The dream of household B-ISDN connections was still a few years away.

The immediate predecessor technology was video telephony over N-ISDN, making use of digital circuits operating at 64 kbits per second, or integral multiples of 64 kbits per second. The standard rate of 64 kbits per second was the ideal for speech telephony, allowing digitized speech to be transmitted without any loss in quality. For video telephony, this rate is about the minimum possible for transmission of any plausible quality. It involves modest sampling of the video signal and compression of the resulting digital data. Both of these factors introduce loss of quality. An international standard from ITU-T called H.261 is used for video telephony compression over narrowband communication channels. It supports rates of $64p$ kbits per second for $1 \leq p \leq 30$, i.e., rates between 64 kbits per second and 2 Mbits per second. The parameter p corresponds to the number of video frames sampled per second — each sample is digitized and compressed to contain at most 64 kbits. At the maximum sampling rate, the quality approximates to that of VHS entertainment video.

For ATM technology, 2 Mbits per second is at the slow end of the possible spectrum. The first standards for widely used ATM channels focused on 25 or 51 Mbits per second, with rates of 155 and 622 Mbits per second in use for trunk connections. Thus, the physical capability for this case study is distinctly superior to first-generation digitized video telephony.

The transmission of moving pictures is not the only issue in video telephony. Accompanying sound is also needed. The digital technology for this is well-understood and well-established, and so does not present any major technical problems. However, in combination with video, there is a problem of ensuring that the audio and video signals are synchronized at the receiver. For example, when a video caller's lips move on screen, the speech heard should match up. This sort of problem is a special case of a more general problem for multimedia transmission. A multimedia communication channel is typically implemented using splitting over separate channels, one for each medium. When the split communication is merged at a receiver, absolute time synchronization is usually needed to reconcile timing differences that arise from the use of separate channels. Thus, some sort of protocol is needed to transfer timing information in addition to the multimedia information.

This is not the only synchronization problem. It only refers to the simplest type of communication, where there is just a single flow of information from one computer to another. More complex types of communications can be built from such communications, only as long as they can be decomposed into independent communications over some collection of channels. For real applications to human communication, this is not sufficient. If two people are involved in a video telephone call, then each can see a continuous picture of the other, accompanied by continuous sound from the other. This gives a reasonable approximation to a conversation between two people in the same place. The main problem comes when more than two people are involved: in a **teleconference**. It is normally the case that each participant receives only one picture with sound at a time, but the problem is that there are two or more possible choices of source.

A solution is to say that the teleconference always has a current focus at any particular time, that is, there is one selected participant who is the source of the pictures and sound that everyone else receives. This has an immediate parallel with message broadcasting networks, where only one computer is allowed to broadcast at a time. There, the computers use a protocol in order to coordinate their multiplexed channels. In this case, some human-oriented protocol is needed, in order to select who is the current speaker in the teleconference. This can range from the totally mechanized, for example, the person generating the loudest sound is chosen, to the totally humanized, for example, each speaker nominates the next speaker to follow.

ITU-T has standardized a collection of protocols to support teleconferencing: its **generic conference control protocols**. These can be used to assist:

- management of the set of computers participating in the conference;
- management of the set of applications and media that constitute the conference;
- control of the current focus of the conference; and
- assignment of a special chairmanship role to one of the participants.

The extent to which such facilities are used depends on the nature of the conference. For conferences involving loosely coupled communications — so-called 'crowds that gather around an attraction' — only limited management of participants is necessary. For more closely coupled communications, where the presence and attention of all participants is important, formal policies must be implemented and/or a chairman must be introduced. Internet protocols to support similar facilities were still at the design stage in 1997. These offer a service broadly aligned with that of the ITU-T standards, but informed by a rather different implementation model: IP-based multicasting, rather than connection-oriented unicasting.

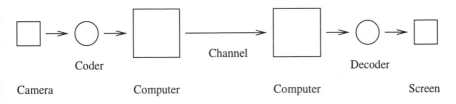

Figure 11.1 Overall video telephony set-up

The above discussion sets the context for this case study. The focus is on the simplex transmission of video from one computer to another via a public broadband network. This is because the aim is to illustrate the implemention methods that lead from an original video signal to the bits that flow along a physical channel. The other issues concerned with multipeer and multimedia teleconferencing are not considered further. However, the importance of these matters cannot be understated. Traditionally, computer communications facilities have offered support for text-based human communication, such as electronic mail, newsgroups and real-time chat. With the advance of higher speed physical connectivity, the communication of speech and video is a reality, and hastening the end of the separation between conventional telephony and digital communications. In the future, further advances can be envisaged, for example, distributed virtual reality worlds supported by computer communications. This is already an active research area.

11.2 THE PROBLEM: INFORMATION, TIME AND SPACE ISSUES

The overall framework for the communication is illustrated in Figure 11.1. The camera and the coder are peripheral devices attached to the sending computer. Their role is to capture continuously the video information and represent it as a stream of binary information. The camera generates an electrical signal that represents the scene viewed by the camera, and the coder converts this into a digital bit stream that can be used by the computer. After the communication of bits between the computers, the process is reversed at the other end. The receiving computer generates a digital bit stream, and the decoder converts this into an electrical signal that causes the screen to display a television picture. The connections between the camera and the coder, and between the decoder and the screen, are not of interest here. However, note that the electrical signal may be analogue or digital. Traditionally, it was analogue, compatible with television and video recorders, but some more modern video equipment has a digital interface.

The essential communication problem is to ensure that the information transmitted between the two computers is adequate to ensure that the moving video picture displayed on the screen is a faithful enough representation of the scene that is being observed by the camera. At one extreme, when the scene is completely static, the problem is much simplified, since no real-time factors need be considered. However, it will be assumed that the scene is dynamic to some extent, which means that the communication must be continuous.

Video information issues

A video camera has a two-dimensional array of regularly spaced sensors, which receive information from the scene that the camera is pointing at. Each sensor detects the intensity of three different wavelengths of light being emitted from one point in the overall scene. These wavelengths correspond to red, green and blue — the primary colours that can be combined in different proportions to give all of the different colours. This type of sensing corresponds to that of the cones in the human eye.

The range of different intensities that the human eye can discriminate is not large, so it is sufficient for a camera to use only eight-bit values to measure the intensity of each primary colour. Thus, the output of each sensor is a 24-bit value: eight bits for each of red, green and blue. The overall output from the camera is a stream of 24-bit values, being the outputs of the sensors when scanned horizontally and vertically. In essence, this scanning is the exact reverse of what a television set does to display pictures.

The quality of the video signal depends on two things. The first is the density of the sensors — the more sensors, the more accurate the representation of the scene in each video frame. The second is the sampling rate — the more frequent the sampling rate, the more accurate the representation of movement in the scene. A rate above 10 frames per second is enough to fool the human eye in most cases; a rate above 20 frames per second is good enough to capture fast movement. From a communications point of view, the larger that density and sampling rate become, the more information there is to be communicated.

As an example, for a 720×576 array of sensors (the resolution used in some television standards), using 24 bits per sensor, and sampling 25 frames per second, there is a raw information rate of around 250 Mbits per second. This is extremely large, even given projected advances in communication technologies. However, the situation can be much improved by the use of video compression techniques. These are implemented by coders and decoders.

A main feature exploited by video compression algorithms is the fact that every single element of a scene does not change between each frame. Once a total representation of a scene at one point in time has been transmitted in one frame, it is adequate only to transmit updates to components of the scene in subsequent frames. Within individual frames, redundancy means that compression is also possible, and also reduction of the level of detail can be used to compress. Overall compression rates in excess of one hundredfold are possible in some cases.

The ITU-T H.261 standard for video compression in video telephony was targeted at N-ISDN transmission rates. It is based upon a logical 176×144 array, sampled up to 30 times per second. The sensing arrangement for any physical video camera has to be mapped to this logical array. There is not a 24-bit red-green-blue intensity sample for each element in the array. Instead, each 2×2 square sub-array of elements has a 48-bit measure associated with it, i.e., an average of 12 bits are used to represent each element. This is an ITU-R (the radio equivalent of ITU-T) standard for representing television signals digitally, and it is used in H.261 as an encoding that is independent of specific video technology.

For this representation, the raw information rate prior to compression is about 9 Mbits per second at the highest sampling rate. The compression algorithm produces at most 64 kbits per frame, so this 9 Mbits per second rate is reduced to around 2 Mbits per second. There is also a higher-quality variant that uses a 352×288 array, i.e., double the size in each dimension. This produces at most 256 kbits per frame.

H.261 was targeted at teleconferencing applications, where the amount of motion between video frames is relatively small. It is not adequate for general-purpose video compression, where there may be rapid changes in the captured scene. The MPEG (Moving Picture Experts Group) standard for video compression, already described in outline in Chapter 2, was designed for general-purpose video compression that ensured high quality. The MPEG standard did not have to observe two major constraints on H.261: a target of an encoding and decoding delay of less than 150 milliseconds, and a target of 64 kbit per second operation. MPEG also deals with the compression of audio signals, and the integration and synchronization of video and audio signals. There are three MPEG standards: MPEG-1, MPEG-2 and MPEG-4. Work began on an MPEG-3, but it was discovered that MPEG-2 was adequate for MPEG-3's target applications with a little tuning, so MPEG-3 was abandoned.

MPEG-1 was the original MPEG standard and, like H.261, was targeted at film-style videos, that is, videos represented frame by frame. It was optimized for CD-ROM usage or applications supporting a compressed rate of about 1.5 Mbits per second.

MPEG-2 was intended to support a wider range of applications than MPEG-1, in particular broadcast television quality video. The main new feature required for this was the ability to cope with the interlaced video signals used in television, in contrast to the frame-by-frame video signals used in film. MPEG-2 was also found to be efficient for applications with higher sampling rates, for example, high-definition television, the original niche of MPEG-3. ISO has adopted MPEG-2 as an international standard, both for video and audio.

Finally, MPEG-4 is targeted at very low bit rate applications capable of *at most* 64 kbits per second. This involves the development of new algorithms that allow the preservation of high quality, using completely different techniques to MPEG-1 and MPEG-2. These centre on using models of the picture information contained in the video signal, rather than a low-level bit representation of the video signal.

The normal parameters used for MPEG-2 are a 720×480 array of elements, sampled 30 times per second. As in H.261, each element yields 12 bits of information. Thus, there is a raw information rate of around 125 Mbits per second. The MPEG-2 compression algorithm reduces this to a maximum of 15 Mbits per second. In typical circumstances, the average rate is around 4 to 6 Mbits per second. For higher quality television, beyond the quality intended for consumer high-definition television, a 1920×1080 array of elements can be used. This increases the raw rate to around 750 Mbits per second, which can be compressed to a maximum of 80 Mbits per second.

The information discussion has focused on the quantity of digital information, measured in bits, that needs to be communicated in order to implement the communication of video information. The details of how the video information is encoded into binary information is omitted, since this would involve a lengthy detour into the mechanics of how the video compression algorithms work. Such details are really the province of the video expert, or the data compression expert, rather than the computer communications expert. What is important from the communications point of view is the type of information to be communicated. Here, this can be treated as a simple bit stream supplied by a specialized client.

Video time issues

Since video is a continuous, real-time phenomenon, there are some absolute time constraints on the communication of video information. If the camera attached to one computer is capturing p frames per second, then the goal is that the screen attached to the other computer should display p frames per second. Thus, there should be a guaranteed frame rate. Latency is also an issue for some communications. If video information is just being transferred from one point to another in isolation, then the delay between frame capture and frame display may not be vitally important. However, for two-way video telephony, or more generally for multi-peer teleconferencing, the delay must be kept to a minimum so as not to hinder the human interaction.

The above comments suggest one obvious way in which to treat the time period of a video communication: as a continuous bit rate communication at a rate that matches the video information rate. For example, if the video camera is generating information at a rate of 30 Mbits per second continously, then a communication channel capable of supporting this rate is provided. This would be a computer communications emulation of what happens with broadcast or cable television. Indeed, this is the standard way of implementing video telephony over N-ISDN, which makes a dedicated 64 kbits per second channel available for the video communication. As seen in Section 7.5.6, the constant bit rate B-ISDN service makes high speed channels of this type available, underpinned by ATM technology. For example, in some experiments with early ATM networks, 2 Mbits per second constant bit rate connections were used to allow maximum-quality H.261 video transmission.

This solution is not the best one, because it ignores the impact of video compression. The information rates quoted above are *maximum* rates, that is, upper bounds on what is required. The actual rates needed at any particular time may be rather smaller, for example, if there is little motion in the scene that is being captured by the camera. Thus, the overall video communication has a natural segmentation over time into sub-communications. Each sub-communication involves a single frame, and has a duration that varies from frame to frame. With H.261 or MPEG compression, these sub-communications do not necessarily begin at regular intervals, corresponding to the video frame rate. In fact, there can be reordering of frame communications, so that later frames are actually transmitted before earlier ones.

If continuous bit rate transmission is requested, then the communication service is unnecessarily making some capacity available. In terms of the service's usage, it is as if a padding sub-communication of dummy bits is inserted between the end of each real sub-communication and the beginning of the next one. This is a waste of the capabilities of the communication system.

For B-ISDN, a better solution is to make use of the real-time Variable Bit Rate service instead. When a variable bit rate connection is established, both a peak required rate and a mean required rate are specified. Here, the first of these would correspond to the maximum rate resulting from the compression process, whereas the second would correspond to the average rate over a lengthy time period. This allows the video communication to proceed with its natural implementation — a series of asynchronous frame sub-communications — but with a guarantee that the maximum absolute rate is available when required.

An optional extension to the MPEG encoding process adds some scope for more subtle behaviour over time. Instead of regarding the representation of each frame as an atomic whole, it can be divided into components that have different levels of importance. Four possible divisions into two components are:

- a lower density sampling array, plus an overlay that adds higher density sampling;
- a lower gradation of sample measurement, plus an overlay that adds a higher gradation;
- a low frame rate, plus an overlay that adds intermediate frames; and
- critical parts of the overall representation, such as header information and motion information, plus an overlay with the less critical parts.

It is possible to have three different levels of importance, by combining these divisions in an appropriate way. In temporal terms, the idea is that the lowest level of information, which contains the most critical information, is given the highest priority in terms of guaranteeing delivery. The higher levels have lower priority, since it is less crucial that their extra luxury content is received.

For modern channels, the most likely cause of information loss is due to congestion, rather than to physical errors. The priority information is used when selecting candidates for dropping. If physical errors are a problem, then more

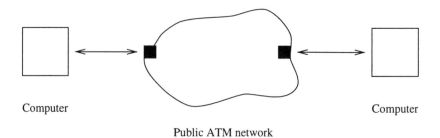

Computer   Computer

Public ATM network

Figure 11.2 Implementation of video telephony channel

powerful error detection or correction codes can be used for the more critical information.

Video space issues

Figure 11.1 includes the basic communication space for this case study: two computers connected by a simplex channel. The further refinement required is to indicate that the simplex channel is provided by a public ATM network. This is shown in Figure 11.2. Note that two channels are shown. Each is a full duplex channel between a computer and an access point to the network. This is in the traditional telecommunications black box style. A computer interacts with the perimeter of the telecommunication network, rather than with its opposite number. The effect of this interaction is that end-to-end communication is carried out by the network. ATM implements this model of network access strictly. In older PSDNs, the X.25 protocol was defined in terms of interactions like this, but really behaved like an end-to-end protocol involving a direct exchange of messages. The computer-network interface is called the **User-Network Interface** (UNI) in ATM parlance. The interface between switches within the network is called the **Network-Network Interface** (NNI).

In practice, this model of communication is rather simplified. Most computers will not have their own direct connection to a public B-ISDN service. Instead, a typical computer will be attached to some sort of computer network, which will supply connectivity to a computer that is attached to the B-ISDN service. With the increasing use of ATM technology, some private computer networks will be based on ATM switching, rather than broadcasting technologies such as ethernet. If so, the typical computer will have through access to the public ATM service via the private ATM service that it is connected to. In the medium term, many organizations will continue to have traditional networks as well as private ATM networks. The standardization of ATM has included the development of standards for how legacy networks, such as ethernets, can be emulated by ATM in order to allow inter-working.

Two different identifier schemes are allowed in public ATM networks, both of them following well-established international standards. One is the E.164 ISDN identifier scheme, which uses 15-digit numbers. These are organized in a geographically-based hierarchy, like normal telephone numbers. The other is the ISO NSAP identifier scheme, as described in Chapter 5. Three of the styles of address allowed within this scheme are permitted, one of them being E.164. The exact choice is not of any particular importance to this case study.

The multiplexing of channels might intrude on a particular video telephony communication. First, there may be multiplexing with other channels implemented using the same physical channel between a computer and the ATM network. This might have the effect of restricting the peak communication rate that can be achieved for a particular communication. Second, there will be many other communications being implemented by the public network at the same time, some of which might interfere with the communication of interest. In principle, since a particular quality of service is guaranteed each time a connection is established through the network, it can be treated as a real black box without worrying what happens inside. However, unless there is an unlikely guarantee of a loss-free, continuous rate connection, the internal state of the network will affect the observed behaviour.

11.3 MANAGEMENT OF AN ATM CONNECTION

In the B-ISDN and ATM specifications, there is a separation between a **control plane** and a **user plane**. The control plane contains protocols used for **signalling**, that is, for the management of connections. The user plane contains protocols used for transferring information over the connections. This type of model is used in N-ISDN as well, and is another reflection of the telecommunications view of the world. In the traditional telephone system, there is signalling communication to control telephone calls as well as the user voice communication. However, the jargon used should not disguise the fact that a familiar time package is in use here. For the connection in this case study, there is an establishment stage, an information transfer stage and a termination stage. Establishment and termination fall within the business of the control plane.

The Q.2931 connection management protocol

The exotically named ITU-T standard Q.2931 protocol is used for the management of point-to-point connections. It is based on the Q.931 protocol used for the same purpose in ISDN. The ATM Forum has defined its own closely aligned but slightly different, protocol — the latest version in 1997 was called UNI 4.0, and it had succeeded the UNI 3.1 during 1996. The differences between Q.2931 and the ATM Forum's protocols are not sufficiently large to be worth describing here. The UNI protocol involves the exchange of messages between a computer and

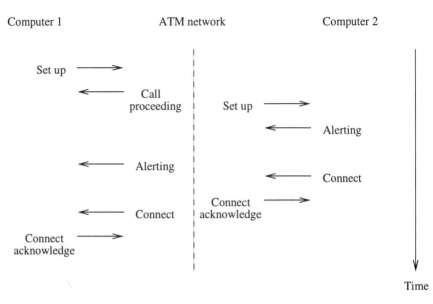

Figure 11.3 Establishment of connection using Q.2931

the ATM network, rather than directly between the two computers. Thus, it adds a certain amount of novelty, compared with the numerous other handshake-style connection management protocols described in other chapters.

The Q.2931 communications use a separate signalling channel between a computer and the ATM network. This channel can transfer messages containing Q.2931 information. The overall Q.2931 communication consists of a series of handshakes that are concerned with the management of one or more user connections. Connection establishment involves five types of Q.2931 message:

- Set up;
- Call proceeding;
- Alerting;
- Connect; and
- Connect acknowledge

(the ATM Forum UNI protocol does not include the Alerting message). The interaction between each computer and the network is essentially a three-stage handshake. This is illustrated in Figure 11.3. In particular, note that handshakes are between computer and network, not between computer and computer. It has been estimated that some connection establishment times might be measured in seconds, which is fairly lengthy compared to the timescales for the communications using the connections.

The Set up message is important, since it contains information about what is required from the connection. There is no negotiation of connection parameters in Q.2931. Each connection request is either accepted as is, or is rejected by the network. The message includes the identifiers of the caller and the callee. Also, there is information about which ATM Adaptation Layer (AAL) is to be used on the connection. Importantly, there is a flow specification required for the connection.

The flow specification was mentioned briefly in Section 7.4.5. It can contain any of the following parameters, which refer to 53-byte ATM messages ('cells'):

- peak message rate: the maximum instantaneous rate at which the user will transmit messages;
- burst tolerance: the maximum number of messages that can be sent back-to-back at the peak transfer rate;
- sustained message rate: the average message transmission rate, measured over a long time interval;
- minimum message rate: the minimum message rate desired by the user;
- message transfer delay: the delay experienced by a message between entry to the network and exit;
- message delay variation: the variance allowed in the message transfer delay; and
- message loss ratio: the percentage of messages lost in the network because of congestion or errors.

The extent to which these parameters are used depends on the B-ISDN service type request: Constant, Variable, Available or Unspecified Bit Rate.

Message delay variation is caused by variable switching and queuing delays within the network. This effect can lead to messages becoming more bunched together, and so artificially increasing the peak message rate. Such an increase can cause problems within the network, or for a user who is respecting the agreed bound on peak message rate when transmitting.

The termination of a connection involves only two different types of Q.2931 message:

- Release; and
- Release complete.

Termination is initiated by one of the computers, which conducts a simple handshake with the network. The network then conducts a simple handshake with the other computer. This is shown in Figure 11.4. In addition to establishment and termination of connections, Q.2931 also has a few other management messages that can be used during the existence of the connection, but these are not of importance to this case study. The ATM Forum UNI protocol has a few other such messages too, but most are different from Q.2931's messages.

Implementation of the Q.2931 protocol

The communication of Q.2931 messages is implemented using the normal facilities for sending messages over an ATM connection. As already described in

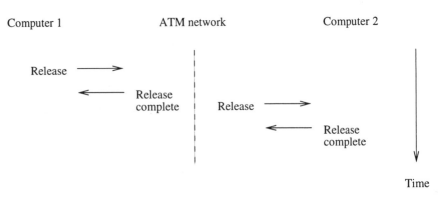

Figure 11.4 Termination of connection using Q.2931

Section 7.5.6, these facilities are supplied using an adaptation layer that civilizes the raw ATM message transmission capability to some extent. There will be discussion of these facilities in Section 11.4, since they are relevant to both signalling and user communication.

Given that an ATM connection is used to implement the Q.2931 communications, a natural question is how this connection is managed itself. The answer is that some connections exist permanently, and so do not have to be established or terminated using a protocol. In this particular case, at least two permanent signalling connections exist between each computer and the ATM network. One of these is used for managing user-to-user connections.

In fact, permanent connections are a general service offered, and users can make use of this service for applications where there is never-ending communication between two points. Indeed, in early experimental ATM networks, this was the only type of connection available, meaning that establishing new connections was a very time-consuming administrative activity for network operators. The availability of Q.2931 changes all that, of course.

The normal ATM service is not completely suitable to support Q.2931 directly, because delivery of messages is not guaranteed. Reliable communication is necessary for connection management. Therefore, an enhanced adaptation layer, the Signalling AAL (SAAL) is used. This is constructed using an extra, lightweight, connection-oriented protocol called the Service Specific Connection Oriented Protocol (SSCOP), which is implemented using the AAL5 adaptation layer. As well as having applications for signalling, the SSCOP protocol can be used as a generic reliable connection-oriented protocol for ATM networks.

Reliable information transfer using SSCOP involves four types of message. Information is carried in **SD** (Sequenced Data) messages, which can have variable length up to 65 535 bytes. Each SD message contains a 24-bit serial number, in addition to the user information. In the case of the SAAL, the user information is a Q.2931 message. Each Q.2931 message has a six-byte header followed by a variable-length information field. The header indicates which connection the message refers to, the type of the message and the length of the message.

Periodically, the sender of SD messages sends a **POLL** (Polling) message to the receiver, to obtain a status report on what SD messages have been received. POLL messages may be sent because a timer has expired, or because a certain number of SD messages have been sent. A POLL message contains the serial number of the next SD message that will be sent, together with a poll serial number. On receipt of a POLL message, the receiver sends back a **STAT** (Status) message containing:

- the serial number of the next in-sequence SD message expected;
- the serial number up to which the sender may transmit (i.e., the upper limit of a sliding window);
- a list showing which SD messages have been received correctly, and which have been lost;
- the poll serial number of the POLL message.

The sender uses this information to advance its sliding window and to retransmit any SD messages that have been lost.

If the receiver detects the loss of an SD message itself, because another one arrives out of sequence, it immediately sends back a **USTAT** (Unsolicited Status) message. This conveys similar information to that carried by a STAT message, the difference being that it was not triggered by the receipt of a POLL message.

From this, it can be seen the SSCOP contains the usual ingredients of a protocol to implement a reliable connection. It is designed to be lightweight and capable of being processed at high speeds. The POLL-STAT mechanism makes the protocol very sender-driven, which means that the transfer of control information is limited to that actually required by the sender. The fact that ATM guarantees the sequencing of messages sent across a connection means that the additional USTAT mechanism can be used in order to increase efficiency further.

In all, there are 15 different types of SSCOP message. Two further contributions to efficient handling are that message lengths are multiples of 32 bits, and that the SSCOP control information is carried as a trailer in each SSCOP message. In particular, the message type is indicated by a four-bit field in the final 32-bit word. The fact that the control information is at the end means that all processing can be done at the time the whole message has arrived, rather than some preliminary processing being done at the beginning and then further processing being done at the end.

SSCOP messages are transferred using the AAL5 adaptation layer. At this point, the implementation follows the same path as the implementation of communications taking place on behalf of B-ISDN users. In this case study, the user communications are concerned with transferring video information.

11.4 USING AN ATM CONNECTION

The purpose of an ATM adaptation layer is to provide agreed rules for segmentation of user communications into sub-communications that involve 53-byte ATM messages. The range of standard adaptation layers was described in Section 7.5.6. In this case study, the AAL5 adaptation layer will be used, since it is appropriate for transferring variable-sized bit sequences, of the type that emerge from video compression algorithms. A practical alternative would have been AAL1, which is designed for continuous bit rate traffic, including video. This would be suitable for uncompressed video or, perhaps, by wastefully transmitting extra dummy information, for compressed video. In fact, this application falls into the territory intended for coverage by AAL2 but, as mentioned in Section 7.5.6, the AAL2 standard has not progressed.

Before describing how AAL5 can be used directly, it is worth mentioning an alternative that involves Internet protocols. In this case study, a direct channel through an ATM network is being used. However, in general, a video communication might involve inter-networking, with a journey through a public ATM network being just one hop on the overall route. If so, then Internet protocols might be inserted between the encoded video information and the AAL5 service.

One possibility is to to use UDP/IP to provide an unreliable connectionless service. There is an Internet standard for implementing IP using the AAL5 service. This involves encapsulating each IP message in an LLC/SNAP message, as seen in Section 9.6. The LLC/SNAP message is then encapsulated in an AAL5 message. One major issue raised by this is that IP is connectionless, whereas ATM is connection-oriented. The effect is that connection management is pushed down from the video communication level to the IP message communication level. Internet standards exist describing how this can be implemented, and also how other problems arising from implementing IP over ATM can be solved.

The AAL5 adaptation layer

The format of an AAL5 message is shown in Figure 11.5. An eight-byte trailer is added to the end of a user message. All AAL5 messages have a length that is a multiple of 48 bytes, so padding is inserted before the trailer if necessary. The reason for using a trailer rather than a header is, as mentioned above in the discussion of SSCOP, to allow processing of messages to be concentrated at the end of reception.

The first two bytes of the trailer are a one-byte User-to-User Indication field and a one-byte Common Part Indicator field. The first of these is available to the users of the AAL5 service for any purpose desired, and will not be used here. The use of the second field is not defined, and so is really just padding. A similar field is used in AAL3/4, but for a reason, unlike in AAL5. The 16-bit length field contains the length of the user message contained within the AAL5 message. Finally, there

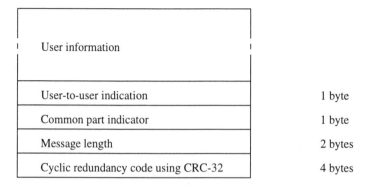

Figure 11.5 Format of AAL5 message

is a 32-bit cyclic redundancy code using the CRC-32 generator polynomial. The CRC is computed over the whole AAL5 message.

The reason that every AAL5 message has a length that is a multiple of 48 bytes is that each message is divided into 48-byte components, each of which is transmitted inside one ATM message. Thus, the AAL5 communication is implemented using a series of ATM sub-communications. The CRC used on each AAL5 message protects against the loss of any of the component parts of the message when the ATM messages are sent through the network.

Earlier, the details of the bit sequences used to represented compressed video information were omitted, through the requirement for specialist knowledge. One interesting point to note is that the MPEG standard information stream consists of a series of 188-byte messages. Two MPEG messages have a total length of 376 bytes, and so fit exactly into a $(376 + 8) = 384 = 8 \times 48$ byte AAL5 message without any padding. Thus, two MPEG messages can be sent using eight ATM messages. This simple mapping of messages simplifies handling by the communicating computers.

Implementation of ATM connections

The general information, time and space features of ATM networking are covered in Section 7.4.5. Here, a little extra relevant detail is included. First, there is a feature of the ATM message that is relevant to supporting AAL5. Second, there is some more detail on ATM connections and the multiplexing of their channels. Finally, there is a brief description of how the transmission of ATM messages is supported by physical channels.

There are two variants of the ATM message format, one used for communications between a computer and an ATM network, the other used for communication between switches within an ATM network. The three-bit payload type field is common to both message formats, and a little more discussion of it is appropriate here.

If the first bit of this field is zero, then the message is carrying user information. In this case, the second bit is used to warn of congestion, as described in Section 7.4.5. The third bit is a user signalling bit, and is preserved as the message passes through the network. It is used by AAL5 to indicate the last message resulting from the segmentation of an AAL5 message communication. The first $(n - 1)$ messages have this bit set to zero, and the last message has this bit set to one.

If the first bit of the payload type field is one, the message is carrying information for network Operation, Administration and Management (OAM). The other two bits indicate what type of OAM information is being carried.

The two variant ATM message formats differ in the length of the Virtual Path Identifier (VPI) which, together with the Virtual Circuit Identifier (VCI), identifies the connection along which the message is travelling. When a new connection is established using Q.2931, the network supplies a new VPI/VCI pair to each computer, for use when sending messages within the connection. Note that the signalling connection used for Q.2931 always has its VCI equal to 5. All user connections have a VCI value that is greater than 31.

In general, a computer attached to an ATM network will have several connections active simultaneously. Even if there is just one user connection active, there is also the permanent signalling connection needed to establish the user connection. The channels used by the different connections have to be multiplexed on to the single physical channel connecting the computer to the network. The VPI/VCI fields of each message allow this multiplexing to be done unambiguously. Within the network itself, there is also multiplexing of channels for communications from many sources over single physical channels.

A variety of types of physical channel can be used to underpin ATM networks, and links to ATM networks. Most of the standard types use the SONET/SDH mechanism for high speed synchronous communication. These mechanisms introduce overheads for control information, as with all protocols. For a 155.52 Mbits per second raw transmission rate, 149.76 Mbits per second is available for ATM messages. For a 622.08 Mbits per second raw rate, 599.04 Mbits per second is available. There are also standards for lower transmission rates, for example, 1 Mbit per second for public ATM networks, and 25.92 or 51.84 Mbits per second for private networks.

When using these synchronous mechanisms, ATM messages are sent continuously. At any time when there is no message that needs to be sent (inevitable, given that ATM is *asynchronous* transmission mode), a special idling message is sent. Thus, this is the point that the asynchronous behaviour becomes synchronous. One feature of the synchronous transmission of ATM messages is that there may be no explicit observable boundary between messages. The bits representing one message may be immediately followed by the bits representing another. In order to cope with this, there is a standard algorithm for determining message boundaries, given a stream of bits representing a sequence of messages.

This algorithm makes use of the header checksum in each message. The idea is to look for valid message headers, that is groups of five bytes that have a valid checksum when they are interpreted as a header. The pitfall is that the checksum can have only 256 different values, so there is a one in 256 chance of a misidentification. However, if the test is repeated over i consecutive candidates, the chance of i misidentifications is reduced to one in 256^i. A value of i between 6 and 8 is suggested, depending on the exact transmission mechanism, which gives an error probability smaller than one in 10^{20}. Once the receiver has become synchronized with the incoming bit stream, the same approach is used to detect when synchronization has been lost. If seven consecutive messages have incorrect header checksums, it is assumed that resynchronization is necessary, rather than that a lengthy burst error has occurred.

One danger with this mechanism for determining message boundaries is that user information within a message may, intentionally or otherwise, contain valid message header formats, and this may confuse the synchronization algorithm. To avoid this, scrambling is carried out on each non-header bit of every message before transmission. This is a variant on the use of bit stuffing to avoid framing bit patterns appearing within user information. Scrambling is used here because it preserves the message size, rather than expanding it.

11.5 OVERALL B-ISDN IMPLEMENTATION

The video communication is implemented by the transmission of bit sequences using a B-ISDN connection existing for the duration of the video communication time period. The effect is that a continuous video signal is transmitted using a continuous bit stream sent over a physical medium. However, there is no direct correlation in time between these two continuous signals. Much asynchronous behaviour lies in between. The information, time and space issues arising in this B-ISDN implementation are summarized below.

Information summary

The information to be communicated using the B-ISDN implementation depends on the encoding and compression scheme used. As one example, the MPEG-2 scheme, used for a video quality suitable for normal broadcast television, produces a maximum of 15 Mbits of information per second, with 4–6 Mbits being more typical on average. This compares with around 125 Mbits of information per second being supplied continuously by the video camera. The MPEG-2 encoding generates fixed-length byte sequences, each 188 bytes long.

The video information is transmitted using the AAL5 adaptation layer. This transmits user messages of length between 1 and 65 535 bytes. An AAL5 message consists of the user information, followed by a trailer containing two under-used bytes, then the length of the user information and a CRC computed over the whole message. The length of the AAL5 message is a multiple of 48 bytes, with padding being inserted between the user information and the trailer if necessary. The CRC can be used for the detection of errors affecting the AAL5 message.

The AAL5 message is transmitted as a series of ATM messages. Each ATM message has a five-byte header, followed by 48 bytes of the AAL5 message. For MPEG encoding, an efficient mapping involves two 188-byte MPEG message communications being concatenated. The resulting 376-byte message can be encapsulated into one 384-byte AAL5 message, which is then segmented over eight 53-byte ATM messages. This means that around 89% of the ATM message information transmitted corresponds to MPEG video information. The rest is AAL5 and ATM control information overhead.

For physical transmission, the ATM message may be directly transmitted as a stream of bits. Alternatively, if a method such as SDH or SONET is used, the bits are further encapsulated into the information-carrying units used by the carrier. In this case, there is a further overhead for control information.

Time summary

The video communication involves the continuous transfer of video information, one video frame after another, at a fixed rate over some time period. This is implemented using a B-ISDN communication with a connection-oriented time package. The connection lasts for the same time period as the video communication. During the connection, there is the transfer of encoded video information. This involves a series of communications with time periods determined by the operation of the video encoding process. The absolute time constraint on each of these communications is that it must be completed before its information is required to maintain the continuous video output at the receiver.

The B-ISDN connection is established by using the Q.2931 signalling protocol between the communicating computers and the ATM network. This is a variable bit rate connection, to take account of the fact that video compression is being used. In particular, this includes a traffic agreement on the peak message transfer rate required, and on the sustained message transfer rate required. For the MPEG example mentioned in the information discussion, 15 Mbits per second (or a little more, to allow for ATM message delay variance) would be suitable as a peak rate, and 6 Mbits per second would be suitable as a sustained rate. This is far more efficient for the network, and cheaper for the user, than asking for a 15 Mbits per second continuous bit rate.

There are several implementation issues concerned with Q.2931 itself. It is a protocol with its own set of message types. It needs a reliable transmission service, and this can be provided by the SSCOP protocol, which is designed for reliable connection-oriented signalling. SSCOP is a sliding window based protocol, but with some non-standard features, to enhance high speed operation. SSCOP is implemented using the same AAL5 service as used for transferring user information. The Q.2931 connection is a permanent connection, which does not itself need to be established or terminated using a protocol.

The ATM connection is implemented using various physical channels. The typical raw information rate of such channels is 155 Mbits per second or 622 Mbits per second for public ATM networks. This seems far more than adequate to support the video communication requirement. Indeed, it seems to remove the need for video compression. However, the catch is that each physical channel is shared by several, perhaps many, channels for other ATM connections. Thus, each new ATM connection is only accepted by the network if there is sufficient spare capacity on the physical channels being used, so that it can be guaranteed that the service requirements can be met.

Neither ATM nor AAL5 supplies a reliable service for the transmission of video information. AAL5 includes a CRC in each of its messages, which allows the detection of damage caused by missing or corrupted ATM messages. When an error is detected, the video information carried by that AAL5 message is lost. The video receiver must be robust enough to cope with some information going missing. Retransmission of lost information is not feasible when the video communication is real time and continuous. However, losses due to physical errors are very unlikely, given modern optical transmission media. Almost all losses will be due to network congestion, but such losses should be within the service guarantee given when the B-ISDN connection is established.

Space summary

The video communication involves two computers, one with a camera attached that is generating video information, the other with a screen attached that is displaying the video information. They are connected by a simplex channel supplied by a public B-ISDN service, and implemented by an ATM network. Each computer has a physical connection to a switch that acts as its entry/exit point to/from the ATM network. Within the network, the channel is implemented by a sequence of physical channels between switches. For each ATM connection, the path through the network is fixed at the time of connection establishment. The network is responsible for multiplexing different channels for ATM connections over each physical channel. Each ATM message sent within the connection is relayed through the network from the sending computer to the receiving computer.

Computers attached to the public B-ISDN service are specified by identifiers, which are used when new connections are established. The address may either be in the standard ITU-T E.164 format used for ISDN addresses, or in the more general ISO NSAP format used for network identifiers in several different types of network. The overall feel of using the B-ISDN service is very similar to that of using the present-day telephone system. The different is that computer messages are transmitted between two points, rather than continuous human speech.

11.6 CHAPTER SUMMARY

Video communication is one important aspect of multimedia communication. Video telephony is a simple special case in which two parties are involved in a conversation. The general problem is to realize multimedia multipeer teleconferencing. This involves the synchronization and control of communication as well as the basic multimedia information transfer. Standards for implementing such communications are under development by both the telecommunications community and the Internet community, not entirely independently. The case study in this chapter focused on just one component of this implementation: the simplex transfer of video information from one computer to another.

Large quantities of digital information are needed to represent directly a video signal to a high degree of quality. Luckily, there are various compression techniques, such as H.261 and MPEG, which can compress the information substantially. This makes the communication of continuous real-time video information more practicable. For example, a straightforward digital representation of a television quality video signal requires about 125 Mbits of information per second, but this can be reduced to as little as 4 Mbits per second in many typical cases.

The B-ISDN service will be an increasingly important public telecommunication service in the future. It is based on ATM cell networking. B-ISDN connections can be established between two computers, in a similar manner to the way that telephone calls are established between humans. The difference is that, rather than there being a continuous flow of information along the connection, there can be a variable rate flow of 53-byte ATM messages along the B-ISDN connection. These messages can carry parts of the compressed representation of a video signal. Guarantees on the delivery rate of messages can be agreed when the connection is established, thereby ensuring that video information arrives in time to be displayed when it is required.

11.7 EXERCISES AND FURTHER READING

11.1 If you have access to a video telephony or teleconferencing facility that is based on computer communications technology, find out its technical specification, and describe how this is related to the technology that supports it.

11.2 Compare ways in which participants in a teleconference might share a multipeer video/audio channel with the ways in which channels are multiplexed in a message broadcasting network.

11.3 Obtain information on the Internet Multimedia Conferencing Architecture, and the Internet Simple Conference Control Protocol (SCCP), both of which were still under development in early 1997.

11.4 If possible, look at the ISO T.120 series standards, which are concerned with telconferencing.

11.5 Try to locate a copy of the ITU-T H.261 standard on the WWW, in order to find out how this sort of video compression works.

11.6 Conduct a World Wide Web search for information on MPEG, in particular on the speeds at which commercially available circuitry can carry out MPEG compression.

11.7 Why are there differences between the ITU-T Q.2931 protocol for ATM connection establishment and the ATM Forum UNI protocol?

11.8 Compare the functions of the Resource Reservation Protocol (RSVP), used to set up connection-style state information in the connectionless Internet, with the functions of Q.2931.

11.9 Consider the list of seven parameters that can be included in a flow specification for an ATM connection. What action might an ATM network take to ensure that it delivers a service guaranteed to be within these parameters?

11.10 Obtain details of any ATM service that you have access to. What are the restrictions on the types of flows that can be specified when establishing a connection?

11.11 In what ways is the SSCOP protocol inadequate to deal with (a) a channel that does not maintain the sequencing of messages; and (b) a channel that can cause duplication of messages?

11.12 Read RFC 1932, which describes a framework for implementing IP using ATM.

11.13 The AAL5 adaptation layer is seen as an improvement on the AAL3/4 adaptation layer, by the computer fraternity at least. Why is this?

11.14 Find out about the Operation, Administration and Management (OAM) side of ATM networks, and so about the use made of ATM messages that have the first bit of the payload type field set to one.

11.15 What are the advantages of using an algorithm to deduce boundaries between ATM messages transmitted synchronously, compared with having explicit marker information transmitted between messages?

11.16 How appropriate would it be to use an SMDS service to support the video communication in this case study, rather than the ATM-based B-ISDN service?

11.17 Find out the current status of any public B-ISDN service offerings in your area, and also throughout the world.

Further reading

The topics of teleconferencing and video compression have a wide literature, ranging from the psychological and sociological factors underpinning teleconferencing behaviour, through to the technical details of video signal processing.

These application-specific matters are not strictly the province of the computer communications literature. B-ISDN and, in particular, ATM are the real computer communications topics of this chapter. An increasing number of specialist textbooks on ATM are now available, all struggling to keep up with the rapid advances in the area. One example is *ATM Networks: Concepts, Protocols, Applications* by Handel, Huber and Schroder (Addison-Wesley 1994). The April 1995 edition of the ACM SIGCOMM *Computer Communication Review* was devoted to ATM.

CHAPTER TWELVE

STANDARDIZATION

The main topics in this case study about standardization are:

- standardization as a human communication activity
- overview of the main standardization bodies for computer communications
- ISO standards, and the ISO Reference Model for Open Systems Interconnection (RM/OSI)
- Internet standards
- ITU-T standards
- IEEE 802 series standards
- proprietary standards: IBM, Digital, Apple, Xerox, Banyan and Novell

12.1 INTRODUCTION

In Chapter 1, the theme of **agreement** was introduced, as a distinctive feature of computer communications. Agreement is concerned with ensuring that communicating computers have a common view of the information that is being shared, the time period in which sharing takes place, and the collection of computers and the channel involved in the sharing. Agreement on the nature of communications at one level of abstraction is essential before these communications can be implemented using communications at some lower level of abstraction. Ultimately, there must be agreement on the means of transferring bits between computers using physical media.

A protocol is a collection of conventions and rules that underpins an agreement on the nature of communications. In general, a protocol has both static and dynamic features. Static features might include the type of information sent between computers, the maximum rate at which information can be sent and the identifier scheme used for computers. Dynamic features might include restrictions on the type of information allowed when the protocol is in a particular state, flow control of when information can be sent and the choice of destination for transmitted information.

In defining a protocol, the static features must be fixed by prior agreement between humans who are going to use the protocol for computer communications. One very important static feature is the definition of how the dynamic features of the protocol are used. Given this human agreement, dynamic agreement between communicating computers is achieved by them using the dynamic features. The result is that communications take place. Defining protocols is an activity that is carried out by external observers of communications, that is, people who have a global picture of what communications achieve.

The subject of this chapter is the way in which humans achieve agreement on the static features of protocols. The first point to note is that protocols do not just exist as bodies of rules that are interesting in their own right. The reason for communications occurring within the agreements policed by protocols is to supply a service. This service is provided to human users or computer applications that need information to be shared by two or more computers. Thus, the role of a protocol is to assist in the implementation of a service.

The components of a protocol are chosen to make appropriate use of the various kinds of implementation methods for information, time and space that have been discussed throughout this book, particularly in Chapters 2, 3 and 5. The precise choice of components, and their detailed definition, are matters for the human agreement process. General ways of packaging implementation methods for time are described in Chapter 4, and ways for space are described in Chapters 6, 7 and 8.

The three case studies in Chapters 9, 10 and 11 all illustrate the use of particular collections of protocols. Beginning with low-level protocols governing the transmission of bits over media, successive layers of protocols enhance this service, until the highest-level protocol can supply a service actually required for a user application. In Chapter 9, the application is fetching a World Wide Web page; in Chapter 10, the application is controlling a factory floor device; and in Chapter 11, the application is transferring a real-time video sequence. Each protocol used carries out one or more implementation tasks, allowing some higher-level communication to be implemented in terms of one or more lower-level communications.

The MAP project, described in Chapter 10, shows a further extension of human agreement. This involves deciding on the collection of protocols that should be used throughout the implementation process, from highest to lowest level. In effect, this is an agreement on a giant protocol, that contains several sub-protocols as components, each contributing some parts of the overall implementation. In the other two case studies, there is flexibility in the mix of different protocols used to achieve the overall implementation. This is a more modular approach to producing a physical implementation of a required communication.

The aim of a more modular approach to implementation can be carried a stage further. This involves making a distinction between a protocol and the service offered by using the protocol. Such a distinction was made when discussing time packages in Chapter 4. For example, a connection-oriented service might make available channels with three types of communication:

- establish a connection between two computers;
- reliably send a user message from one computer to the other; and
- terminate the connection.

This service could be supported by various protocols, for example, the Internet TCP, or the ISO TP4 transport protocol, or some privately devised protocol. The service abstraction means that a user of the service does not need to know what protocol is being used to implement it. This idea is familiar from the world of structured computer software design, in particular from the use of object-oriented programming.

When protocols are designed, it may or may not be the case that a precise description of the supported service is designed first. In some cases, the service might just be defined as 'everything that the protocol makes possible'. Again, this situation is familiar from software engineering. It is regarded as good practice to have a requirements specification before producing software. However, for various reasons, some of them good reasons, this is not always the case. Specifications and software may evolve in parallel, or specifications may be produced after the software. Sometimes, specifications may never be produced.

The complication that arises with communication services and protocols, compared with computer software, is that more than one computer, possibly a vast number of computers, are affected — by definition. This means that all of the humans involved with these computers have to be in agreement, even if this is only silent or grudging agreement. If not, then the protocols cannot be made to work.

With software, in general, there is at least a prospect of making it work on a single computer, even if no one else agrees with its specification or implementation.

Thus, standardization of services and protocols is a complex issue. For example, where protocols implement communications spanning the world, worldwide agreement is needed, so that computers can use the protocol, regardless of their locations or owners. In general, agreements on a protocol must take account of individuals' viewpoints, manufacturers' viewpoints, governments' viewpoints, in fact the viewpoints of any person or organization with an interest in making the protocol work. The way in which standardization is carried out is the subject of this chapter.

In some cases, standardization happens by stealth. An individual, manufacturer or organization devises a new protocol that becomes widely used because of its attractiveness, its availability, or pressure from its creator. Gradually, the protocol becomes a *de facto* standard, as even those who are reluctant become forced to conform, in order to continue communicating with the large user population. This is not a phenomenon unique to computer communications. It is familiar in the computer world from such things as operating systems, programming languages and application packages. Often, *de facto* standards are given an official seal of approval by being turned into official standards by standardization organizations, either without change or with some generalization and improvement.

Some standards exist entirely within closed populations, for example, single organizations or users of particular proprietary communication systems. Although not impacting on the entire computer communications community, some of these standards are of technical interest in their own right, or are of interest because their user community is large. In some cases, these specialized standards become adopted for more general use, either by escaping from their closed environments or by forming the basis for more widely used standards.

This chapter is mainly concerned with the principal organizations involved in official standardization. However, where appropriate, notable *de facto* standards or proprietary standards are mentioned. Of course, official standardization is not unique to computer communications, so some of the organizations are concerned with a much wider range of artifacts. After an outline of the main standardization bodies, there is a description of the main computer communications standards that have emerged from these bodies.

12.2 STANDARDS BODIES

The official standards bodies concerned with computer communications come from several different perspectives. First, there are the general purpose national or international standardization bodies. Then, there are the more specialist national or international bodies. These cover areas such as electronics, telecommunications and information technology. Finally, there are the Internet-related bodies, which solely focus on computer communications in the Internet community. In addition, there are various professional and manufacturer organizations that are involved in standardization work. The next sections describe the main bodies of interest. Most of these feature in earlier chapters, since they are responsible for various protocols of interest.

12.2.1 International Organization for Standardization (ISO)

The International Organization for Standardization (ISO) was established in 1947, and is a worldwide federation of national standardization bodies from around 100 countries, one body from each country. Its mission is to promote standardization, with the aims of facilitating international exchange of goods and services and of fostering international cooperation. Its work results in international agreements that are published as International Standards. The first ISO standard was published in 1951, and was entitled 'Standard reference temperature for industrial length measurement'.

The intricacies of international standardization are illustrated by ISO's name itself. Most English-speaking people regard ISO as an acronym for 'International Standards Organization'. However, ISO stresses that its short name is not an acronym, rather it is a word derived from the Greek $I\Sigma O\Sigma$, meaning 'equal'. This word is acceptable in the three official languages of ISO — English, French and Russian — whereas the acronym is not. Of course, the choice of these three languages shows a further example of international agreement and compromise.

The member bodies of ISO are national standardization bodies, each being the 'most representative of standardization in its country'. For example, national bodies include SAA from Australia, AFNOR from France, JISC from Japan, BSI from the UK and ANSI from the USA. The actual work of ISO is carried out within a decentralized hierarchy of around 2700 technical committees, sub-committees and working groups. The major responsibility for administrating each committee is taken by one of the national bodies.

Standardization work in the field of Information Technology, which includes computer communications, is carried out by a committee called JTC1 — Joint Technical Committee 1. This should not be confused with the TC1 committee, which is concerned with standardization of screw threads. The 'joint' in JTC1 refers to the fact that the committee's work is carried out jointly with another standardization body, IEC (International Electrotechnical Commission). IEC was founded in 1906, with the object of promoting standardization in the fields of electrical and electronic engineering; this is the only field that ISO does not cover. Since 1987, ISO and IEC have collaborated through JTC1 to take joint responsibility for standardization in information technology — a further example of international agreement, this time on standardization procedure. ISO and IEC also cooperate with the Telecommunication sector of the International Telecommunication Union (ITU-T), which is described in the next section.

In early 1997, the JTC1 committee was directly responsible for 316 standards, and involved in a total of 1178 standards. It had 19 sub-committees, three of which are the most relevant to computer communications:

- JTC1/SC6 : Telecommunications and information exchange between systems;
- JTC1/SC21 : Open systems interconnection, data management and open distributed processing; and
- JTC1/SC25 : Interconnection of information technology equipment.

although several other sub-committees also impact on the area. The three named sub-committees were responsible for 148, 195 and 29 international standards respectively, and had five, five and three working groups respectively.

The three main principles guiding the development of ISO standards are: consensus; industry-wide; and voluntary. The need for a new standard usually arises from an industry sector, which makes a national member body of ISO aware of the need. This body then proposes a new work item to ISO and, if the need for the new standard is recognized and formally agreed, the standardization process begins.

First, a working group of experts devise one or more **Working Drafts** (WD), leading up to a **Committee Draft** (CD). This is circulated to the members of ISO to build a consensus in favour; this need not imply unanimity, but means general agreement without sustained well-founded opposition. If necessary, revised committee drafts may have to be produced. When a broad consensus is achieved, a revised version is issued as a **Draft International Standard** (DIS). Finally, if at least two-thirds of the ISO members that participated actively in development of the standard vote in favour, and no more than one-quarter of all voting members vote against with supporting technical reasons, an **International Standard** (IS) is issued. Every IS is reviewed at least every five years in order to ensure that it remains up to date.

For some standards, where there are very different national positions, this process can take many years to conclude, and the resulting international standard may be watered down or riddled with compromises. For existing mature standardization documents produced by others, there is a fast track procedure whereby such a document can proceed immediately to DIS status. Reaching international agreement on standards is usually hard in itself, but it is important to realize that the agreement reached may not be the best technical standard. Although the participants in ISO work are volunteers, most are employed or funded by an organization with a special interest in the area. These include manufacturers, vendors and governments. In addition, academics are often involved, and even such apparently neutral people often have their own personal technical agendas. Thus many ISO standards are the result of compromise between different special-interest groups.

12.2.2 International Telecommunication Union

The International Telegraph Union was founded in 1865, and acquired its present name, the International Telecommunication Union (ITU), in 1934. It became an agency of the United Nations in 1947. ITU is an inter-governmental organization for the development of telecommunications and, in particular, it develops standards to facilitate the interconnection of telecommunications systems on a worldwide scale. This is carried out by ITU's Telecommunications sector (ITU-T). Until reorganization took place in 1993, the terrestrial telecommunications standards work of ITU-T was carried out by CCITT, an acronym for its French name, *Comité Consultatif International Télégraphique et Téléphonique*, and older ITU-T standards are still often referred to as CCITT standards. Until 1993, radiocommunications standards work was done by CCIR (also a French acronym) but this was absorbed by ITU-T also. Meanwhile, a new Radiocommunications sector (ITU-R), was created to carry out the work done by CCIR on the efficient management of the radio-frequency spectrum in terrestrial and space radiocommunications.

In early 1997, ITU comprised 186 member countries, together with 363 other members drawn from public or private telecommunication operators, broadcasters, industrial or scientific companies, and regional or national organizations. ITU is funded by financial contributions from the member countries and organizations. The scale of contributions ranges from 1/16 of a unit to 40 units for countries; and one-fifth of this scale for organizations. The intention is that the contribution reflects the relative richness of the country or organization, the 1/16 and 1/8 rates being reserved for the least developed countries.

The duties of ITU-T are to study technical, operating and tariff questions and to issue recommendations on them, with a view to standardizing telecommunications on a worldwide basis. Note that only *recommendations* are issued — they are not binding standards. However, ITU-T recommendations are usually complied with by national organizations, in order to guarantee connectivity on a worldwide scale. ITU holds a telecommunication standards conference every four years (1996 was one of these years), at which draft recommendations are approved, modified or rejected. A programme of future standardization work is also approved. Until the 1988 conference, recommendations could only be approved at the four-yearly conference. However, as part of major streamlining to speed up the standardization process, new procedures were introduced to allow standards to be adopted as soon as they were ready, after a ballot of members.

The main work of ITU-T is not carried at its conferences, but rather via a decentralized hierarchy as in ISO. There are 15 different **Study Groups** of ITU-T, each considering a list of Questions over a four-year period. The Questions represent the programme of work for each Study Group that was decided by the previous standards conference. In turn, Study Groups are sub-divided into Working Parties, which are further sub-divided as necessary. By considering Questions, Study Groups formulate recommendations. The topics for standardization are classified into 33 different areas, some more relevant to computer communications than others. For example, five of the areas are: ISDN; B-ISDN/ATM; data networks; data services; and OSI. The last of these areas is introduced later in Section 12.3.2.

The resulting recommendations are categorized into different series, each identified by a letter of the alphabet. Recommendations are then numbered within the series. For example, the X.25 recommendation is one of the X series, which is concerned with public data networks. This is the most relevant series to computer communications. Most of the X series recommendations are developed in cooperation with ISO/IEC, and are also ISO/IEC standards. The V series, which deals with data communication over the telephone network, including modems, is also relevant. Six others that are sometimes encountered as the distinction between computer communications and telecommunications blurs, are:

- E series: telephone and ISDN networks;
- H and J series: transmission of non-telephone signals;
- I series: N-ISDN and B-ISDN;
- Q series: switching and signalling; and
- T series: telematic services.

There are 16 other series, covering all the letters of the alphabet except W and Y.

Like ISO standards, ITU-T recommendations are the results of compromise between different countries and, in particular, between the principal telcommunication operators (e.g., Public Network Operators) in different countries. Thus, recommendations may take a long time to be decided upon, and the final version may not be the best technical solution. However, ITU-T is a productive organization, compared with some other arms of the United Nations, and it plays a very important role, given the internationalization of the world's telecommunication system.

12.2.3 Internet Society

The process of standardization in the Internet community is distinctly different from the processes in ISO/IEC and ITU-T. A remark of Dave Clark is often quoted:

> We reject kings, presidents and voting.
> We believe in rough consensus and running code.

This was particularly true until 1989, when standardization in the Internet world was very informal. Standards were discussed and announced by the Internet Activities Board (IAB). This was a small group of around 10 active Internet researchers, each heading a task force on a topic of interest. IAB was set up by, and reported to, the American Department of Defense and the National Science Foundation, who were providing most of the funding for the Internet at that time. The standards appeared as documents in the Request for Comments series. These were, and are, technical memoranda, freely available in machine-readable form. The topics covered by RFCs range from standards, through information bulletins, to April Fool's Day jokes. One of them, RFC 2026, issued in October 1996, describes the Internet standards process.

In 1989, the Internet was growing rapidly, and was no longer just the domain of researchers. Equipment manufacturers and vendors had an increasing interest in how the Internet developed, and so IAB was reorganized to allow broader participation in standards making. The main act was to create the Internet Engineering Task Force (IETF). IETF is responsible for identifying problems in the Internet, proposing solutions and making recommendations on standardization. It is divided into eight functional areas: Applications, Internet, Network Management, Operational Requirements, Routing, Security, Transport and User Services.

Each area has one or two directors, and has several working groups. Headed by a chair, each working group has a finite lifetime, and works to achieve a specific goal. In early 1997, there was a total of 86 active working groups. There is no formal membership of working groups or, indeed, of IETF itself. Anyone on the relevant electronic mail list, or who attends meetings, can participate. In line with Clark's second sentence above, working groups strive to achieve rough consensus, and then build running code to test the consensus and evolve it into an Internet standard. There is no formal voting, and disputes are resolved by discussion and demonstration.

IETF is part of a larger organization. In 1992, the Internet Society (ISOC) was formed, as an international organization concerned with the growth and evolution of the worldwide Internet. In particular, ISOC provides an institutional home, and financial support for, Internet standardization activity. Also in 1992, IAB became associated with ISOC, and was renamed the Internet Architecture Board. It is responsible for overseeing the process used to create standards, including the creation of new IETF working groups, and it acts as an appeal board for complaints about improper execution of the standards process.

The standards process is administered by the Internet Engineering Steering Group (IESG), which is composed of the IETF area directors and the chair of IETF. The ISOC-IAB-IESG structure is the formal structure that makes Internet standardization work. However, the technical work is done under the auspices of IETF, with wide participation.

The standardization process follows a similar sequence to that of ISO standards, although the actions required for progression are somewhat different. First, working documents are announced and disseminated by IETF as work in progress, with a maximum lifetime of six months. Then, if a proposal joins the standards track, it is published in the RFC series. There are three standardization stages:

- Proposed Standard: a complete and credible specification with demonstrated utility;
- Draft Standard: multiple, independent, inter-operable implementations exist, and work well from limited operational experience; and
- Standard: the real thing, with demonstrated operational stability.

The lifetime of a Proposed Standard is between six months and two years, and the lifetime of a Draft Standard is between four months and two years. Standards can stay for ever, or may be made Historic if no longer required. The Standards are published in the RFC series, but also have their own STD sub-series within the RFCs.

ISOC has formal relationships with ISO and ITU-T, to facilitate cooperation on standardization. However, the Internet standardization process differs from those of ISO and ITU-T in several significant ways. First, wide participation in the formulation of standards is possible. For those people who are not active participants, documents are freely available for inspection. Second, working groups operate by considering all possible ideas in an initial phase, but then choosing and developing a particular solution. The choice made is not revisited at a later stage. Third, there is an emphasis on producing demonstrations of the object being standardized. This contrasts with hammering out a specification, and then worrying about how to implement it. The overall impact of the Internet process is that standards can be created relatively quickly, reflecting rapid advances in technology. The resulting standards have tested implementations and, moreover, are documented in a freely accessible form.

12.2.4 Professional and manufacturer bodies

There are many organizations around the world that are involved in standardization activities which impact on computer communications. Most have historic roots in other areas — electronics, computers or telecommunications — but have become involved as their technologies become part of computer communications. This section contains a list of eight bodies that are encountered relatively often, and is intended as a brief guide to the reader. In most cases, standards emerging from these organizations proceed to become international standards issued by ISO/IEC, ITU or ISOC. The Institute of Electrical and Electronics Engineers (IEEE) is particularly notable for its local area network standards, adopted by ISO, and these are discussed more fully later in Section 12.6.

ATM Forum

The ATM Forum was founded in 1991 and, by 1997, included more than 750 companies representing all sectors of the communications and computer industries, as well as a number of government agencies, research organizations and users. The objective of the ATM Forum is to accelerate the use of Asynchronous Transfer Mode (ATM) products through a rapid convergence of inter-operability specifications. The ATM Forum aims to work with standards bodies such as ISO and ITU-T, selecting appropriate standards, resolving differences among standards, and recommending new standards when existing ones are absent or inappropriate. Thus, a particular part played by the ATM Forum is in accelerating the development of ATM standards ahead of the slower-moving ITU-T, which is responsible for the international ATM and B-ISDN recommendations. Unfortunately, this sometimes leads to incompatibilities between the two organizations' standards, which is undesirable.

Electronic Industries Association

The Electronic Industries Association (EIA) is an American national trade association representing electronics manufacturers. It also represents the American communications and information technology industry in association with the Telecommunications Industry Association (TIA). Both of these associations have roots in organizations formed in 1924. EIA standards cover ground both above and below computer communications. Above, there are standards for electronic information interchange between software tools. Below, there are standards for the physical transmission of bits over electronic media.

ECMA

The European Computer Manufacturers Association (ECMA) was formed in 1961. The members of ECMA are European computer and communication equipment manufacturers. Its purpose is to develop standards for information processing and telecommunications systems, in cooperation with national, European and international organizations. In addition, it encourages the use of standards, and assists in promulgating standards. Since 1961, ECMA has developed over 200 standards, many of which have been accepted as a base for European and international standards. This reflects the fact that ECMA standardization work is fairly far-sighted, and takes technological trends into account. To more truly reflect the activities of ECMA, its name was changed in 1994 to just its initials: ECMA — a European association for standardizing information and communication systems (with no new acronym).

European Telecommunications Standards Institute

The European Telecommunications Standards Institute (ETSI) was established by the European Community in 1988. Its mission is to set standards for Europe in telecommunications, and to assist in standardization for broadcasting and office information technology. The aim is to accelerate technical harmonization in Europe in order to enable a pan-European telecommunication infrastructure with full inter-operability. In addition, ETSI contributes to worldwide standards making. ETSI is a forum that brings together all interested parties — network operators, service providers, manufacturers, governments, users and the research community — with the intention of achieving consensus on requirements. Thus, the membership of ETSI is rather broader-based than that of ITU-T, but by definition is mainly concerned with European interests.

Frame Relay Forum

The Frame Relay Forum, like the ATM Forum, was founded in 1991. It has a similar remit, namely encouraging the use of a particular communication technology. However, as the forum's name suggests, the technology is frame relay, rather than ATM's cell relay. The Frame Relay Forum does not formulate standards itself. Rather, it takes existing national and international standards for frame relay, and then creates **Implementation Agreements**. These specify exactly how the standards should be applied, in order to achieve inter-operability between the products and services of different carriers and vendors. This approach is possible for frame relay, because most areas are already well-agreed and standardized. In contrast, the ATM Forum operates in a far less well-defined area, and so contributes to standards making itself.

Institute of Electrical and Electronics Engineers

The Institute of Electrical and Electronics Engineers (IEEE) is a worldwide professional society, which exists to promote the development and application of electrotechnology and allied sciences. Although a worldwide society, IEEE is often seem as being an American organization, since it is based in the USA. As one of its many activities, IEEE is involved in standardization in the electrotechnical field. Confirming an American orientation, IEEE is a member of the American National Standards Institute (ANSI). The most well-known IEEE standards in the computer communications area are its 802 series standards for local area networking, which are covered extensively in Chapter 6. Most of these are subsequently adopted by ISO as standards. IEEE is also active in the development of standards for global and national information infrastructures.

National Institute of Standards and Technology

The National Institute of Standards and Technology (NIST) is part of the American government's Department of Commerce. It issues Federal Information Processing Standards (FIPS) for equipment that is sold to the American government — the largest purchasing organization in the world. This makes the NIST a significant influence. Rather than devising new standards, FIPS are concerned with stating which existing national or international standards must be respected. The best-known NIST FIPS was the Government Open Systems Interconnection Profile (GOSIP), first issued in 1987 and revised in 1990, which defined implementation agreements on the use of ISO protocols in communication systems. Similar profiles were adopted by many other governments. The assumption was that ISO protocols would dominate, but this did not foresee the rise of the Internet, despite the fact that the American government was funding it. Thus, bowing to the inevitable, a move towards ISO protocols was not enforced and, indeed, the NIST is now involved in investigating moving from ISO standards to Internet standards. In addition, NIST is concerned with helping industry consortia and standards groups to improve the technical quality of, and to speed the development of, selected proposals for advanced networking technologies and distributed multimedia protocols.

World Wide Web Consortium

The World Wide Web Consortium (W3C) started in 1994, and is an industry consortium that develops common standards for the evolution of the World Wide Web by producing specifications and reference software. Its membership is only open to organizations or companies, not individuals, but its products are made freely available for all to use. The W3C is a global organization which, in early 1997, was jointly hosted by MIT in the USA, INRIA in France and Keio University in Japan. It maintains an activity list of current technical directions. These are classified into three general areas: User Interface; Technology and Society; and Architecture. W3C has no formal links with Internet standardization efforts, although some W3C team members participate in IETF working groups. W3C standards may be submitted as candidates for Internet standardization, but this is not a prerequisite for their issue, in order to ensure rapid promulgation of new work.

12.2.5 Summary of standardization bodies

The different standards bodies covered in this section illustrate the range of different mechanisms for achieving human agreement on computer communications. A main point is that traditional ways of achieving standardization, involving protracted processes within rigid hierarchies, conflict with the world of computer communications, where technological progress is swift and the culture is informal. Bodies such as the ATM Forum and the W3C illustrate one extreme, where special-interest groups have been formed to ensure rapid formulation of standards for a particular technology. The products of such bodies are likely to become fast track standards for traditional bodies, such as ISO and ITU-T, in the course of time.

The Internet standardization process is interesting, because it shows an originally very informal process that has now been formalized to acquire authority in international standardization. The hierarchical structure, and phases of standardization, are very close to those of a traditional organization such as ISO. The difference is that the mode of operation is very different: open participation at the lowest levels of the hierarchy, and the progression of standards by demonstration of implementations, rather than by international wheeler-dealing. This is significantly different from ISO, where members of working groups represent national positions (or, at least, some kind of national majority opinion) or ITU-T, where the interests of national telecommunications organizations predominate. The effect is one of achieving standards that are more technically up to date, but might be rather more special-case than desirable. However, there is still the option of devising new, improved standards fairly quickly.

The bottom line is that standards are necessary, since they formalize the human agreements necessary to make computer communications possible. However, there is room for multiple standards for particular objects, preferably along with some means of making them inter-operable. The important thing is that delays in standardization do not hold back progress, and that standards can be implemented to produce efficient and effective products.

The remaining sections of this chapter cover most of the significant computer communications standards. First, ISO standards are considered in Section 12.3. In fact, the most significant ISO standard is one that describes an architectural model for communication systems. This has dominated many people's thinking, often in a very negative way, since it was finally issued in 1984 after several years of deliberation. After this, by way of contrast, Internet standards are considered in Section 12.4. Here, there is a modest architectural model, but one that plays a background role to the protocol standards. After the ISO and Internet 'whole model' sets of standards, there is a brief account of important standards that have been originated by ITU-T and IEEE, in Sections 12.5 and 12.6 respectively. Finally, Section 12.7 of the chapter departs from formal standardization bodies, and covers some major proprietary standards. These are often closely related to formal standards, but represent instances where one manufacturer has developed its own self-consistent products to pursue commercial advantage or to take a lead in making new technology available.

12.3 ISO STANDARDS

12.3.1 Structure

ISO standardization efforts centre around the Reference Model for Open Systems Interconnection (RM/OSI, or just OSI), which is standard number ISO 7498. This model supplies a framework for the development of protocol and service standards. The model consists of seven layers, each of which contains related functions that contribute to the implementation of a communication system. Some layers contain sub-layers within, adding further structure. Together, the functions in the layers allow the communication needs of user applications to be implemented using physical communication media. That is, they include the various agreement and implementation mechanisms discussed in this book.

A large amount of new jargon surrounds RM/OSI, much of it invented deliberately to ensure international agreement on the nature of the model, so avoiding possible ambiguities that would arise from the use of traditional terminology. Where possible, this jargon is avoided here, to avoid confusing the reader.

The basic idea is that each standardized protocol falls within one layer (or sub-layer). That is, the facilities that the protocol implements fall within the range of functions ascribed to one particular layer. Further, each standardized service falls within one layer. That is, the service offered corresponds to making available

User applications

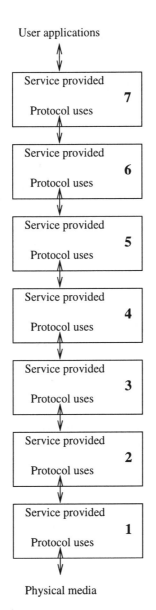

Physical media

Figure 12.1 RM/OSI layering of a communication system

the products of a set of facilities within the scope of one layer. A service offered within one layer can be used to implement protocols in the next layer above. This process is illustrated in Figure 12.1. Thus, RM/OSI is an architectural framework used for classifying protocols and services. ISO, and other organizations, define protocols and services that fit within the guidance of this framework.

The thorny problem with devising such a model is in deciding how many layers are apt, and then which functions are grouped together within each layer. One extreme, effectively rejecting such a model, is to have one layer, containing all possible functions. ISO followed certain principles in deciding on the seven layers of RM/OSI, including among others:

- do not create so many layers that the task of describing and integrating the layers is made difficult;
- create boundaries at points where the service description can be small and the number of interactions across the boundary is minimized;
- create separate layers to handle functions that are manifestly different;
- collect similar functions into the same layer;
- enable changes of functions or protocols within a layer without affecting other layers; and
- create boundaries at points where past experience had proved successful.

The final principle listed is a significant one. Past history, and the need to satisfy vested interests in existing technology, played a large role in the shape of RM/OSI. A particular influence in the lower layers is the fact that ISO anticipated most computers being interconnected by worldwide public data networks developed by ITU member countries. A general influence is the model used for IBM's Systems Network Architecture (SNA), discussed in Section 12.7, which was already in existence.

The **application layer** of RM/OSI is the seventh and topmost layer, and provides services to users of the OSI environment. It contains specialized functions such as file transfer and terminal handling, as well as more general functions such as remote procedure call and concurrent transaction handling. The most general function is one to allow associations to be created between applications on different computers, that is, connections for sharing information.

The **presentation layer** is the sixth layer, and provides services to the application layer. Its implementation task is centred around information representation. The main function is to negotiate the syntax used for transferring information, given the abstract syntax used by applications to represent information. That is, to establish an agreement on how transmitted bytes are used to represent higher-level information.

The **session layer** is the fifth layer, and provides services to the presentation layer. Its implementation task is concerned with the management of time periods and decomposition of time periods. The functions are concerned with coordinating and synchronizing a dialogue between two parties. This includes enforcing simplex or half duplex channel behaviour, and allowing the marking of synchronization points within the dialogue.

The **transport layer** is the fourth layer, and provides services to the session layer. Its implementation task is centred around constructing an end-to-end reliable connection-oriented time package between two computers. The functions include flow control, error detection and correction, and sequencing. The need for the functions varies depending on the quality of the service provided by the layer below.

The **network layer** is the third layer, and provides services to the transport layer. Its implementation task is concerned with the use of networking to create channels needed for the communication space of the service user. The functions are those concerned with routing and relaying within switching networks. There are several sub-layers, in particular, a higher one dealing with multi-protocol inter-networks formed from other networks and a lower one dealing with single-protocol switching networks formed using physical channels.

The **data link layer** is the second layer, and provides services to the network layer. Its implementation task is centred around constructing a time package of appropriate quality between two computers linked by a physical channel. The functions can include flow control, error detection and correction, and sequencing. The need for the functions depends on the service needed by the layer above. There is a lower sub-layer, containing functions needed for multiplexing channels on to the physical multipeer channel in broadcasting networks.

The **physical layer** is the first and bottommost layer, and provides services to the data link layer. It makes use of a physical transmission medium. Its function is to provide electrical or mechanical procedures that allow bit transmission over the medium.

Figure 12.2 shows the classical ISO diagram of the protocol dialogues in RM/OSI. It includes three computers. Two are end systems, which are conducting an application layer dialogue. The third is an intermediate system involved in message relaying, and it conducts network layer dialogues with the two other systems. At this point, an immediate relationship with the MAP protocol architecture described in Chapter 10 is obvious. The protocols used in MAP fit neatly into each of the seven layers of RM/OSI. This is, of course, no coincidence, since the designers of MAP were careful to follow ISO standards throughout.

Another impact of the layered model is on the vocabulary used in computer communications, where the same concept may have different names, depending on which layer it appears in. For example, the term 'message' is used throughout this book for a unit of information that is communicated. ISO chose the neutral term 'data unit' as a general term. This is embellished to **protocol data unit** (PDU) for a message communicated within a protocol, and **service data unit** (SDU) for a message passed across a service interface. These are further embellished by layer, to give names like **TPDU** for a transport protocol message. In practice, special case names are often used. Thus, for example, a 'frame' is usually a data link layer message, and a 'packet' or 'datagram' is usually a network layer message. Another example of this is that a 'bridge' is a data link layer switch, and a 'router' or 'gateway' is a network layer switch.

Just as similar objects occur in different layers, so there are similar functions. In particular, several of the layers include functions corresponding to two particular implementation methods: one for time and the other for space. Segmentation and concatenation may occur within some layers, with communications requested

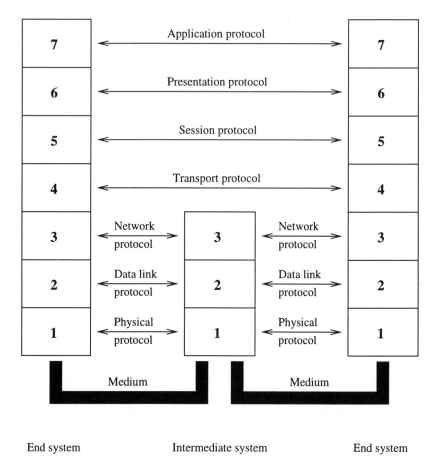

Figure 12.2 Protocol dialogues in RM/OSI

through the layer's service being split up or joined together over different time periods. Splitting and multiplexing may occur within some layers, with channels requested through the layer's service being split up or joined together over different spaces.

A final point to note about the model is that the services offered by one layer may be used by one or more processes involved in implementing the layer above. Each such process makes use of the services of the lower layer through a **service access point** (SAP), to use the ISO jargon. To allow this, each SAP has a different identifier within the layer offering the service. This is the derivation of the term 'NSAP identifier' used to describe ISO standard network identifiers at various points in the book. The NSAP identifier is the identifier of a SAP for the Network layer of a particular computer system.

The sub-layering within the network layer and the data link layer was not present in the original standard, but was added to deal with the rise of inter-networking in the first case, and broadcast networking in the second case. The original model is perfectly fitted to the traditional case where the network layer service is provided using a single-owner switching network formed from physical point-to-point channels. A further influence from this type of networking was that the original model envisaged connection-oriented services and protocols at each layer. Connectionless services and protocols were added later as amendments, and were not integrated fully with the model until a revised version was standardized in 1994. However, the two types of operation are still separated in the model, and inter-working using both types is not possible. A final point is that the model only covers unicasting and, even in the 1994 version, has no support for multicasting or broadcasting.

These comments indicate why RM/OSI has been controversial. The lower layers are best suited to describing connection-oriented protocols over single-owner switching networks supported by reliable connection-oriented services over physical links. This is in complete contrast to the philosophy of the Internet. A further controversy exists over the three upper layers: session, presentation and application. Many argue that a more natural scheme would be to merge these three layers into a single application layer, particularly given the first two of the layering principles listed previously. There is also the practical issue that few uses have been found for the session and presentation layer functions — MAP is just one example where the layers perform no significant functions. A final reason is that the Internet manages without such distinctions, a fact which helps to encourage anti-OSI sentiment in that community.

Given the controversy over the model, it is reasonable to ask what effect it has had on practical computer communications activities. First, it has formed the basis for the development of ISO standard services and protocols, discussed in Section 12.3.2. In this respect, the role of the model has been one of guidance, rather than compulsion. Interestingly, the 1994 revision of RM/OSI strengthened the role of the model, so that protocols must respect the requirements of the model in order to be compliant with RM/OSI. However, all of the ISO standard protocols in 1994 violated the model in some respect(s), and so became non-compliant, strictly speaking. This strengthening moves RM/OSI still further away from a common viewpoint, especially in the USA, that RM/OSI should not be a standard at all — it should just be a source of guidance.

The idea that RM/OSI should just exist as a tool to advise on the architecture of communication systems, has led to one of the model's major impacts. It underpins the structure of numerous textbooks on computer communications. These have successive chapters, beginning at the physical layer and proceeding to the application layer. This book differs from the RM/OSI centred view, on two major counts. First, the general principles of the subject can be found in many different contexts, in different layers. Second, a significant number of communication protocols do not map neatly into particular RM/OSI layers, and so authors of

RM/OSI-based books spend significant space discussing how best to classify such protocols. The true answer is that the classification scheme is not adequate. This is not surprising, since it was designed as a means of classifying ISO protocols, not all protocols ever invented.

Finally, an undesirable but unintended feature of RM/OSI has been its impact on implementations of communication systems. Some early implementations of the ISO protocols were organized very strictly on a layer by layer basis. That is, there were seven independent sub-systems, with interfaces between sub-systems for adjacent layers. This leads to great inefficiency, particularly in the three upper layers, which are best implemented as a whole. The important point to realize is that RM/OSI is intended as an abstract model for the development of protocols and services, as part of the standardization effort. It is not intended as an implementation model. As long as the protocols are implemented to send the right messages to the outside world, the internal organization can be chosen to have any form that is convenient and, preferably, is of high quality.

12.3.2 Services and protocols

There are a large number of ISO standards covering the different layers of RM/OSI. This section contains a brief summary of these, by layer. Essentially, this is just a list of key standards, rather than an attempt at describing their details, in order to give a flavour of the extent of ISO's work. Some of the standards are covered in more detail at relevant points in the text. As one descends to the three bottom layers, the standards become far less systematic. This is because various historic standards are being accommodated, and also because ISO is not the main body creating standards at these low levels.

Application layer

Standard ISO 9545 describes the internal structure of the application layer. ISO makes a distinction between services that can be used to support different applications, which are called **Common Application Service Elements** (CASE) and services that are used to support particular applications, which are called **Specific Application Service Elements** (SASE). In this section, the actual ISO standard numbers are not included, in order to spare the reader from a numerical deluge. The five main CASE service and protocol standards are for:

- Association Control Service Element (ACSE), connection-oriented and connectionless;
- Commitment, Concurrency and Recovery (CCR);
- Remote Operations Service (ROS);
- Remote Procedure Call (RPC); and
- Reliable Transfer Service (RTS).

The first of these is for managing associations between application processes, and is seen in Section 10.3 as part of the MAP protocol set. The other four implement particular handshake-based time packages that may be of general use to applications. CCR, ROS and RTS are all discussed in Chapter 4, as are the general principles applicable to RPC. RTS is actually a combination of ACSE with some of the session layer service primitives, rather than something independent.

A fairly wide range of SASE service and protocol standards has been defined by ISO, and this includes:

- File Transfer, Access and Management (FTAM);
- Virtual Terminal (VT);
- Job Transfer and Manipulation (JTM);
- Directory Service (DS);
- Message Oriented Text Interchange System (MOTIS);
- Manufacturing Message Service (MMS);
- Remote Database Access (RDA); and
- Transaction Processing (TP).

These are fairly self-explanatory, since they refer to well-known computer applications (MOTIS covers electronic mail, along with other types of text messages). The standard for MMS is included in Chapter 10 as part of the MAP protocol set. In addition to these particular applications, and other specific applications, ISO is also developing a reference model for **Open Distributed Processing** (ODP), which is an analogue of RM/OSI, to be used as a model of the distributed applications that make use of the services of RM/OSI.

Presentation layer

ISO 8822 covers the standard presentation service, ISO 8823 covers the standard connection-oriented presentation protocol, and ISO 9376 covers the standard connectionless presentation layer protocol. The connection-oriented protocol standard features in the MAP protocol set. The Abstract Syntax Notation 1 (ASN.1) standard is not layer specific, but is very relevant to this layer, in that it is often the subject of presentation protocol negotiations. Its specification is standard ISO 8824, and its coding rules are standard ISO 8825.

Session layer

Standard ISO 8326 covers the session service, ISO 8327 covers the standard connection-oriented session protocol, and ISO 9348 covers the standard connectionless session layer protocol. Note that, as for the presentation layer standards, the connectionless service definition was grafted on as an addendum to the original connection-oriented definition. However, the two different protocols have different standard numbers. This policy also applies to the transport layer. The connection-oriented protocol standard features in the MAP protocol set.

Transport layer

ISO 8072 covers the standard transport service, ISO 8073 covers the standard connection-oriented transport protocol, and ISO 8602 covers the standard connectionless transport protocol. The transport service and protocols are described in Chapter 4. There are five different classes of connection-oriented service, which cater for different types of service offered by the underlying network layer. The transport protocol for the Class 4 service is seen in Section 10.3 as part of the MAP protocol set.

Network layer

Standard ISO 8648 describes the internal structure of the network layer, and ISO 8348 is the standard network service. ISO 8473 is the connectionless inter-network protocol, which is described in Section 8.3.3 and is also used as part of the MAP protocol set. The connection-oriented network protocol is lifted from the ITU-T X.25 standard, and is standard ISO 8208. Note that this is the highest layer at which ISO did not define its own protocols. As well as the recycling of the X.25 protocol, ISO 8473 was very strongly influenced by the Internet IP protocol. There are also standard protocols for the exchange of routing information. ISO 9472 is used between a user computer (end system) and a switch (intermediate system), and ISO 10389 and ISO 10747 are used between switches, the latter when the switches are in different administrative domains.

Data link layer

ISO 8886 covers the standard data link service. ISO 3309 is a fairly old standard for the message format used in High level Data Link Control (HDLC) procedures, and there are various other standards related to HDLC procedures. These include the ISO 7776 standard, which is for HDLC procedures compatible with the ITU-T LAPB procedures. There is no single standard dealing with particular HDLC-based protocols. The data link layer standards also include the local and metropolitan standards, which are mostly directly taken from the IEEE standards described later in Section 12.6. The naming convention is that IEEE standard 802.x is numbered ISO 8802-x, where x is between 1 and 6. The standard ISO 9314 covers the Fibre Distributed Data Interface (FDDI) style of MAN.

Physical layer

ISO 10022, a latecomer on the service standards scene, covers the standard physical service. There are miscellaneous ISO standards dealing with very specific physical arrangements. However, this area is largely the province of ITU-T. Some components of the LAN and MAN standards have a more natural home in the ʰal layer, but it is not worth attempting to rigorously classify them between ʳ and the one above.

12.3.3 Management

Communications management was not included in the original 1984 RM/OSI standard, since there was not general agreement on where it should fit. In 1989, the standard ISO 7498-4 was added as an extra part to RM/OSI, to cover the Management Framework. This has guided the development of the numerous ISO management standards, in association with standard ISO 10040 of 1992, which gives an overview of OSI systems management. The content of ISO 7498-4 has now been superceded by the content of the various more detailed standards mentioned below.

The activity of system management itself is treated as an application. It makes use of standardized application layer facilities. The model used for systems management is object-oriented. Each system component that is to be managed is represented by a **managed object** that is stored in a **Management Information Base** (MIB). System components include computers, switches and protocols, in fact, any hardware or software system elements. Each managed object is characterized by:

- the current state of the object;
- the management operations that can be applied to the object;
- the behaviour exhibited in response to operations applied; and
- the events that can be signalled by the object.

The definitions of particular managed objects are not part of the ISO management standards. The standard ISO 10165 covers the management information model in general, including guidelines for defining managed objects. Then, appropriate objects are developed by the different groups working on the various standard protocols and system components within RM/OSI. For example, ISO 10737, ISO 10733 and ISO 10742 are standards for management information in the transport layer, network layer and data link layer respectively.

In addition to the information base standards, there are also standards for system management functions. The multi-part standard ISO 10164 covers a different **system management function** (SMF) in each part. An SMF is a collection of services that can be performed on the managed objects which are relevant to the particular management function. Each SMF falls within one or more overall management categories:

- accounting management;
- configuration management;
- fault management;
- performance management; and
- security management.

In 1997, the process of developing SMF standards was still continuing. The first standards appeared in 1992, for alarm reporting and security alarm reporting, and others emerged in later years. A total of 23 different parts of ISO 10164 were in the standardization process.

The communications requirements of the system management functions are provided by the ISO Common Management Information Service (CMIS) and the ISO Common Management Information Protocol (CMIP) service and protocol standards, which lie within the application layer of RM/OSI. CMIS supplies a general management messaging facility that can be used to interact with remote objects. Facilities include the creation and deletion of instances of managed objects, retrieving or changing the attributes of managed objects, requesting an operation to be performed by an object, and receiving notifications of events from objects. CMIP is supported using the ACSE and ROS common application layer services, ROS being used to supply a request/response messaging facility.

The CMIS and CMIP standards are stable, and pre-date the various emergent MIB and SMF standards. They have been used to form the basis of some special-case network management systems, pending the official ISO standards becoming stable. Also, these standards have been used over the Internet TCP/IP protocols, not just over ISO protocols. There is a specialist forum, the Network Managment Forum (NMF), which is a grouping of users and suppliers with the remit of encouraging and accelerating the use of ISO standards, among others, for the management of networked information systems. The NMF does not develop new standards, but just issues recommendations on how existing, and new, standards should be used in practice. These are known as the **OMNI***Point* specifications.

12.3.4 Other standards

There are many other ISO standards for computer communications, apart from services and protocols for each RM/OSI layer and overall system management. Security, like management, was a major issue not addressed in the original RM/OSI standard, but was later added as an extra part. Standard ISO 7498-2 is concerned with the Security Framework. The principles for security are contained in three particular later standards:

- ISO 10181 on the Security Framework;
- ISO 10745 on the upper layers security model; and
- ISO 13594 on the lower layers security model.

These three supercede ISO 7498-2, and were at various pre-standardization stages in early 1997.

Another standardization activitiy is concerned with how services and protocols are actually defined. Traditionally, this is done using natural language, with the aid of some figures and diagrams. However, this can lead to very indigestible standards — as evidenced by many of the ISO standards documents

— and, more worryingly, inconsistent or incomplete standards. **Formal Description Techniques** (FDT) are formal languages suitable for describing services and protocols. There are two ISO standard FDTs, each of a rather different character.

ESTELLE is an FDT that is based on an extended finite state machine model. It is based upon the programming language Pascal, with additions to allow the specification of state machine-like behaviour. This is useful for protocols, which are often described in terms of their effect on some global state information. The alternative, **LOTOS**, is a rather more abstract FDT that is founded on theoretical work in the areas of algebraic specification of data types and algebraic calculi for describing concurrent systems. The intention is that services and protocols can be described in a more abstract way that does not force, or suggest, particular implementation methods. ESTELLE and LOTOS specifications have been produced for a variety of the ISO standard services and protocols. However, the natural language specifications are still the ones usually referred to in practice, with the FDT versions being the province of specialists.

As well as accurately specifying standards, a further problem is ensuring that implementations actually match the specification. ISO also has standards for **conformance testing**, that define how implementations can be tested for conformance with particular protocol standards. **TTCN** (Tree and Tabular Combined Notation) is a further standardized FDT, that can be used for specifying test suites for conformance testing.

Finally, the production of standardized **profiles** is an activity that has led to numerous standards. A profile is a set of combinations and options from various standard protocols that fit together in order to implement some required functions in terms of others. That is, it is concerned with how standard protocols are actually used. The intention is that implementations which both conform to the same profile will definitely be able to inter-work. This is the same idea as that behind the MAP project described in Chapter 10. Essentially, MAP is just a large profile to allow inter-working of factory floor devices. The NMF's **OMNI***Point* specifications are further examples of profiles drawn from international standards.

12.4 INTERNET STANDARDS

12.4.1 Structure

The preceding dozen pages devoted to the various ISO standards is just thin coverage of a vast array of standards related to computer communications. An awareness of the topics standardized is useful to give guidance on the issues of importance in practical communications. However, as already mentioned, the fact is that manufacturers have been slow to adopt many of the ISO standards. As a result, certain niches have been filled by other *de facto* standards available earlier. The Internet standards have been a particular rival in this respect. These standards arrive with tried and tested implementations, a result of the philosophy

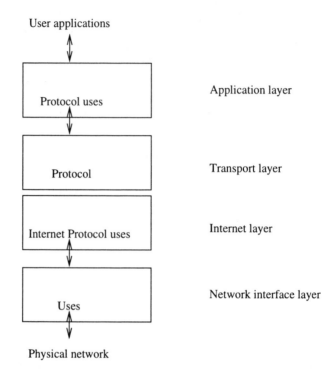

Figure 12.3 Internet model layering of a communication system

of demonstrating new ideas in action before standardizing them. This contrasts with the ISO tendency of having lengthy deliberations over proposed standards, and only exposing the results to implementation after standardization is complete.

Just as ISO has a reference model, so there is an Internet reference model of sorts. One difference is that, while RM/OSI was designed before the standards that fit within it, the Internet reference model was invented after the key standards that fit within it. In other words, the Internet model follows from the protocols used, not the other way round. Again, this illustrates the principle of doing things before standardizing them. The other significant difference between the ISO and Internet reference models is that the Internet model is distinctly simple. In fact, it could be argued that there is little real benefit gained from introducing it explicitly, except to allow a direct contrast with the ISO approach.

The Internet model is illustrated in Figure 12.3. It has four layers, in contrast to the seven layers of RM/OSI shown in Figure 12.1. The heart of the model is formed by the transport layer and the internet layer. This corresponds to the transport layer and the inter-network sublayer of the network layer in RM/OSI. Note that these two central layers are shown as adjacent, rather than separated by a service interface. This reflect the close relationship between the standard transport protocols and the Internet Protocol (IP). In the original scheme of things, these layers were one, with the Transmission Control Protocol (TCP) as the transport protocol and TCP/IP being the combined layer protocol. However, realizing the need for an unreliable connectionless service as well as a reliable connection-oriented service, the User Datagram Protocol (UDP) was added as an alternative transport protocol. At this point, the split-layer model became relevant.

The top layer of the model is the most varied. It corresponds to the top three layers of RM/OSI, with the rather superfluous session and presentation layers being absorbed by the application layer. The application layer of the Internet model contains protocols to support particular applications. These are similar sorts of applications to those included as ISO standards, for example, file transfer, virtual terminals, electronic mail and system management. The application layer protocols make use of a service offered by the transport layer. The UDP service offers unreliable transport of a message containing a byte sequence. The TCP service offers a reliable bidirectional channel transporting byte streams.

The bottom layer of the model is rather slim, and corresponds to an RM/OSI network layer sub-layer that allows an inter-networking protocol to make use of the networking services offered by the constituent networks. In the Internet model, this layer is responsible for supplying a means of transmitting IP messages over particular network types. The Internet model is not concerned with any levels of implementation below this. Networks are taken as the implementation primitive, in contrast to RM/OSI, where raw physical media are the primitive. This is consistent with the *raison d'être* of the Internet: supplying an inter-networking service over whatever networks are available. This does not involve standardizing networks or their implementations. Rather, it means coping with all the standards that do emerge from other bodies.

Note that the Internet model does not single out the notion of *service* as a matter for standardization. Only protocols are standardized, so the service available is implicit from the capabilities of the protocol. Any abstraction of service from protocol is left as a matter for particular implementations.

The attraction of the Internet model is that it gives a perfect description of the standardizations matters that are of interest to the Internet community. There is no need to warp or mould concepts to fit, as is sometimes the case when trying to map communication systems into the RM/OSI model. This is no surprise of course, since the Internet model followed the definition of its components, and so is a perfect fit. However, the special-case approach is the major shortcoming of the Internet model, in that it is not really suitable as a general-purpose model. The idea of a central transport/inter-networking layer is a common feature of virtually all communication systems, and so this is a useful feature. As normally presented, IP is a special-case feature of the Internet model, but the layering still works if a different protocol is inserted instead. However, an attempt at an all-embracing model should also address the issues involved in individual networks and on physical channels.

In terms of practical Internet standardization, the model is not a major factor to consider, except to the extent that the various protocol standards can be classified as being within the different layers. A more immediate matter is that standards are classified as being in one of three categories:

- Required: must be implemented by an Internet system;
- Recommended: should be implemented by an Internet system; and
- Elective: may or may not be implemented by an Internet system.

The distinction between 'must' and 'should' is that 'must' means an absolute requirement, whereas 'should' means that there may be valid reasons in particular circumstances not to do something or to do something differently, but that the full implications of doing so must be fully understood and carefully considered. The idea behind an elective protocol is that, if anything is to be implemented in the area covered by the protocol, then the elective protocol must be used. In some cases, there is a choice of more than one elective protocol for a particular area — examples of this include electronic mail and message routing.

In January 1997, there were 49 Internet standards. Of these, four were required, 20 were recommended and the rest were elective. Three earlier standards had been relegated to historic status. There were 41 draft standard protocols and 191 proposed standard protocols. All of this information was gleaned from Internet standard number STD 1, which describes the standardization process and contains the list of standards and their status. It is one of the four required Internet standards.

Another of the four is STD 2, which gives lists of assigned numbers that are used with the different Internet protocols. For example, these include the IP version numbers, ICMP message type codes and the well-known port numbers for TCP and UDP. All of the assigned numbers are managed by the Internet Assigned Numbers Authority (IANA), which is a further standardization arm of the Internet Society.

The third required standard is STD 3, which specifies the requirements for communications software on computers attached to the Internet. This includes details of how the system must behave at each of the layers of the Internet model. Together, STD 1, STD 2 and STD 3 supply a framework within which the other standardized protocols fit. There was another standard, STD 4, which was like STD 3, except for Internet switches rather than attached computers, but this has been consigned to historic status.

12.4.2 Protocols

Internet and transport layers

The fourth required standard is STD 5, which specifies the Internet Protocol (IP). This is the protocol at the heart of the Internet, that gives it a distinctive character. The Internet Control Message Protocol (ICMP) is also a required part of this standard. The Internet Group Multicast Protocol (IGMP) falls within the standard as well, but not as a requirement. This is because multicast switching within the

Internet is still at an experimental stage. However, multicasting will become a recommendation in the future, as the field matures.

The two standard transport protocols are the User Datagram Protocol (UDP), which is STD 6, and the Transmission Control Protocol (TCP), which is STD 7. Together with the IP standard, these two standards form the central core of the implementation of the Internet. It is possible to use other transport protocols over IP, but UDP and TCP are the only standardized protocols — albeit recommended, rather than required, protocols. TCP is the more senior partner of the two transport protocols, since it is used to support most of the commonly used applications. This is why the TCP/IP label is often applied as a description that captures the essential flavour of the Internet.

Application layer

Recommended standards exist to cover the most common applications used on the Internet:

- Terminal protocol (TELNET); also, six other recommended standards covering TELNET options (plus 33 other elective standards for options);
- File Transfer Protocol (FTP);
- Simple Mail Transfer Protocol (SMTP);
- Format of electronic mail messages (RFC 822);
- Network Time Protocol (NTP);
- Domain Name System (DNS);
- Mail routing and the domain name system;
- Simple Network Management Protocol (SNMP);
- Structure of management information;
- Management Information Base (MIB) for TCP/IP internets; and
- Echo protocol (ECHO).

These are standards STD 8 to STD 17 and STD 20, respectively. The first six applications are mentioned elsewhere in the book at various points. The next three are part of the Internet's system management standards, which fill a similar niche to ISO's management standards. Finally, the ECHO protocol is a simple application that is based on either TCP or UDP. As its name suggests, it just sends back any message received from another computer. This facility is useful for debugging and measurement. The standard is interesting for being a mere 16 lines long, excluding the title. No ISO standard is known to approach this level of compactness.

The Internet management standards have a lot in common with the ISO standards. Originally, they were developed as a short-term measure until corresponding ISO standards were finalized. However, this was done in a way that ensured broad compatibility. As the word 'simple' in the name of the SNMP protocol may suggest, the Internet standards are rather less elaborate. They are still rather lengthy though, as Internet standards go. The concept, and the terminology, of the Management Information Base (MIB) is shared by both sets of standards, as is the idea of manipulating a MIB remotely.

The SNMP protocol is analogous to ISO's CMIP protocol, but rather simpler. That is, it is a simple protocol for performing remote operations on a MIB. SNMP messages are represented using the Basic Encoding Rules of ASN.1, which is unusual for Internet protocols. Then, message transfer is implemented using the unreliable connectionless service of UDP. This is in contrast to CMIP, which is implemented using the reliable handshake service of ROS. The difference is because there is no standard Internet transaction protocol, and TCP was felt to be too heavy duty for SNMP to use.

A standard MIB for TCP/IP internets is defined to contain variables relevant to the management of an Internet component system. It contains objects to represent physical interfaces, as well as objects to support key protocols, such as IP, ICMP, TCP and UDP, as well as SNMP itself.

Network interface layer

As would be expected, most of the network interface layer standards relate to how IP messages should be transferred over different types of computer network. Standards STD 36, STD 39 to STD 49, STD 51 and STD 52 cover, respectively, FDDI, ARPANET, wideband networks, Ethernet, Experimental Ethernet, IEEE 802, DC networks, Hyperchannel, ARCNET, serial links (SLIP protocol), NET-BIOS, Novell IPX, point-to-point links (PPP protocol) and SMDS. Many of the networks listed are mentioned elsewhere in the book; details of the others are not of immediate relevance here.

In addition to specific network interfaces, there are MIBs designed for the management of different types of network. The only such MIB that was a full standard in March 1996 was an Ethernet MIB, but a large collection of other MIBs were at the draft or proposed standard level.

The Address Resolution Protocol (ARP) and the Reverse Address Resolution Protocol (RARP) complete the collection of network interface layer standards. ARP is used to map an Internet identifier to a physical network identifier. RARP is used by a computer to obtain its own Internet identifier from a server on the same physical network. The message format used for both ARP and RARP contains a field indicating the physical network type, the values used being allocated by IANA as assigned numbers. This is the only tailoring to particular network types. Physical network identifiers are also carried, but are preceded by a length field, to make the protocol usable whatever the identifer scheme used. The identifiers are interpreted as being within the identifier scheme of the specified physical network type. For example, IEEE 802 networks have the type value 6, and the length of the identifiers may be either two or six bytes.

12.5 ITU-T STANDARDS

In many cases, ITU-T standards are harmonized with ISO standards. Of these, some standards originated with ITU-T, and were adopted by ISO. Others originated with ISO, and were adopted by ITU-T. The remainder were developed jointly. This section is concerned only with those ITU-T standards that are often encountered in their own right. These fall within two particular areas of relevance that are standardized by ITU-T:

- physical channels and networks; and
- applications related to telecommunications.

Thus, in RM/OSI terms, the first area falls within the bottom three layers and the second area falls within the top layer. ITU-T includes the RM/OSI standards as part of its own series. Thus, standard X.200 is the RM/OSI model, X.700 is the OSI management framework and X.800 is the OSI security framework.

The bottom three layers of RM/OSI model are the same as the three-layer model used by ITU-T in its X.25 standard for the interface between a computer and a Packet-Switched Public Data Network (PSPDN). This standard is the norm for such networks, with the ISO numbering rarely used. X.25 includes two possible physical layer standards: X.21 or X.21*bis*. X.21 is a standard for physical channel to a digital network. X.21*bis* is a standard for an analogue network, as in the traditional telephone system. The intention was that X.21 would quickly become the most used, with X.21*bis* being a transitional standard based on the analogue standard V.24, mentioned below. There are two possible data link layer standards, both HDLC variants. The more modern one is the Link Access Protocol Balanced (LAPB), while the older one is the Link Access Protocol (LAP) — which is not balanced, in that there is a master-slave relationship over the interface. The network layer standard does not have a separate name, just being known as the X.25 packet protocol. There is also an inter-networking standard X.75 that covers the interconnection of X.25 networks. The identifier scheme is covered by the X.121 standard, which defines the use of 14 decimal digit identifiers in X.25 networks.

The other best-known X series standards are concerned with applications. The XXX (X.3, X.28 and X.29) triple of standards is very long-standing, and covers a Packet Assembly/Disassembly (PAD) facility, which means a facility for a fairly dumb user terminal to interact with a computer over a PSPDN. This provides a service of a similar type to that provided by the Internet TELNET protocol.

Apart from XXX, the two more recent series of application standards that are well-known in their own right are X.400 and X.500. X.400 has formed the basis for the ISO MOTIS application, which includes electronic mail as a special case. One particular component, X.409, has evolved into Abstract Syntax Notation 1 (ASN.1), which finds many uses as a way of encoding high-level types in terms of bit sequences. X.500 has formed the basis of the ISO directory service, which is used to underpin many applications that require mappings of names to objects. One example is the mapping of human-friendly names to computer-friendly addresses.

The V series standards feature most frequently when discussing modems to allow computer communications over normal telephone lines. The V.24 standard covers the interface between a computer and a modem. It originated from the EIA standardization body, and is best-known by the EIA standard name: RS-232-C. The X.21*bis* standard, which is allowed at the physical layer of X.25, is a very slight variant of V.24.

Other V series standards describe the behaviour of a modem, that is, how it modulates a carrier signal to represent binary information. Early standards are V.21, V.22 and V.22*bis*, which allow transmission rates of up to 300, 1200 and 2400 bits per second respectively. More recent standards are V.32, V.32*bis* and V.34 (called V.Fast in a preliminary proprietary version), which allow rates of up to 9600, 14 400 and 33 600 bits per second respectively.

There are also standards for improving the quality of the transmission. The V.42 standard covers error-correcting procedures, and it can be used together with V.42*bis*, which covers data compression procedures. The latter standard is what allows modems to be advertised with transmission rates that appear to exceed the maximum rate of around 30 kbits per second that results from Shannon's channel coding theorem.

Other ITU-T standards of interest come from more general telecommunications, rather than specifically computer communications. For example, from the applications side, there is the H.261 standard for video encoding, mentioned in Section 11.2, or the T.4 and T.6 standards for facsimile encoding, mentioned in Chapter 2. Also from the 'telematics services' T series is the T.120 series covering audio-visual teleconferencing, mentioned in Section 11.1. Finally, there is the T.400 series covering Document Transfer Access and Manipulation (DTAM), which is at the centre of an ISO standardization exercise not mentioned earlier: the Open Document Architecture.

From the physical side, the most important other standards are those concerned with ISDN — both N-ISDN and B-ISDN. From their underlying technologies, the former has spawned frame relay standards and the latter has spawned ATM cell relay standards. Most of the standards are in the I (ISDN) series, but those connected specifically with signalling are in the Q series. The I.320 standard gives a reference model for N-ISDN, and I.321 gives a reference model for B-ISDN. The distinction between transmission of user information and of signalling information means that the models not only have layers, but also have **planes**.

The B-ISDN model is illustrated in Figure 12.4. There is a user plane and a control plane, corresponding to the two types of transmission, and also a management plane for network management. The management plane has two parts: one for layer management in the user and control planes, and the other for overall management of the user and control planes. Some authors attempt to map the layers of the model into RM/OSI layers, but this seems fruitless since there is no direct mapping. The functions of the layers of the B-ISDN model are:

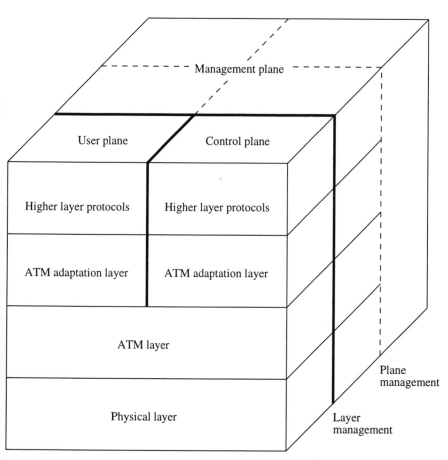

Figure 12.4 ITU-T B-ISDN ATM reference model

- Physical layer: transporting bits and ATM messages (cells) over a physical medium;
- ATM layer: switching and multiplexing of ATM messages over channels; and
- ATM Adaptation layer: offering service based on ATM message transmission.

The physical layer is divided into two sub-layers, one concerned with bit transmission, the other concerned with ATM message transmission given bit transmission. The ATM Adaptation layer is also divided into two sub-layers, one concerned with segmentation of communications into ATM message sub-communications, the other concerned with implementing the service offered. The various ITU-T standards for B-ISDN and ATM fit within this reference model, as do the standards developed by the ATM Forum. The latter standards are closely aligned with those of ITU-T, but are usually more promptly agreed and issued.

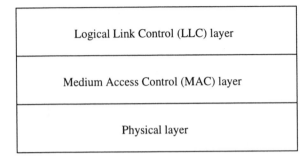

Figure 12.5 IEEE layering for LAN standards

12.6 IEEE 802 STANDARDS

The IEEE standards of interest for computer communications are those in its 802 series, covering local and metropolitan area networks. The first standard in the series, IEEE 802.1, gives an overview of the standards and their architecture. An extension to this standard, also adopted as an ISO standard, deals with bridges between different types of IEEE standard networks.

The longest-established and best-known three IEEE 802 standards are the 802.3, 802.4 and 802.5 LAN standards, which are described in detail in Chapter 6. These fit within a three-layer model for LAN implementations that is shown in Figure 12.5. This model is appropriate for message broadcasting networks based on a physical broadcast medium.

The physical layer is concerned with the physical medium, and how it is used to transmit bits. The Medium Access Control (MAC) layer is concerned with multiplexing broadcast channels on to the shared multipeer channel. The 802.3 ethernet, 802.4 token bus and 802.5 token ring standards cover both the physical and MAC layer protocols. For example, the 802.3 physical layer standard includes details of different cabling schemes, such as 10 BASE 5, 10 BASE T, etc., along with the bit encoding scheme and the restrictions on length and number of attached computers. The MAC layer standard covers the CSMA/CD technique for medium access control. There are various extensions to these standards, the most significant being 802.3u, which covers 100 Mbits per second ethernet operation, and 802.3z for 1000 Mbits per second operation.

The Logical Link Control (LLC) layer is covered by the IEEE 802.2 standard. This is independent of which protocols are used at the physical and MAC layers. 802.2 defines three types of service, each with an associated HDLC-based protocol to implement it:

- LLC1: unacknowledged connectionless;
- LLC2: connection-oriented;
- LLC3: acknowledged connectionless.

The third of these is the least common, and was added at a later stage to the original standard.

The 802.2, 802.3, 802.4 and 802.5 standards were adopted by ISO, with some modifications, as ISO 8802-2, 8802-3, 8802-4 and 8802-5 respectively. The IEEE 802.6 standard for a DQDB-based MAN was also adopted by ISO, as standard ISO 8802-6. This standard has a three-layer structure that is roughly the same as the model used for the LAN standards. However, in the top layer, the service offerings are different from those in the 802.2 standard.

In addition to the above mature standards, which result from the work of committees numbered 802.1 to 802.6, there are other standards, and work in progress, resulting from other committees. In summary, these are:

- 802.7: recommended practice for broadband LANs;
- 802.8: recommended practice for fibre optic LANs and MANs;
- 802.9: integrated services LAN;
- 802.10: inter-operable LAN and MAN security;
- 802.11: wireless LAN;
- 802.12: demand priority 100 Mbits per second LAN;
- 802.14: cable television broadband LAN.

The 802.9, 802.11, 802.12 and 802.14 committees have produced standards for new types of LAN or MAN. The 802.9 standard is for **isoEthernet** — an integrated voice and data LAN that supports both a 10 Mbits per second ethernet and 96 64 kbits per second N-ISDN channels. The 802.12 standard is the 100 BASE VG-AnyLAN standard, discussed in Section 6.4.6. A small point to note is that there is an ISO 8802-7 standard, for slotted rings. This is not related to any IEEE 802.7 work, but is adopted from a ring developed as part of research work at Cambridge University.

12.7 PROPRIETARY STANDARDS

To conclude this chapter, some important proprietary standards are briefly reviewed. These had their origins before, first, ISO standards looked like dominating the computer communications scene and then, second, Internet standards actually started to dominate the scene. All of the standards considered have a layered model to underpin them. This reflects the need to implement in stages, rather than as a whole. Although the models used can be roughly fitted to the RM/OSI model or the Internet model, there is no particular benefit in doing so in any strict way. It is enough to observe that the models involve applications, which are supported by networking, which is supported by physical media.

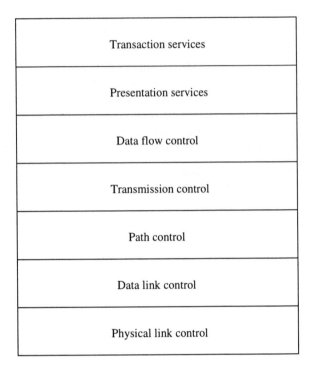

Figure 12.6 SNA layering of a communication system

12.7.1 IBM System Network Architecture

IBM introduced its System Network Architecture (SNA) in 1974, in an attempt to rationalize the hundreds of differing communications products that it had already launched. SNA facilitates the interconnection of computer equipment from IBM and other suppliers. The structure of SNA was a major influence on the design of the RM/OSI model, in particular, the notion of layering. SNA has seven layers, like RM/OSI, most of them with very similar roles. The layers are illustrated in Figure 12.6. The model is entirely connection-oriented — there are no connectionless facilities. The bottom five layers correspond very closely to those of RM/OSI. A slight confusion arises because SNA uses the word 'session' to refer to a connection at the transport layer. Originally, the top two layers shown were one layer. The subsequent split into two has a rough correspondence with RM/OSI, although the presentation layer can include a wider range of information transformation facilities, such as compression.

Apart from its overall model, one of the biggest gifts of SNA to the communications community has been its Synchronous Data Link Control (SDLC) protocol used in the data link control layer to provide the reliable transfer of information across a physical point-to-point channel. SDLC formed the basis for ISO's High-level Data Link Control (HDLC) family, which in turn has led to ITU-T's LAP and LAPB protocols and IEEE's LLC protocols.

In the original versions of SNA, the structure of networks was very centralized, with a hierarchy of equipment. Mainframe host computers were at the top level, with specialist front-end processors handling communications below them. Then, in lower strata, there were controllers for terminals and other peripheral devices and, finally, the terminals and peripherals themselves. The assumption was one of decreasing power and intelligence when moving down the hierarchy. Communication between hosts and terminals, in a master-slave manner, was possible, as was communication between hosts. However, direct communication between terminals was not allowed.

With the advent of cheap and powerful personal computers, the hierarchy began to creak, since there was no reason to treat these computers as dumb terminals. However, they could not be regarded as mainframe computers either. At first, networking of personal computers was outside SNA. Token rings, the LAN type invented by IBM and favoured thereafter, supplied the physical connectivity between personal computers. Then, a connection-oriented session service was supplied using the NETBIOS interface for the IBM Personal Computer (PC) and its clones. A standard, but unpublished, protocol was used to support NETBIOS on IBM LANs. An extended version of NETBIOS called NETBEUI appeared later and, for it, the protocol was also published to allow inter-working over LANs. In 1985, support for personal computers was brought into SNA as well, allowing direct communication between terminal-style devices, without the need for mainframe computers and their front ends to act as intermediaries. The new architecture is known as Advanced Peer-to-Peer Networking (APPN).

SNA is still widely used in IBM environments. However, to allow inter-working with other computers, IBM also supports both ISO and Internet protocols in its products. The range of physical media is also wide, embracing more modern technologies such as frame relay and ATM, as well as ethernet LANs in addition to token ring LANs.

12.7.2 Digital Network Architecture

The Digital Equipment Corporation (DEC) Digital Network Architecture (DNA) had its beginnings in 1975, around the same time as SNA. The motivation for its creation was similar to that of SNA, namely to provide a way of interconnecting Digital computer equipment. It did not begin with a large legacy of Digital communication products, but rather evolved with Digital's computers as they progressed from being fairly simple stand-alone minicomputers to powerful networked computers. While DNA is the architecture, Digital's facilities are better known by the main product name: DECnet. From 1983 until 1996, DECnet Phase IV was available, but it has been supplanted by DECnet Phase V, which is referred to as DECnet/OSI by Digital. As the name suggests, this supports all of the ISO

OSI protocols, as well as the proprietary protocols supported in earlier versions of DECnet. The Internet protocols can also be used with this product.

In traditional DNA, the centrepiece protocol was the Datagram Delivery Protocol (DDP), which supplied an unreliable connectionless service between computers attached to a DECnet network. The Network Services Protocol (NSP) could then be used to supply a reliable connection-oriented service on top of DDP. The role of DDP and NSP is similar to that of CLNP and TP4 in the ISO world, or of IP and TCP in the Internet world. Below these, a variety of physical media are allowed in DNA. Above them, DNA has a session control layer which, in turn, can be used to support user communications. Thus, the overall DNA model is fairly similar to the Internet model, except that a session layer is separated out from the overall application layer.

12.7.3 AppleTalk

When Apple Computer introduced the Macintosh computer, networking was seen as being very important, and so the AppleTalk network architecture was devised, to support communication between Macintoshes. The initial version of AppleTalk was only suitable for fairly small networks, so was replaced by AppleTalk Phase 2, which worked for larger networks in organizations with large numbers of Macintosh computers. The specifications for AppleTalk were published by Apple, in order to encourage other vendors to produce products based on AppleTalk.

Like DECNet, a protocol called the Datagram Delivery Protocol (DDP) is at the core of AppleTalk. It is not the same as the DECNet protocol with the same name, but is for the same purpose: supplying an unreliable connectionless service between two computers on an AppleTalk network. There is no TCP-style transport protocol directly above DDP. However, there is the AppleTalk Transaction Protocol (ATP), which is a reliable request-response protocol. Above both of these, in a session layer, there is the AppleTalk Data Stream Protocol (ADSP), which does supply a reliable connection-oriented service. Alongside this is the Printer Access Protocol (PAP), a connection-oriented protocol for supporting connections between clients and servers, originally used only for printers, but now of more general applicability.

A range of physical media can be used to support DDP. In addition to the usual technologies, a further alternative is Apple's proprietary LocalTalk technology, which is a contention bus running at 230 kbits per second, and suitable for short distances and small numbers of computers. Different applications make use of the AppleTalk facilities. The underlying intent was to fully integrate networking into the desktop environment of the Macintosh, to allow sharing of resources such as files and printers.

12.7.4 Xerox Network Systems and descendents

The Xerox Corporation introduced the Xerox Network Systems (XNS) protocols around 1980, particularly to support office applications using a variety of communication media. In common with DECNet, AppleTalk and the Internet, there is an unreliable connectionless protocol as the centrepiece. It is called the Internet Datagram Protocol (IDP). It supports the Sequenced Packet Protocol (SPP), which gives a reliable connection-oriented service, and the Packet Exchange Protocol (PEP), which is a semi-reliable request-response handshake protocol. These protocols populate the centre two layers of the XNS model. Below them, there is a layer for handling the physical communications medium, with no specific protocols being mandated. Above them, there is a layer with various application protocols, for file access, printing and naming.

The most important feature of XNS has been its influence on other standards. Its Routing Information Protocol (RIP) became the most-used interior gateway protocol in the Internet. Further, because XNS was one of the earliest available protocol suites, it has been adopted and modified by several other manufacturers. Two major examples are both concerned with networked operating system products. First, there is Banyan VINES, an operating system based on Unix, which has the VINES Internetwork Protocol (VIP) at its heart. This supports transport services which, in turn, support a remote procedure call protocol which is used for the server-client interactions for distributed operating system purposes.

Second, there is the more famous Novell NetWare, which originated for client-server relationships between Personal Computers, but has since been implemented on various other types of computer. In 1996, NetWare was renamed 'IntraNetWare', reflecting this proprietary standard being positioned to fill a niche where interaction with other standard facilities may not be necessary. NetWare has the Internet Packet Exchange (IPX) protocol in the middle. This supports the Sequenced Packet Exchange (SPX) protocol, which is based on the SPP protocol of XNS, and provides a reliable connection-oriented service. In turn, IPX and SPX are used as a basis for the NetWare applications. They can also be used to support NETBIOS, to allow inter-working with PC applications which use that alternative IBM standard.

12.8 CHAPTER SUMMARY

Agreement between humans on the protocols used for computer communications is a prerequisite for achieving agreement between computers when communications take place. The process of achieving human agreement occurs through standardization, an activity applied to many areas where widespread harmony is desirable. Some standards emerge from practical use: a particular product becomes dominant, and gradually becomes a standard for all users. Other standards emerge from official standardization bodies, which exist as fora for interested parties to try to reach agreement.

For computer communications, one significant standards body is the International Organization for Standardization (ISO), which is an umbrella body for national standards organizations, one per country. ISO standardized the Reference Model for Open Systems Interconnection (RM/OSI), which forms a conceptual framework for other ISO protocol and service standards. The nature of the RM/OSI model has always stimulated debate, and some of the standards developed within it have not become widely used in practice. However, the layered philosophy of RM/OSI is generally accepted as a good way to structure communication systems, even if ISO may not have chosen the best set of layers.

An alternative, specialized, body is the Internet Society (ISOC), which is responsible for developing standards for Internet protocols. Although the formal structure of ISOC standardization is not unlike that of ISO, the mechanisms are distinctly different, allowing wide participation in standards formulation and a philosophy of demonstrating implementations before fixing standards. This results in standards that are immediately usable, and have support in a dynamic networking community. They are then well-poised to become *de facto* standards outside the Internet.

In between the generalist and specialist organizations are standards bodies for related areas, particularly telecommunications and electronics. One particularly important body is the International Telecommunications Union (ITU), part of the United Nations, which develops standards for both low-level physical transmission and networking, and also high-level applications for human communication. Another significant body is the Institute of Electrical and Electronic Engineers (IEEE), a professional body, which develops standards for local and metropolitan area networking.

Finally, there are proprietary standards developed by manufacturers, principally to support communications between their own computers, but sometimes becoming, or influencing, more general standards. The most famous proprietary architecture is IBM's System Network Architecture (SNA), which influenced the design of ISO's reference model. This can be used alone to interconnect IBM computers, but can also inter-work with other computers using ISO or Internet protocols. Other examples of proprietary architectures include Digital's Digital Network Architecture (DNA), Apple's AppleTalk and Xerox's Xerox Network Systems (XNS). The last of these has spawned other notable descendants, including Banyan VINES and Novell NetWare.

12.9 EXERCISES AND FURTHER READING

12.1 Give examples of everyday objects that have become standardized, and indicate whether the standardization was designed, or happened through dominance of usage.

12.2 Consult the ISO WWW page (URL `http://www.iso.ch`) to find out more about the International Organization for Standardization and its operation.

12.3 Check whether you have access to ISO standards documents. These may be available under the banner of your own country's ISO member organization, for example, BSI in the United Kingdom.

12.4 Consult the IEC WWW page (URL `http://www.iec.ch`) to find out more about the International Electrotechnical Commission and its operation, and how it cooperates with ISO.

12.5 Consult the ITU WWW page (URL `http://www.itu.ch`) to find out more about the International Telecommunication Union and its operation.

12.6 Consult the Internet Society WWW page (URL `http://info.isoc.org`) to find out exactly how IAB, IESG and IETF interact in order to formulate Internet standards. Also read RFC 2026, which describes the Internet standards process.

12.7 Find, and consult, WWW pages for other standards bodies mentioned in this chapter: ATM Forum; EIA; ECMA; ETSI; Frame Relay Forum; IEEE; NIST; and World Wide Web Consortium.

12.8 Compare the four WWW URLs given above, and also the URLs you found for the WWW pages in the previous exercise. Do they allow you to deduce anything about the nature of the organizations?

12.9 Obtain further details of the ISO RM/OSI model. For each layer, comment on how well it fits the full list of layering principles used when formulating RM/OSI.

12.10 Do you think that there is a case for having separate Session, Presentation and Application layers in RM/OSI?

12.11 For the two case studies considered in Chapters 9 and 11, investigate how well the protocols described fit within the ISO RM/OSI model.

12.12 Read the paper 'The (Un)Revised OSI Reference Model' by Day, which appeared in *Computer Communication Review*, October 1995. Do you agree with the author's feelings on recent RM/OSI developments?

12.13 Gather information on how widespread use of the various ISO application layer standards is.

12.14 What was the underlying rationale for creating five different varieties of the ISO standard connection-oriented transport protocol?

12.15 Construct a list of all the different System Management Functions that have been, or are being, standardized by ISO. What management category (or categories) does each fall into?

12.16 Find out about the activities of the Network Management Forum.

12.17 Learn more about the ESTELLE, LOTOS and CHILL formal description techniques.

12.18 If you have access to examples of ESTELLE and/or LOTOS specifications of ISO standard services or protocols, compare the merits and demerits of these specifications with the corresponding natural language specifications.

12.19 Give further examples of profiles constructed from sets of ISO standard protocols.

12.20 Compare and contrast the ISO RM/OSI model and the Internet reference model.

12.21 Read the latest version of Internet standard STD 1 (for example, the March 1996 version was RFC 1920), in order to discover the details of the Internet standardization process. How many Internet standards are there today?

12.22 Consult RFC 1157, which describes the first version of SNMP, and then RFC 1441, which introduces the second version of the Network Management Framework.

12.23 Search an index of RFC documents, to discover how many different MIB types are described. Note the range of different communication mechanisms that are covered.

12.24 If you have access to an (elderly) PSPDN using a PAD, compare the facilities offered by the ITU-T XXX standard and the Internet TELNET standard.

12.25 Look for examples of non-X series ITU-T standards that appear in computer communications contexts.

12.26 Discover the dates at which the various IEEE 802 standards were first released, and compare these with the approximate dates in which the various LAN technologies were first invented.

12.27 Discuss the influence of the SNA layered model on the ISO RM/OSI model.

12.28 If you have any information about IBM computer installations using SNA networking, investigate the extent to which the original hierarchical network arrangement still applies.

12.29 For any personal computer that you have access to, find out details of its networking arrangements (if any). Does it rely on proprietary protocols, formally standardized protocols, or a mixture of both?

12.30 Do you think that there is a future for proprietary standards for computer communications?

12.31 Compare the protocols that humans use for their communications on standardization issues with the protocols that computers use for their communications in general.

Further reading

The main further reading for this chapter are actual standards documents, or product specifications in the case of proprietary standards. The Internet Society is the shining example, in that all Internet standards and discussions on standards are freely available for consultation. ISO, ITU-T and IEEE standards documents must be purchased, and are often extremely expensive, when compared with the price of textbooks. Until recently, the ISO RM/OSI model dominated many general textbooks, although some are now migrating to the Internet reference model. Specialist textbooks on network managment are available, as this become an increasingly important topic. There are several textbooks on SNA, reflecting its very significant position among proprietary standards. Some research has been carried out on how standardization processes work in practice, and how their goals are reached. Results of this research are available in the academic literature.

INDEX

Note: Underlined page numbers denote main, or defining, occurrences.